ALSO BY MASHA GESSEN

Blood Matters

From Inherited Illness

to Designer Babies,

How the World and I Found Ourselves

in the Future of the Gene

MASHA GESSEN

Mariner Books · *Houghton Mifflin Harcourt*

Boston   New York

First Mariner Books edition 2009

www.hmhbooks.com

Portions of this book have been published in slightly different form in the following: "Mutations" in *Granta*, 2004; "A Medical Quest" in *Slate*, 2004.

Library of Congress Cataloging-in-Publication Data
Gessen, Masha.
Blood matters: from inherited illness to designer babies, how the world and I found ourselves in the future of the gene/Masha Gessen.—1st ed.
p.   cm.
Includes bibliographical references and index.
1. Medical genetics—Social aspects.   2. Human chromosome abnormalities—Diagnosis—Social aspects.   3. BRCA genes.
4. Genetic counseling.   I. Title.
RB155.G475   2008
616'.042—dc22   2007036751
ISBN 978-0-15-101362-3
ISBN 978-0-15-603331-2 (pbk.)

Text set in Adobe Caslon and Priori Sans
Designed by Liz Demeter

Printed in the United States of America

DOC 10 9 8 7 6 5 4 3 2 1

Contents

PART I

The Past

Chapter 1

MY MOTHER'S
FATAL FLAW

I spent the day of August 21, 1992, driving to a mountainous desert town whose name, in the scorching heat and fine dust, was a seductive mockery: Palm Springs, California. I had embarked upon the most Californian of endeavors, an editorial retreat for the Los Angeles–based magazine where I worked. I ate dinner with my colleagues at a bland Mexican restaurant. I had two margaritas, talked more than I usually did, and told a story that left me vaguely uneasy, as I always feel when I talk about my mother: I cannot talk about her without telling lies. I do not remember what I said, but it was something complimentary, even prideful, I think, and though I loved my mother and was proud of her, talking of her in that way, with all that had gone wrong between us, was most certainly a lie.

I woke up at four that morning, in the bedroom of a rental bungalow, with a wave of nausea pushing its way up to my burning throat. I stumbled to the bathroom, drank from the tap and threw cold water on myself, washing my face and head clumsily, then looked at my bloated face in the mirror and wondered how two margaritas could have done this to me. I went back to bed and next opened my eyes at a few minutes before seven, without a trace of a hangover but with a sudden wakefulness I could not fight. With hours to kill before the meetings began, I tried going out for a walk in the desolation of Palm Springs, considered a swim in the kidney-shaped pool, and finally went back inside the bungalow intending to read some magazine submissions. I spread them out on the coffee table and, before starting, picked up the phone and dialed my parents in Boston. I was checking in at least daily back then and knew they would be awake—they were three hours ahead. These considerations were background noise; I had picked up the phone without pausing to think, just getting one of my daily chores out of the way while I had time to kill.

A strange male voice answered the phone.

"Papa?" I asked, knowing that it was not.

"Hold a minute," the man said nervously, and a moment later my father came on the line.

My mother was dead. The man answering the phone was a policeman who had come to fill out a report, which, as it turned out, was a necessary part of letting someone die at home.

My mother had been diagnosed with breast cancer two years earlier. By the following summer, it had already spread to her bones, and then it got to her liver and killed her.

My mother had last woken up at seven that morning—four o'clock in California, when I had first awakened—and asked for ice cream. Her liver was failing. Her throat must have been burning up. She died a few minutes before ten. That was the moment I had bolted awake for the second time, the bizarre toxic symptoms of three hours earlier mysteriously gone, and my inextricable physical relationship to my mother proven to me for the first time in my conscious life—at the moment hers ended.

～

The second time the physical relationship proved itself was on January 28, 2004, at a coffee bar in Cambridge, Massachusetts—an accidental location I shall avoid in the future, much as I have avoided revisiting Palm Springs. I was sitting at a small square table, trying to fix my ailing laptop, when my cell phone rang and a professionally sensitive woman's voice said, "I am returning your call. The results of your tests have arrived. And there is a change." She paused. "In the BRCA1 gene, there is deleterious mutation." She paused. "I'm sorry."

BRCA stands for "breast cancer." BRCA1 and BRCA2 are two genes known to play a role in the development of breast and ovarian cancer. The caller was a genetic counselor informing me that my mother had passed on to me a mutant gene. I was surprised. I was shocked. I should not have been. I had gone to get tested, I had known enough to know that I was a likely candidate for the mutation, but I was convinced that I was negative. Even if my mother had been a carrier—I could not know, because she died before the gene was discovered—I had only a 50 percent chance of having inherited it. That night in Palm Springs had taught me nothing: I was certain I was immune to my mother's physical legacy.

Something had gone wrong between me and my mother, something so profound and so old that I find it difficult to describe. There was no tragic fight, no horrible misunderstanding. For as long as I can remember, we simply felt like strangers, not particularly intimate ones except by virtue of circumstance: We happened to live together. Nothing between us was ever unconditional, not even our physical proximity. I left home at fifteen.

My mother died before we had had much chance to claim our tiny islands of common ground: before I wrote anything she—also a writer and a translator—would have enjoyed reading, before I translated my first book, using what I had learned from watching her work, and before I too became a mother. When I started writing professionally, she said proudly, "My genes have won out." I remember being surprised, and silently dissenting: I did not doubt my mother's gifts, but I never believed they were also mine. I counted on more—and less. My mother was a more talented writer, a more diligent reader, and a more enterprising student. She was also handicapped by a desperate fear of people, and that fear could turn routine communication into a feat of heroism. She died at forty-nine, still gifted but not accomplished: Even if by external measure she could be considered successful, she felt anonymous and overlooked. I think that long-ago conversation with my colleagues in Palm Springs had in fact concerned my mother's career achievements, and this was why it had left me so uneasy. I knew the fear too, but of necessity I learned to get out and make my way among people early, and I had always thought that this was why I had done well with barely half of her gifts. I had assumed I was simply better at living than she had been. And even though, like all daughters of mothers who die young, I had a difficult time visualizing myself past a certain age, I had always, without really thinking about it, assumed that I would make better of what I had, and for longer, because I am not as afraid. I thought my gifts were my own, making me free from her legacy altogether. Then I found out that I got everything from her, including the flaw that killed her.

I learned the basics of the story of my flaw. I carry a genetic mutation that kills women early—earlier and earlier with each generation—through breast and ovarian cancer. "My" gene was identified about two years after my mother died and ten years before I tested positive for the mutation. It seems to be a gene that works as a tumor suppressor—unless it is damaged, as it was in my mother's case and is in mine. The hereditary roulette works as follows. For most people, the genome consists of one pair of sex chromosomes (XX in women and XY in men) and twenty-two pairs of chromosomes called autosomal, plus mitochondrial DNA, which is something of a separate story. The autosomal chromosomes contain two copies of every gene, one inherited from the mother and one from the father. The BRCA1 gene resides on chromosome 17. Those born with a mutation have one normal copy and one damaged one. A child born to a parent who has a mutation has a 50 percent chance of inheriting it. If a female child inherits a BRCA1 mutation, her lifetime risk of breast cancer may be as high as 85 percent, and the risk of ovarian cancer may go up to more than 50 percent. For some reason, probably having to do with the environment or diet or lifestyle, these days women with the mutation are getting the cancers at an earlier age than their mothers' and grandmothers' generations.

If a fetus inherits two bad copies of a BRCA gene—one from each parent—it will not be viable. A girl baby who is born with only one defective copy of the gene will not develop cancer as long as the other copy is functioning. But when the "good" copy also suffers a mutation—as, it seems, will happen in most cases—cancer will develop, and the disease may be more aggressive than in people without such mutations. A male child with the mutation may also eventually develop breast cancer, but this happens far less frequently. The risk of cancer goes up steadily with age: about 20 percent by age forty, 40 percent by age fifty. Rarely do women under thirty develop the cancer, and the chances of cancer only pass the 50 percent mark at around the age of fifty-five. So throughout human history, a woman would most likely become sick after she had given birth to and raised her children. For modern women, particularly western Jewish professional ones who have children later, the mutation may bring cancer before the child-rearing years are past.

Mathematically, women are just as likely to inherit this breast-and-ovarian mutation from their fathers, but they are not as likely to suspect that they have it unless their mothers have been stricken. Because this sort of genetic testing is not routine (although Israeli geneticists are debating the possibility of population-based screening for these mutations), women who discover they carry the mutation often come from matrimonial cancer dynasties: Great-grandmothers, grandmothers, mothers, and sisters in every generation have had the cancers. These women are terrified of having daughters. Some of these women hate their mothers.

That these mutations were discovered first among Jewish women is probably largely, though not entirely, an accident. Jews are an obvious choice for the study of genetics: They make up compact populations certain to share many genetic traits. So do Icelanders, Scandinavians, and a large number of other ethnic groups, but they are not as frequently found near large medical-research centers. It may also be that Jewish medical researchers have chosen to study familiar communities. Finally, the position mainstream Judaism takes on most key issues of the genetic age has allowed this kind of research—including research on prenatal and preimplantation testing, gene therapy, and stem-cell research—to flourish in Israel.

My generation, making radical and underinformed decisions, may be lucky to be the guinea pigs—or not. In the past ten years a few thousand mostly Jewish, mostly midlife women, mostly in the United States and Israel, have gained the kind of knowledge humans are unfamiliar with having. I had my fortune told by a genetic counselor at a high-tech medical center in Boston.

～

My daughter was born nine years after my mother died. I gave her my mother's name: Yolka (in my daughter's case, a diminutive for Yael). She fashioned it into a story of life. "You know," I heard her say at the age of three. "Before, I used to be a small baby. Then I was born. Now I have grown and become a girl. Before, I used to be Grandmother Yolka, but I died."

The calm simplicity of that story pleased me, but I worried now about what I had passed on. My overconfidence about my own hereditary fortune affected my daughter too: When I gave her my mother's name, I was certain she would take only the best. Women who know they have a cancer-gene mutation are, I have discovered, rarely so cavalier. I have talked to young women who would rather adopt a child with an entirely unknown health history than risk passing on the possibility of breast cancer. I have talked to older women who are wracked with guilt over having passed on the mutation. I have talked to a woman who hates her eighty-four-year-old mother for being cancer-free while she and all her sisters, who inherited the mutation, have developed cancer. They focused their anger on the mother's refusal to be tested for the mutation—as though it would have made a difference—and when she finally relented, and tested positive, they rejoiced in the chance to lay blame.

With a disease as unpredictable as cancer, the opportunity to blame an actual person is an unexpected temptation. Sometimes mutations can be traced through generations, based on the history told by death certificates, obituaries, and fears passed on from mother to daughter. I have talked to women who have seen their mothers, grandmothers, aunts, sisters, and cousins struggle with cancer and succumb in a pattern that becomes familiar. When I learned about my own mutation, I knew only that it went back to my mother, who got it from her father. (He was killed in World War II at the age of twenty-two, but his sister later developed ovarian cancer, pinpointing their branch of the family as the culprit.) Then the trail was lost, as it often is, since in chains of mutation carriers generations seldom overlap by much.

The first time I went to a gathering of women who carry the breast-and-ovarian-cancer mutation, I found myself looking hard into their faces, searching for familiar features. Unlike, say, people with HIV or multiple sclerosis, we share more than a similar condition, parallel concerns, and identical hopes: We have common blood. The mutation I carry is called a "founder mutation," which means it goes back to a single event: one person who was born with a mutant gene and passed it on to his or her offspring, who passed it on and on. The fact that my particular mutation is found among both

Ashkenazi and, albeit less often, Sephardic Jews means the "founder" was a very distant ancestor, someone who lived long before the split into Ashkenazim and Sephardim.

That is why one can refer to these mutations as "Jewish" without incurring the wrath of the politically correct, as long as they can make the leap—especially difficult for Americans—to thinking of Jewishness as an ethnicity rather than a religion. In February 2006 I went to the first-ever public gathering of women who carry BRCA mutations, a conference at a cancer center in Florida. The conference included networking sessions, for which sign-up sheets were posted on a bulletin board. One of the sign-up sheets was titled "Latino mutations." No one signed up. After a while the woman who had posted it crossed out the heading and replaced it with "Ashkenazi mutations"—and got better results. The same mutation that I carry is found, with relative frequency, among Catholics in the American Southwest. Not only is it the same mutation, but the general genetic neighborhood in which it is located looks similar, pointing to the probability of a common "founder." The Hispanic carriers of this mutation may be what have come to be known as "crypto-Jews"— the descendants of Spanish Jews who were forcibly converted to Catholicism before leaving Spain and eventually making their way to the New World. The woman who created that sign-up sheet told me she believed she was probably of Jewish ancestry herself—a belief she had acquired since testing positive for the mutation. The cutting-edge science of the twenty-first century has a way of turning one's thinking about blood, religion, and disease positively medieval.

~

When I learned of my mutation in January 2004, I was thirty-seven, my daughter was two, and her brother—who is adopted—was six and a half. I had been thinking of having another child. My cumulative risk of breast cancer at that moment was roughly 14 percent, to be adjusted up to 16 the following year and 18 the year after that.

I was, in the absurd argot of the trade, a "previvor": not yet diagnosed with cancer but with a high risk of getting it. I went to see a genetic counselor.

Those who learn that they carry a mutation like mine are immediately admitted to the cancer caste. I found myself carrying a cancer center's patient card, walking past a wig-and-prosthesis shop on my way to see my doctors, and retracing my mother's steps down the hospital corridors—still hoping that in my version, the same genes would add up to a better life, and a longer one.

The genetic counselor, her head permanently cocked to one side in a show of sympathy, suggested there was only one way out of the cancer ward. She advised me to cut out my ovaries. She said I might also consider removing my breasts. I was still using them to feed and comfort my daughter then. Breast milk had turned out to be the magic potion of motherhood: It nourished my daughter during her frequent illnesses when she was a baby, and it could cure frustration, anxiety, or a stubbed toe once she became a toddler.

I spent weeks reading medical studies and doing frantic arithmetic, careening from one option to another. In the end I decided to proceed slowly. I weaned my daughter. I managed to convince her that big girls do not drink from the breast. It took a couple of weeks and then she took to holding my breasts—before she went to sleep, when she woke up, or for comfort when she hurt herself or felt insulted. Every time I cuddled her, I worried: How could I get rid of them when she needed them? The argument that ought to have trumped them all—that any trauma was worth it if it meant having me around—did not convince me. What worried me, gnawed at me to the point where I felt a stabbing pain in my chest—my breast— was the fear of losing the physical connection with my daughter that I never remembered having with my mother. It turned out it had been there all along, as sure as the fact that I bolted awake the moment she died three thousand miles away, but I had never felt it. Touching her did not give me comfort.

There is a black-and-white picture of me at the age of perhaps two months. My mother is holding me, her lips buried in the black down on my head, and she has that look of being desperately tired and profoundly in love, a look—or, rather, a state—I recognize. This is the last evidence that we had this connection. When I walked into her hospital room at Boston's Brigham and Women's Hospital in August 1990, she pointed with her chin toward the flatness under her gown and said, "I fed you with that breast." She sounded like she was saying something one said in such situations. I mumbled something noncommittal, which is to say, unfeeling: I felt nothing.

Fourteen years later, I sat at a large table with women who carried my mutation. The hospital where my mother had her mastectomy and where I now went for my breast MRIs was across the street. The dozen women around the table dined on takeout and talked about cancer as a social worker looked on. She was there to help us talk to one another and also, this being a research institution, to observe how this newly isolated breed—women who knew they carried a cancer-causing mutation—would behave when placed together. I tried not to draw attention to myself as I looked around the room. I looked at noses, eyebrows, hands. The women looked nothing like me, but many of them had the thick postchemo hair, the brittle postchemo skin, and the protruding breast prostheses that reminded me of my mother. Other than that, our genetic commonality was invisible.

My daughter will go on in the world with feet and eyebrows that are replicas of mine, a stubbornness just like mine, and the habit— my habit—of scrunching up her face when doing something that requires great concentration, such as coloring within the lines. Most important, she will carry with her the memory of me, perhaps even the physical sense of me. That physical awareness is the essential element of security. Whatever trace of my mother I carried—even if it was the mere knowledge of her existence—had kept me from feeling mortal as long as she was alive. When I awoke on the morning of her death, I felt a fear that has not left me since. For months after I learned of my mutation, I would think about this in the sleepless early mornings, when my daughter pressed her hot heels into the small of my back, and I knew I was the only thing that protected her from the cold wind of fear and freedom that came into the room through the open balcony door. Then she would tap me on the shoulder and ask me to turn around so she could hold my breasts.

⁓

One of the women who carried a BRCA mutation started a Web site to help women share information about risks and possible remedies. I found myself checking the message boards several times a day. Women talked about surgeries, drugs, and screening procedures. The level of detail and the depth of knowledge were astonishing: Even as patient communities go, this was an extraordinarily educated and motivated one.

Whenever I pulled out my cancer-center patient card—whether to schedule an appointment through my regular health center or to park in a hospital lot—I received a recognizable sympathy look, the sort that says, "So young. So tragic. So frightening." I felt a bit like an impostor: I did not have cancer. I had come to realize that my ad-mission to the cancer caste, conditional as it was, was but a tempo-rary substitute for the yet-unrecognized group of which I had become a member. It was not quite a secret society, but it did live, invisibly, by a set of rules different from those of the rest of the world. In this society, people become privy to the information contained in their genes and reshape their bodies—and their fates—accordingly. The rules by which this society lives are an approximation, albeit a very crude one, of the rules by which my daughter's generation will run its life.

Understanding my accidental role in this rehearsal of the future was part of what made me want to write this book. But there was something else, too. The frontier women who were cutting off their breasts to spite their genes were just the most visible and vocal expression of the transformation in our understanding of ourselves. I belong to a generation that grew up believing we were shaped by love, care, or lack of it—or perhaps even the number of books on our parents' bookshelves. But we will go to our graves believing that it is a combination of letters in our genetic code that determines how we get there, and when. Our concept of the stuff we are made of will have undergone fundamental changes. I got a glimpse of that when I was looking around that room at my fellow mutants, and again and again in the process of writing this book, as I looked at myself, my biological daughter, and my adopted son. I was transported to a new era, a future that will rest on a different understanding not only of what causes things to go wrong in human beings but of what makes a human being in the first place, and what connects any one of us to any other.

Chapter 2

THE FOUR MOTHERS
OF JEWS

Once you join a secret society, you find fellow members everywhere. After I wrote a series of articles on my mutation for *Slate* I began to get letters from long-lost friends, friends of friends, and people who thought we should be friends. Women whom I met professionally would confess their genetic information to me. In a colleague's office in New York, I was pulled roughly aside to hear, "I am BRCA, too"—or was it "BRCA2"? Even though there were only two of us in the office, for a moment I had a hard time identifying the source of the whisper, so sharply did it break with the context of the conversation we had been having. Once I got my bearings, I realized I was staring at my interlocutor's chest, which was perfectly flat.

For the most part, I welcomed the communication: The same impulse that moved people to write or call led me to respond enthusiastically. Each new flesh-and-blood acquaintance made my predicament seem less extreme, and many shared useful information. I also found myself relentlessly looking for clues that would make sense of my situation. One of my newfound friends mentioned that her family hailed from the Polish city of Bialystok. My paternal grandmother was born in Bialystok, and I found myself wondering if I had inherited the mutation from my father, and he from his mother, who was alive and cancer-free at eighty-three. This would be good news if it were true: If she could carry the mutation and not get sick, my odds of developing cancer would be ratcheted down. The mechanism by which my mutation causes cancer is unclear, but some families seem to do better with it than others. Whether this is because of genes or environment is impossible to say: Families tend to share both.

But, of course, the Bialystok lead was a red herring. Bialystok was a major center of European Jewish life before World War II. Statistically, 1 percent of the Jews of Bialystok would have carried the mutation. It seems my friend's family had it and my paternal grandmother did not.

A brighter ray of hope appeared when my maternal grandmother—eighty-six and complaining only of arthritis—told me that her mother had had her ovaries removed before the age of fifty. Why? "She had an inflammation." Back when my great-grandmother was ill, *cancer* was a word rarely uttered: In the family it may well have taken on the name "inflammation." In some ways, this imagined lineage for my mutation had even greater appeal. My grandmother was a very fit, very active, and unfailingly happy person—very much unlike my mother but, I liked to think, rather like me. It must have been my grandmother's skill at extracting joy from life, combined with her physical activity—she had recently given up her bicycle but still insisted on cross-country skiing in winter—that allowed her to beat the genetic odds. I was also possessed of something of a manic personality and exercised obsessively. Had the ability to defeat genetic fate skipped a generation? The idea fit my internal understanding of genetics perfectly. I should note that, inasmuch as they are genetically determined, traits do not, of course, skip generations:

They are inherited directly or not at all. Still, the idea that exercise and fitness could moderate the risk of cancer even in mutation carriers was not entirely without merit; it is just that the adjustment was small, and my statistical risk of developing cancer would remain many times that of someone who did not carry an altered gene.

I stopped chasing after rays of hope after Natasha, my mother's first cousin, got the test. I had called her when I was about to be tested, to check which kind of cancer had killed her mother—yes, it was ovarian—and got in touch again a few months after getting my own results. Natasha's test came back positive for the same mutation as mine, proving that in our family's case the simplest and most obvious solution fit the puzzle best. The mutation must have gone back to my great-grandfather, who died of some sort of cancer in his sixties. That death too may have been related to the mutation, which raises the risk of all cancers, including prostate, pancreatic, stomach, and colon—though the risk is not nearly as high as the risk of breast and ovarian cancer in the women who carry the mutation. My great-grandfather broke the fifty-fifty rule by passing the mutation on to both of his children: his son, who was killed in the war at the age of twenty-two, and his daughter, who died of ovarian cancer at fifty-two. Each of them had a child, both of whom in turn inherited the mutation. They were my mother, who died at forty-nine, and Natasha, who was apparently healthy at fifty-three, which was wonderful but hardly statistically significant. The evidence seemed to say that I came from the sort of family where female mutation carriers did develop cancer—and tended to die of it.

~

Natasha had been tested at Hadassah, the large teaching hospital that sits on a mountain overlooking Jerusalem. For a major international medical and research center, the accoutrements seem a bit too modest: The offices are small, the waiting areas virtually nonexistent, the furniture scuffed, the ceilings peeling, and the handouts badly xeroxed. To get to the genetics clinic, one has to pass a metal detector, present one's bag for inspection, and wind through a couple of labyrinthine hallways. A plain white sheet of paper lists the conditions for which prenatal tests are offered. The list is split into three categories: conditions frequently found among Ashkenazi Jews, among Sephardic Jews, and among Yemenites. The last category contains just a few disorders, the Sephardic list comprises perhaps a dozen, and more than half the sheet of paper is taken up by tests available to the Ashkenazim. Michal Sagi, the young woman who ran the genetics program at Hadassah when I interviewed her in 2005, said, smiling, "Some of the Yemenites look at this sheet and say, 'Ah, the Ashkenazim are so sickly!'"

We both laughed. The Ashkenazim are just better studied than other Israeli populations, and other Middle Eastern population groups appear to have many common genetic conditions of their own. Still, the question of why the Ashkenazim are so sickly has occupied scientists for some time.

There are several genetic conditions that are unusually common among Ashkenazi Jews. Most of these are recessive traits, meaning that a person with a single altered copy of the gene has no symptoms but has a fifty-fifty chance of passing the gene to a child. A child who inherits two mutant copies—a one-in-four probability if both parents happen to be carriers—will get sick. Two explanations for the prevalence of these recessive traits compete for primacy in science. One explanation is selective advantage: The theory suggests that a single copy of the affected gene may offer protection against another disease. The best-known case of selective advantage is the genetic mutation that causes sickle-cell anemia in a person with two copies of the mutant gene but offers protection against malaria in people with just one copy. The same may be true of "Jewish diseases": a person with two copies of a "bad" Tay-Sachs gene, for example, will die in childhood, but perhaps a person with one copy will enjoy protection against tuberculosis and therefore survive longer than many of his compatriots and have a chance to have more children, thereby ensuring that the gene gets passed along.

The other possible explanation for the high incidence of recessive conditions is bad luck. The professional term is one of the most evocative in all of science: *genetic drift.* The theory says, in essence, that not everything evolves in accordance with the laws of natural selection: Sometimes things just happen. The smaller the population, the more likely things are to just go awry. The classic explanation involves a coin toss. If you toss a coin a thousand times, you are virtually guaranteed to see a fifty-fifty split between the number of times it lands heads up and tails up. But if your sample is drastically smaller—if, say, you toss a coin ten times—you are likely to see a statistically lopsided picture: seven heads, for example, versus three tails. By the logic of genetic drift, a mutation can take hold in a population simply because the population is very small. Say it originates with a few founder families, or even just one. If the person or persons with the mutation do not have much success reproducing, the mutation will get washed out. But if they happen to be fertile, or rich, or both, they may manage to have many children who will grow to adulthood and make up a significant group—half of them carrying the mutation—within the small population.

A good analogy for the concept of genetic drift is what happens to surnames. Surnames are inherited patrilineally, which means that generally only sons will carry on a family's last name. Since the chances of having a boy child are roughly fifty-fifty, the chances of passing on one's surname are equivalent to the chances of passing on a genetic trait. My great-great-grandfather had eight children, all of whom carried the last name Gessen. Two sons died young, two others never had children, two daughters had children who carried their husbands' last names, but my great-grandfather had four children named Gessen, and one of his sisters had an out-of-wedlock son who was also a Gessen. In other words, the last name suffered some losses in that generation but still had a decent chance to survive in the population. My great-uncle—the out-of-wedlock Gessen—had epilepsy and decided, in accordance with Jewish tradition (if not with genetic logic) forbidding epileptics from marrying lest the illness be passed on, never to have children of his own. He married a woman who already had children, whom he raised as his own. He died surrounded by grandchildren and even a great-grandchild, none of whom carried the name Gessen. My great-grandfather's two sons, on the other hand, not only survived World War II but had six children between them, three of them male. On the whole, this was not great going: The number of Gessens in this generation grew by just one person. In the following generation, my uncle and my father's cousin had two daughters each: The chances that these girls will carry on the Gessen name are negligible. My father had three sons, and his sister had an out-of-wedlock son who got the Gessen name, which makes a total of four Gessen males in my generation, but so far all of them have failed to reproduce. As it happens, the only person in my children's generation who is poised to pass on the surname is my son, Vova, who is adopted and got his surname the way one cannot get a gene: by court order. If the Gessen name were a genetic trait, it would be on its way out. At the same time, I have a couple of dozen cousins descendant from the Gessen line; if things had gone a little differently in terms of sex distribution and reproductive patterns, Vova could have four or five dozen cousins named Gessen, rather than none. This is the essence of genetic drift.

~

The most frightening of the so-called Jewish diseases are Niemann-Pick, Canavan, Gaucher's, and Tay-Sachs. The last is the best-known and one of the most common: One in twenty-seven Ashkenazi Jews is a carrier. A baby born with two copies of the mutant gene lacks an enzyme necessary for breaking down fatty substances in the brain and nerve cells. These substances build up, and around the age of six months begin blocking the functioning of the baby's brain and central nervous system: a baby that seemed healthy suddenly stops smiling, turning over, and grasping toys. The child degenerates, becoming blind, paralyzed, and totally unresponsive, and usually dies before the age of five.

One in thirty-seven Ashkenazi Jews is a carrier of the gene for Canavan disease. Affected babies lack a different enzyme, which leads to the buildup of a particular acid that in turn destroys myelin, or "white matter," in the brain. Symptoms appear between three and six months of age; the baby becomes developmentally delayed and may also go blind. The children die before the age of ten.

Niemann-Pick is another enzyme deficiency that leads to the buildup of a fatty substance—in the brain, liver, spleen, and lymph nodes. Babies become blind and disfigured and usually die before the age of two. One in seventy-five Ashkenazim is a carrier.

Gaucher's is also an enzyme deficiency that leads to the buildup of another fatty substance, which can cause symptoms ranging from mild to severe, including anemia, fatigue, and bleeding. Unlike the other enzyme deficiencies, it is not universally fatal, and treatment is available. One in fourteen Ashkenazi Jews is a carrier of the Gaucher's mutation.

None of these conditions is unique to Jews, but they are more common among the Ashkenazim than in any other population. The fact that they share some features—all are enzyme deficiencies, and all but Canavan are what are called lipid-storage diseases—has long led scientists to suspect that this is no accident. Neither bad luck nor genetic drift can explain the emergence of different mutations (each of the conditions can result from one of several) that affect different genes but cause similar processes to occur in the body. The explanation must lie with selective advantage—but what sort of selective advantage could it have conveyed?

In 2005 three Utah researchers published a study suggesting that Jews are so sickly because they are so smart. Ashkenazi Jews have the highest average IQ of any ethnic group that has had its IQ measured as a group. That does not mean that there are no Ashkenazi Jews of average or below-average intelligence. Indeed, the difference in mean IQ is not huge: The average Ashkenazi IQ is 108 to 115, while the average IQ in general is, by definition, 100. Again, this does not mean that all Ashkenazi Jews are smarter than all gentiles. What it does mean is that, proportionately, there are many more very intelligent people among Ashkenazi Jews than among other groups. This high general intelligence is frequently realized in measurable intellectual achievement. Ashkenazi Jews, who constitute less than 3 percent of the U.S. population, account for 37 percent of the winners of the U.S. National Medal of Science, 25 percent of American Nobel Prize winners in literature, and 40 percent of American Nobel Prize winners in science and economics. Ashkenazi achievement in the visual arts and architecture, on the other hand, does not stand out—and Ashkenazi Jews tend to score slightly lower, on average, than other Europeans on tests of spatiovisual ability. Indeed, the gap between Ashkenazi IQ and spatiovisual scores is one of the odd features of this ethnic group.

Current scientific wisdom, often couched in much additional information aimed at masking the study authors' discomfort, is that intelligence is, to a large extent, genetically determined. Studies of siblings reared together and apart, studies of twins, and studies of adopted children and their biological and adoptive parents show that the genetic component in determining intelligence is strong—and that it gets stronger as a child ages, suggesting that inherited ability is more important than all the early development strategies contemporary parents can produce (or fail to produce). How exactly intelligence is passed on from generation to generation is an open question; many scientists are working on genetic models of intelligence, often focusing on genetic causes of mental retardation, on the assumption that this trail may lead them to understanding the genetic mechanisms of intellectual ability in general. But if intelligence is inherited like genes—or like surnames—then the explanation for Ashkenazi Jews' high average IQ may also be either selective advantage or genetic drift.

The Utah researchers argued that it was selective advantage. Before the diaspora era, they pointed out, written accounts made no mention of Jews' unusual intelligence, but once in exile, Ashkenazim tended to find themselves in financial and managerial occupations. Those who excelled had significantly more success reproducing— and thus more children who survived to adulthood and themselves went on to reproduce—thereby passing on whatever intellectual ability allowed the parents to succeed in their middleman occupations. Presumably, these were verbal and mathematical abilities, while spatiovisual aptitude played no role in the Ashkenazi model of success. That would explain the contemporary Ashkenazi Jews' lopsided test scores, the researchers argued.

This was not exactly a new theory. The debate on the origins of Jewish success dates back at least to the turn of the twentieth century—with future fascist propagandists arguing the biological line alongside Jewish theoreticians. The Jewish historian Joseph Jacobs attributed the Jews' intellectual leadership to what was then widely known as the "germ-plasm." He wrote, presciently, "There is a certain probability that a determinate number of Jews at the present time will produce a larger number of 'geniuses'... than any equal number of men of other races. It seems highly probable, for example, that German Jews at the present moment are quantitatively (not necessarily qualitatively) at the head of European intellect." Therefore, he concluded, "the desirability of further propagation of the Jewish germ-plasm is a matter not merely of Jewish interest."

But the Utah researchers went further. They argued that selection for intelligence also explained the Jewish diseases. Those who carry only one copy of the genes that cause one of the lipid-storage diseases have an altered chemistry affecting the brain and the central nervous system, and the effects of this change may actually facilitate learning. To bolster their argument, the researchers used the case of Gaucher's disease—the only one of these illnesses that does not always kill people before they reach adulthood. A list of 322 adult Gaucher's patients—basically all the affected adults in Israel—showed that a disproportionately high number were engaged in intellectually challenging professions such as the academe, the sciences, engineering, law, and medicine. Five of the patients were physicists. On the other hand, these numbers were disproportionate when compared to the general U.S. or Israeli population, where the smart Ashkenazim do not constitute a majority, whereas the Gaucher's patients were Ashkenazim virtually by definition. The researchers did not compare the Gaucher's patients' occupational profile to a general picture of Israeli Ashkenazim— whether because such statistics were not available, because the comparison would not have been as dramatic, or because they were getting into politically charged and potentially distasteful territory, I do not know.

They did, however, discuss two other diseases. Torsion dystonia shows up as painful muscle spasms that can spread and become so severe that sufferers are confined to a wheelchair and unable to dress or feed themselves. The disease is dominant—only one copy of the gene is required for symptoms to appear—but only 30 percent of carriers show symptoms. Roughly one in nine hundred Ashkenazim is a carrier, and carrier status seems to correlate with a higher IQ. Nonclassic congenital adrenal hyperplasia is a recessive disorder in which an enzyme deficiency causes a hormonal imbalance that can manifest itself in symptoms ranging from negligible, such as hirsutism in women, to severe, such as infertility and the failure to develop breasts. One in five Ashkenazi Jews may be a carrier, and most of them have higher-than-average IQ.

But that is not all. The researchers argued that mutations like mine—disorders of DNA repair that cause cancer or untreatable anemia (a recessive condition known as Fanconi's anemia) or a number of physical changes and increased susceptibility to infections and cancers (a recessive disorder called Bloom syndrome)—may also be related to intelligence. Perhaps, they suggested, the same deficiency in the BRCA gene that makes it ineffectual at preventing cancer may encourage neural growth and thereby enhance intelligence.

The paper had all the hallmarks of a promising scientific theory: However wacky the premise seemed, the story was well told and internally coherent. The authors admitted their theory was unproven, made their case, and suggested further investigation. A perfect study would involve many pairs of siblings who were heterozygotes—genetically different—for the genes implicated in Ashkenazi diseases. If it could be demonstrated that the siblings who carried a deleterious mutation were on average more intelligent than their unaffected brothers or sisters, the we-are-so-sick-because-we-are-so-smart theory would be all but proved.

The media loved the study. I liked it, too, in part because it affirmed my new world view: It seemed there was a basis for my sense that I inherited my mother's intelligence, her verbal ability, and her cancer gene as a package. But, probably like most people, I found the discussion of Jewish intelligence uncomfortable, the way an inappropriate remark that elicits giggles at a dinner table can make one uncomfortable. It is not that I do not think that Jews are smarter, on average, than other people. Indeed, thinking that way is part of my cultural heritage: My grandmother, a principled cosmopolitan, always told me in a stage whisper that Jewish intelligence and the envy it engendered were the cause of anti-Semitism. But you could make that argument and then fall back upon cultural explanations, such as a tradition of literacy and a premium placed on education, even if you secretly suspected that Jews were just born smarter. So what made me uncomfortable about the Utah paper was what always makes people uncomfortable about genetic research: It takes private facts and makes them a matter of public discourse. This is the essence of obscenity.

~

Less than a year later an Israeli study purported to debunk the natural selection theory and to establish genetic drift as the cause of all the Ashkenazi illnesses and genetic advantages. The researchers claimed to show that nearly half of all contemporary Ashkenazim were the descendants of just four women who lived some fifteen hundred years ago. That settles the argument because, the logic goes, if Ashkenazi started that small, anything could happen. When a population starts with just a few people—or, at the extreme, with one woman—geneticists call this the "founder effect," meaning that the entire population or a large proportion of it will carry the genetic traits that happened to characterize that one person or small group of founders.

I went to see the authors of the study in Haifa. Doron Behar, listed first among twenty authors—the spot usually reserved for the graduate student who does the bulk of the research—had defended his Ph.D. dissertation less than a year earlier ("a recent founder event," he joked), and his skyrocketing career in genetics had been a matter of drift.

Doron was a critical-care doctor at Rambam, the large medical center right next to the Port of Haifa. In an age when doctors the world over are criticized for cookie-cutter care, critical-care doctors exemplify the assembly-line approach. Unlike general practitioners and most specialists, they do not get to know their patients. They come in when the patient's condition is dire and exit if he improves. In the absence of familiarity and communication, they use standard measurements as their guide—and they cannot help noticing that just as the same trigger causes reactions of varying severity in different people, so the same treatment can be more or less effective. Doron Behar wondered what sort of knowledge might allow a doctor to administer a test that would tell him which approach would work best for the person with systemic inflammatory response syndrome who has just been wheeled in. The term "personalized medicine," just coming into use as Behar was developing his interest in genetics, usually refers to complex, often long-term procedures, such as chemotherapy for cancer or medicating a patient for a major surgery, but it stands to reason that a doctor involved in the least personalized practice of medicine would want to find a shortcut to knowing his patients.

In addition to being a critical-care doctor, Doron Behar was also an inveterate people-watcher and a shameless picture-taker. The walls of his home office were covered with pictures of faces seen in China, Tibet, India, and elsewhere—the faces of strangers, ad nauseam. Behar was a man obsessed. His being a Jew, and an Israeli, made his obsession perhaps less socially suspect than it might otherwise have seemed. He really wanted to know what it is that causes the members of one tribe to have necks, noses, and skin color markedly different from the members of a tribe who live half a mile away. He also wanted to know what causes people to develop medical reactions of varying severity in response to identical triggers. And he talked about this a lot.

He happened to talk about this to Karl Skorecki, the head of the nephrology department at Rambam. Skorecki's own foray into genetics, quite accidental, happened about a decade earlier. One day in the midnineties Skorecki let his mind wander in the synagogue. Just at that moment a Cohen was called to the Torah reading. Skorecki was a Cohen himself; his mind still not focused on the service, he wondered what it was that connected all the Cohanim. Being a religious man and a learned man, he knew: The Cohanim are a priestly caste endowed with a set of particular privileges in synagogue and burdened with a set of restrictions in ordinary life. The Cohen lineage is paternal: The son of a Cohen is a Cohen and his son is a Cohen, and so on. All the Cohanim are sons of Aaron, brother of Moses. If that is the case, thought Skorecki, then he, a Jew of Ashkenazic roots, and the man of apparently North African extraction who was just then making his way to the front of the congregation, would carry the traces of their common forefather. And they would carry them most evidently where all men store the traces of their fathers and their fathers' fathers: in the DNA of the Y chromosome.

Skorecki, who had no research background in genetics, contacted an Arizona geneticist named Michael Hammer, who had distinguished himself in Y-chromosome research, and told him of his idea. It was a perfect research hypothesis: simple and obvious but for the fact that no one had thought of it before, and eminently verifiable. Hammer loved it, and together they conducted a study showing that, indeed, a majority of Jewish men who have been told by their fathers that they are members of the high priesthood are the descendants of a single male. Whether that man was named Aaron and had a brother named Moses, the DNA will not tell us—or, it will not tell us definitively. The signs did match. To determine when the common ancestor had lived, the researchers looked at the level of relatedness—or, rather, the degree of difference—among the Y-chromosome DNA of his various descendants. The difference is measured in mutations. Using estimates of the average time it takes a mutation to occur, the researchers calculated that the common ancestor had lived 106 generations, or roughly 2,650 years, ago, give or take a few hundred years. In other words, the man may have been alive during the exodus from Egypt. Furthermore, genetic signs place the father of most Cohens (Cohns, Kagans, Koches, Cheneys, and others) unambiguously in the Middle East.

The study of genetic heritage combines science that is staggeringly precise with what are at best decent estimates and respectable probabilities. Contemporary equipment allows scientists to decipher segments of DNA and compare them to each other. Some of these segments, which are shared among many individuals, are called haplotypes (short for "haploid genotypes"), a sort of DNA signature of a given population. The greatest variety of haplotypes is found in Africa, which establishes Africa as the common homeland of all humankind: People have been there the longest—roughly 150,000 years, it is believed—and in that time have achieved the greatest diversity. As people migrated to other parts of the world, and as some of them then migrated again, different haplotypes came to be predominant in different places. This knowledge has engendered a number of projects, both commercial and nonprofit, that aim to place a person's ancestors by looking at his or her haplotypes. But the results are mere probabilities: If a given haplotype is seen frequently in, say, the Middle East, infrequently in Europe, and never anywhere else, then there is a greater probability—but no certainty—that the person's ancestors hail from the Middle East and a slight possibility that they lived in Europe.

Time estimates are even less precise. To calculate the number of years that have elapsed since a group's common ancestor lived, geneticists look at the diversity within the group, essentially counting the number of mutations that have appeared in a particular segment of the genome. They estimate the number of generations required to acquire that many mutations, then multiply that number by twenty-five to get the number of years. There are two fallible assumptions here: the number of generations required for a mutation to appear, which is based on current statistics but can be skewed or imprecise, or inapplicable to the particular mutations in question; and the equation of a generation with 25 years. If the number of years in a generation is adjusted down from 25 to 15 years—an obvious biological probability given that Jewish girls, for example, could be as young as twelve when they started having children—the number of years that have elapsed since the common Cohanim ancestor will shrink from 2,650 to 1,590.

Consider families you know. Take mine. My father was twenty-two when I was born; his mother was just twenty when she had him.

I have two uncles who are roughly my age; their parents were in their forties when they were born. I am thirty-five years older than my youngest brother; my father was fifty-seven when this son was born. Twenty-five years is a bit more than a generation lasts in my particular lineage and a lot less than it does in the case of my brother or my uncles. Whether the twenty-five-year estimate gets any truer when spread out over many generations is debatable: My great-great-grandfather, for one, did not start having children until after he was discharged from the czar's army around the age of sixty; other men in his lineage also tended to live a long time and have children late. Which brings us to the problem of sample size: the smaller the sample—and sometimes a haplotype grouping examined in a particular paper includes just a couple of people—the larger the "confidence interval," or the margin of error. It is not unusual to see a table in a genetics paper that lists a number like 1,500—say, years to common ancestor—followed, in parenthesis, by a confidence interval of 1,000, meaning that the person may have lived 2,500 or 500 years ago. The numbers are all blinks in genetic time, but in our understanding of history they are huge.

Still, the magical quality of studies like Skorecki's overshadows their imprecise nature. The media loved his study; his colleagues—for Skorecki suddenly found he was a star in the field of genetics, and geneticists were his colleagues—loved it. Scientists generally love a question well asked, and this is why Skorecki's study appealed to them. And everyone loves a question well answered, which is why the media, and the public, ate it up. Skorecki himself thought his question was "parochial," as he put it, and the study would not draw much attention. Instead, it generated endless press and continued to have reverberations—and, it seems fair to say, to draw fans—nearly ten years later.

It made for a great story. On the one hand, it proved something quite improbable: that people from entirely different parts of the world, people who look completely different—just as Skorecki, fair-skinned, tall, and slim, must have looked completely different from the North African Cohen who gave him the idea for the study that morning at the synagogue—can be related. At the same time, this incredible discovery confirms, and is confirmed by, the most believed story of all time, which is the story contained in religious texts. It also offered a chance at the reconciliation for which the modern world yearns: the agreement between hard science and religious tradition. Here was scientific evidence that the Cohanim, as a group, existed and that they were who they said they were.

"I always give this example," Skorecki said when I interviewed him at Rambam, in an eleventh-floor corner office overlooking the Mediterranean shore in Haifa: a military base, the port, blue-gray water, and a chunk of Lebanon in the not-too-great distance (in about a month, Lebanese rockets started falling within yards of the building). "If you have a male who gave his son a secret word and he transferred this secret word to male offspring only, and so on and so forth, and we go into the future a hundred generations, to the extent that the secret is kept, any male on the planet who knows this secret word, their Y-chromosome DNA markers should be closer, more related, than the Y-chromosome DNA markers of any two males on the planet. That's the idea. The secret word is an oral tradition. It's a simple idea."

It was an imperfect analogy. The word in Skorecki's example had built-in protection against falsification: It was a secret. To appropriate the tradition, someone who was not a descendant of the original secret holder would have to guess the word. To pose as a Cohen, all one had to do was claim to be one. Not all Cohanim have Cohen or a variation of it as their last name: Skorecki does not. And Jewish history is ripe with examples of broken lineages, migrations, and other opportunities for impostors. The fact that in most cases a Cohen is a Cohen is a Cohen is, in its way, an affirmation of truth and honesty, which seem to have won out through a hundred generations of humans.

Using DNA evidence, one could try to corroborate any story in the world: All you have to do is ask the right question. Skorecki and his colleagues have been adamant about refusing to help anyone check Cohanim claims through DNA; in any case, rabbinical authorities are not about to institute DNA checks of an oral tradition— though the possibility of doing so will certainly be discussed over and over in the coming years and decades. But anyone who is uncertain of his heritage—say, a Jew raised in the Soviet Union, where the Jewish oral tradition was interrupted—can have his DNA tested for the Cohen modal haplotype through a commercial company.

One story that people have longed to confirm and develop for centuries is the story of the ten lost tribes of Israel, which were said to have been taken away by Assyrian armies some seven centuries before the current era. That was when they disappeared from the written account. Could their progeny be found now, through DNA? Possibly.

A number of small populations in different parts of the world have legends linking them to the ancient Hebrews. One such group is the Lemba, a population that has remained endogamous—has married within the group—despite being scattered throughout southern Africa. The Lemba circumcise their newborn males, keep one day of the week holy, and do not eat pork. None of these traditions is unique to Judaism, but generations of the Lemba have maintained a story of Jewish descent, and have been known to neighbors as the "black Jews." A closed population with mysterious roots, the Lemba long attracted the attention of geneticists, and a 2000 study in a scientific journal reported that their claim of Jewish provenance seemed to be true. Not only did many of the Lemba carry a haplotype found predominantly among Semites—which might as well have meant that they were of Arabic descent—but one of the Lemba clans had a very high frequency of the Cohen modal haplotype identified by Karl Skorecki and Mike Hammer.

The discovery confirmed the Lemba oral tradition. That dove-tailed nicely with a rare occurrence: a burst of Jewish proselytizing among the Lemba. Strictly speaking, Jews are not allowed to seek converts. But the Lemba—at least some of them—had long be-lieved themselves to be Jewish in some sense, and had practiced some rough approximations of Jewish religious customs. So proselytizing among the Lemba might be interpreted as reaching out to less-observant Jews in an attempt to make them observe more of that tra-dition, and this is certainly allowed. In the three years following the discovery of the Cohen haplotype among the Lemba, at least three rabbinical missions from the United States brought the Lemba books on Jewish tradition, prayer shawls, and other gifts to celebrate their affirmed Jewishness. It seems the Lemba elite, or at least some of it, welcomed the missions and began something of a Judaic revival.

One thing the Lemba did not do, however, was rush to emigrate to Israel—something they certainly could have tried to do, since Is-rael, in accordance with the Law of Return, grants citizenship to all persons who are born Jewish or have a Jewish parent or grandparent or convert to Judaism. Had the Lemba tried to immigrate en masse, Israel, as a state, would have had to take a position on the relation-ship between DNA information and ethnic or religious identity. In-deed, Israel is quite likely to face this discussion sooner or later, both because it is one of a few countries that grant citizenship on the basis of ethnicity and because Jewish DNA is arguably the best-studied DNA in the world. Plus, there are still at least nine lost tribes out there.

~

The most obvious way to try to match DNA information against history is to try to perform genetic tests on the remains of people who lived many centuries before us. "The problem with ancient DNA is, it is in pieces. It is like reading fragments of words instead of complete sentences," Marina Faerman told me over coffee in a cafeteria at Hadassah, the Jerusalem medical center. She was a slight woman with wispy dirty-blond hair and large gold-rimmed glasses. A physical anthropologist, she had been casting about for something to do to gain a foothold in Israeli science after she immigrated from Russia in 1990. A couple of her newfound colleagues suggested she try to tackle ancient DNA: The work would be tedious, but the pay-off would be publications of notice. She teamed up with several local researchers, including a charismatic senior geneticist-hematologist named Ariella Oppenheim, and began deciphering the DNA of remains found in archaeological digs.

In 1998 they published a paper on determining the sex of infanticide victims found in a dig in Ashkelon. The remains of about a hundred babies had been found beneath a late–Roman Empire bathhouse. Because the babies were newborns, showed no sign of disease, and had been disposed of in what seemed to have been a gutter, and because historical records indicate that infanticide was practiced in that culture, they assumed the babies had been killed soon after birth. They collected their left femurs (this was done to avoid testing the same individual twice) and set out to extract their DNA to see if they had been boys or girls. They should have been girls, for infanticide was known to have been practiced more frequently on girl babies, but most of them turned out to have been boys. Upon consulting with historians and archaeologists, they postulated that the bathhouse may have been a bordello whose staff disposed of their babies but kept some of the girls to rear as courtesans.

That was a neat trick, and Faerman and Oppenheim went on to perform more of these. One of their papers is called "From a Dry Bone to a Genetic Portrait: A Case Study of Sickle Cell Anemia." This time the DNA they were testing was contemporary, but they had even less information than if they had received the tissue sample from a dig. All they knew about the bone fragment was that it came from a person who had had sickle-cell anemia. A genetic test confirmed the diagnosis. A look at the Y-chromosome DNA (the person had a Y chromosome, which made him a male) showed that his father's family came from a Bantu heritage (the Bantu actually comprising about four hundred ethnic groups in Africa), while his mitochondrial DNA placed his maternal ancestors in West Africa—possibly Nigeria. This combination of heritages pointed to the Caribbean as the man's own likely place of origin. Only after creating a DNA sketch of the man whose tiny bone fragment they had tested did the researchers obtain his medical history: He turned out to be a black Jamaican-born male who had died of sickle-cell anemia in the United States. The researchers were so taken with their ability to derive the relevant facts of sex, provenance, and diagnosis from a tiny tissue sample that they published a paper about it.

In fact, they had already performed even more impressive feats: They had used bone fragments to diagnose people who had lived many centuries before. Oppenheim, a hematologist, was particularly interested in blood disorders and had spent many years studying beta-thalassemia, a recessive condition that results in severe anemia and bone deformities. Babies born with beta-thalassemia become weak and lethargic at around the age of three months and, if the disease is allowed to run its course, often die by the age of one. But if patients are given regular blood transfusions—every few weeks starting in infancy and continuing through life—they can have a normal life span. Beta-thalassemia and other thalassemias are the world's most common genetic disease, with more than 100 million carriers alive today and more than 120,000 affected individuals born annually. Because thalassemias are common in areas that have been affected with malaria, the current theory is that carriers—people with only one copy of a gene with a thalassemia-causing mutation—are better-protected against malaria than those who do not carry an abnormal gene. There are very few carriers among Ashkenazi Jews but about 20 percent of Kurdish Jews have the mutation, and some Arab communities in Israel have a carrier rate of more than 10 percent. There are, however, only a couple hundred people with beta-thalassemia living in Israel, apparently because births of affected individuals are prevented through prenatal screening and therapeutic abortions or even earlier, via preimplantation screening: When both people in a couple know they are carriers, they may opt for in vitro fertilization, during which only unaffected embryos will be implanted.

In a dig in Akhziv, in Israel, archaeologists found the remains of a child who likely died of anemia two to five hundred years ago: The child's bone deformities suggested the diagnosis. Ariella Oppenheim's team performed a DNA analysis and confirmed that the child was a homozygote for one of the roughly hundred mutations that lead to the disease. The odd thing was, the child had lived to be about eight years old—usually victims of the disease cannot make it past infancy in the absence of regular transfusions. The DNA analysis (performed separately on three different fragments of the skull, to ensure that the researchers were testing the remains and not some intruding matter) suggested the answer.

Symptoms of beta-thalassemia do not show up in newborns right away because their blood contains fetal hemoglobin, which the mutation does not affect. Sometime in the first year of life, adult hemoglobin, which is actually a different substance, takes the place of fetal hemoglobin. People with beta-thalassemia cannot produce normal adult hemoglobin, which is why they get weak and their bones stop developing normally. But some of them carry another gene variant, which causes their bodies to continue producing fetal hemoglobin. This was why the boy from Akhziv had lived to the age of eight.

Here was another cool genetic study. Not only were the scientists able to diagnose a long-dead person, they were able to use centuries-old remains to confirm a contemporary clinical hypothesis: The fetal-hemoglobin-producing gene variant actually ameliorated the effects of beta-thalassemia. It would have been impossible to confirm this hypothesis experimentally: No person with beta-thalassemia could be allowed to go without transfusions in order to observe the course of his disease.

Oppenheim's team had more fun with beta-thalassemia mutations. In another paper, coauthored with Canadian researchers, they reported the simultaneous discovery of the same rare beta-thalassemia mutation in Ashkenazi families in Jerusalem and Montreal—and the consequent discovery that the families were probably related. When two people share not only a particular mutation but also the variants of genes in its particular neighborhood, this sort of mutation is called "identical by descent" and points to the existence of a shared ancestor. The two thalassemia families traced their roots to the Pale of Settlement, where their ancestors appeared to have lived in towns 250 miles apart about 150 years ago. The genetic quest turned into a genealogical one: The families, and the researchers, were now looking for a single common ancestor (they had not found one by the time they published a paper in the journal *Human Mutation*).

In another study, Oppenheim suggested looking at the many beta-thalassemia mutations found in Israel as a sort of historical map that recorded various migrations to Israel: Here was a mutation that hailed from Kurdistan, there one that came from Turkey, one from Yemen, one from India, and so on. This was the sort of project in which ancient DNA could prove very useful: It could show who came to Israel, when and where from. The tools for analyzing the material were there, but one thing was missing: a genetic picture of the current Israeli population, which would provide a frame of reference.

～

So Oppenheim, Faerman, and their colleagues took DNA samples from various local groups and went to work trying to answer a few simple earth-shattering questions. Faerman listed them for me: "What is the connection among different populations of Jews? What is the connection between Jews and Arabs? And where did they all come from?" Did they have a hypothesis? "What hypothesis could we have had?" bristled Faerman. "There is the Bible, there is the Tanakh."

That hypothesis proved pretty much on target. By comparing DNA samples from Israel with DNA samples from Muslim Kurds now living in Kurdistan, the researchers concluded that the current population of Israel hails from there, where the Tanakh—the Hebrew Bible—indicates Abraham came from. They also found that Jews and Arabs are indisputably related, though, for instance, the Cohen genetic signature is found almost exclusively among Jews. They also found that Ashkenazi Jews, who constitute the Israeli elite, are perhaps a bit more closely related to Kurdish Jews, whose social standing places them at the bottom of the population heap, than they are to the Sephardic majority. This gave Oppenheim particular pleasure. She was one of those very rare Israeli Jews whose personal history in the region goes back more than three centuries, when some of her family moved there. Her grandfather wrote poetry in Arabic, and her other grandfather was a dentist in Arab villages. She had always been privy to the little-known fact that Jews and Arabs are brothers, and she hoped, briefly, that their study would open the eyes of her compatriots.

Oppenheim, a professor emeritus at the Hebrew University in Jerusalem by the time I interviewed her in her very cluttered office at Hadassah (where she, being emeritus, went every day solely for the pleasure her work brought her), was a smart woman. A very smart woman. She had the intellectual capacity to know that scientific evidence and facts in general have no impact on ethnic enmity; it was just that these DNA studies, reaching as they do into the depths of our molecules, have a way of enticing magical thinking. Anyway, she said, "Then I thought more about it, and I realized in many families brothers many times fight. So these genetic relationships are no guarantee of good relations. I also found that the ideologies of people don't change and they just take the facts to fit the ideology. For example, one person said that it's fantastic because now we can convert all the Palestinians to Judaism, sort of prove to them that they belong to the Jewish state."

There was a different sort of sensation hidden in that 2001 study. "We found that between 10 and 15 percent of Ashkenazi Jews, from different samples—we took Ashkenazi Jews from the United States, from Africa, from Israel, from many different locations, and it all was quite consistent—about the same percentage appears to have originated from a gene flow that came from Eastern Europe–Central Asia, somewhere in that region," said Oppenheim. "Gene flow" is just what it sounds like: the flow of genes into a population, from the outside. Gene flow happens, for example, when people begin to intermarry. The researchers traced this genetic signature to the first millennium, to someplace, they thought, in Eastern Europe. Did that mean that the Ashkenazi foremothers, recently settled in Europe, may have mixed with the locals? Possibly. But Oppenheim's group proposed a more intriguing explanation. They thought the pattern they saw in the Y-chromosome DNA samples they analyzed suggested that the incursion into the Ashkenazi population occurred through the introduction of several men who were related to one another. Perhaps, they thought, they were a group who converted to Judaism. Perhaps they were the Khazars.

The *Encyclopedia Judaica* says the Khazars were "a national group of general Turkic type" who lived and battled their neighbors somewhere in the Caspian–Caucasus region between the seventh and tenth centuries and then converted to Judaism. The thing the *Encyclopedia Judaica* does not tell us is whether the Khazars ever actually existed, and hence whether they really had a king and whether the king, along with all his dignitaries, converted to Judaism. That is what legend tells us. And most Jewish population geneticists find this legend unspeakably alluring. This may be because it is like genetic drift: It can be used to explain a great many things. But more likely, this is because it dangles before the population geneticist the ultimate research carrot: the possibility of proving something that has not yet been proved.

~

Karl Skorecki and Doron Behar, whom Skorecki had now drafted as a geneticist, also stumbled upon something unexpected when they tried to replicate Skorecki's Cohanim finding with a study of the Levites. Membership in the Levite caste, too, is passed from father to son, and Skorecki initially expected the lineage to be even more strongly evident—because, he explained to me, becoming a Cohen impostor would have carried more benefits than becoming a Levite impostor. Levites are those male descendants of Jacob's son Levi who are not Cohanim. During some of Jewish history, they enjoyed many of the same benefits and privileges as the Cohanim without some of the restrictions (such as restrictions in marriage, for example). When the researchers started looking at the Y-chromosome DNA of contemporary Levites, they saw that about half those Levites who were Ashkenazi carried a distinctive genetic signature that pointed to a common male ancestor who lived a mere thousand years ago, give or take a thousand years. "We have an event," said Skorecki. "I think DNA has provided a glimpse into a probable historical event, some minor event perhaps, which has left a remnant today, which I don't think could have been uncovered in any other way. Maybe one day someone will find some archival record that matches this, but I think it would be very difficult." For the purposes of genetic research, "an event" is anything that causes changes in the gene pool. A rape can be an event, as can a conversion, a migration, or a pogrom. But what was this particular event? Could it have been the Khazars?

The scientists try to be cautious, but the specter of the Khazars haunts Jewish population genetics. "I cannot argue with somebody who is telling me, Look, we know that the Khazarian king lived in Eastern Europe," Behar tried to explain to me, pointing out that this was not his argument exactly but just an argument he would be hard-pressed to refute. "I mean, I'm telling you what people told me. We know that the royal family converted to Judaism, okay? If I was the king, I wouldn't want to become an ordinary Israelite, I would want to become a Levite or something. So is it possible that our day's Levites are the great-great-great-sons of one Khazarian nobleman or king who converted to Judaism? It's an explanation. We can't do much to approach the genetics, because we don't have anyone who says, 'I'm the descendant of Khazarians' today. It's possible? Yes. Can I prove it? No. Should we be cautious? Yes. We are not coming out with the title, 'The Levites Are Descendants of Khazarians.' We don't think it's appropriate."

In fact, a couple of villages populated with people who call themselves Khazars exist in Lithuania and in the Crimea. They do not, however, practice Judaism, and they have never agreed to be tested genetically. But if they did, and they did not show genetic similarities to the Levites or other Ashkenazim, this would do nothing to prove or disprove the Khazar "event" theories. So the scientists try to keep their Khazar theories to themselves. Only Ariella Oppenheim was either brave or careless enough to put the possible Khazar explanation in print in a scientific paper. Well, she and a group of Harvard doctors who reported, in the mid-1990s, that Jewish and Iranian patients with a ghastly skin disease called pemphigus vulgaris had similar genetic traits.

~

DNA evidence is tricky that way: It seems that, outside the courtroom, it can serve only to corroborate, never to disprove. "I talked to one historian," Oppenheim told me, "who said she was so disappointed because they thought we were going to reveal something completely new that they never thought about, never knew about, and here we were going to prove it by genetics. But we did not, and I feel very good about it. If we found something that was not in recorded history, I would not feel quite good about it, quite uncertain about it." Her team's work showed, in essence, that Jews are Jews and are pretty much who we say we are. Nice, but not a historical sensation.

Doron Behar and Karl Skorecki's 2006 study showing that half of all Ashkenazim are the descendants of four women was neither a historical nor a scientific sensation. We all know—or we claim to know—that all humans have a single common mother. We also know that any two people in the world have common ancestry—a forefather and a foremother whom they share. That is the science. History is also fairly straightforward: Ashkenazi Jews were a small group that settled in the Rhine Valley around the seventh or eighth century. The small group grew to around eight and a half million at the turn of the twentieth century and more than ten million before World War II, in which about six million were killed, and doubled again by the turn of the twenty-first century, reaching about eight million people. With this kind of dramatic expansion—and with what we already knew about Jews, by and large, being Jews, it should not have surprised anyone that so many Ashkenazim could be shown to have come from so few.

The scientists took eleven thousand blood samples and used them to isolate genetic signatures passed from mother to daughter to daughter (the mitochondrial DNA, unlike the Y chromosome, is passed on to offspring of both sexes, but a son's children will inherit his wife's mitochondrial DNA, so this effectively is a woman-to-woman thing). They compared the DNA of Jewish women to samples from what geneticists call the "host population"—the titular nations of the European countries in which Ashkenazi Jews have lived. It turned out that roughly three and a half million people, or 40 percent of Ashkenazi Jews now living in Europe, Israel, and North America, were the descendants of four women who had lived one thousand to fifteen hundred years ago, give or take a thousand.

"I had discussions with my coauthors about where we were going to send this paper," Karl Skorecki told me. "I was saying, 'This is parochial, this will not be of great interest to the larger scientific community, not so much to the larger community.' They were saying, 'This is a finding which is absolutely stunning!' I was saying, 'It's an interesting finding, but I don't think it's startling or stunning.'" Skorecki had used the same word—*parochial*—when he described the question posed by his Cohanim paper, which remains one of the most noticed genetic studies of our time. And he was wrong this time too. The study, published in the *American Journal of Human Genetics,* was one of those very rare scientific papers that make it to ABC News, MSNBC, and *USA Today.*

Why? "It's not an abstract concept," suggested Skorecki. "It's not a theoretical woman: These were real living, breathing women."

That was an obvious exaggeration. They had not lived or breathed in a thousand years, give or take as much. But Skorecki had a point. Knowledge of how many women there were, and when, gave scientists minimal grounds for theorizing who the women were. "It sounds like such a family had to have some sort of economic advantage," said Behar. Previous studies attempting to link mitochondrial DNA to physical attributes that would have given the women a selective advantage—health, beauty, and extraordinary fertility would be three such traits—had failed. So the edge had to be socioeconomic. Perhaps our great-great-great...grandmothers were rich and this was why their daughters survived to marry well and have lots of daughters of their own (in which case the genetic-drift theory and the selective-advantage theory are, in fact, casually reconciled). Perhaps, like the Orthodox Jews of today, they followed a tradition of having many, many children. "It must have been something in the beginning," said Behar. The beginning, he theorized, would have been enough: After a few (three, six, maybe ten) generations, the lineages would have gained enough of a foothold in the population to continue to propagate.

So, many of us had a wealthy, nurturing great-great-great... grandmother. All she left us was her DNA signature. After the study came out, a *New York Times* reporter sent away for a DNA cheek-swab test kit from a commercial company (Skorecki's original coauthor, Michael Hammer, and Behar both work with the company) and published a breathless article titled "Love You, K2a2a, Whoever You Are." K2a2a was one of the four genetic signatures the researchers identified.

There is a thin line between theoretical knowledge and not-so-theoretical knowledge. In 2001 the Oxford geneticist Bryan Sykes published a bestselling book called *The Seven Daughters of Eve,* claiming that all modern Europeans are descendant from seven women (other researchers later commented it may have been more like a dozen). Sykes also decided to name the alleged seven daughters, or mothers. The one whose signature began with K he chose to call Katrine. Thanks to Behar and Skorecki's research, she had actually acquired enough distinguishable features for us to be able to say Katrine was an unlikely name. A Hanna or a Sara she may have been, or a Rakhel. Perhaps that is the magic trick of studies like these: Somehow, they make it all more specific, so the abstract foremother of all yields to the specific four mothers of us.

The question is why we are so taken with this kind of specificity. Why do small numbers—four women fifteen hundred years ago, as opposed to a dozen women tens of thousands of years ago—hold such sway? "I don't know what is it in us humans that makes us want to learn about similarities," Doron Behar said, trying to answer my question why, being a medical doctor and the father of three small children, he chose to pursue a second, challenging, wildly time-consuming, and often tedious career as a geneticist. "I don't know what makes us dig in the ground to look for the ancient—I don't know. There is no easy answer for this, because I can't really say that if you will dig in the ground and find a cave then you will get this, this, and this benefit." Humans are driven to look for continuity. On some level, it may be part of our eternal search for immortality. No one feels fully mortal as long as his mother is alive. No one dies fully as long as his children go on living. And perhaps if our great-great-great...grandmother's DNA is maintained in our flesh and blood, that makes her, suddenly, not quite so dead, and us not quite so doomed.

~

"I know that genealogy is the second most popular hobby in the United States," mentioned Behar. This is one of those frequently repeated factoids that seem to bear no relationship to any statistics, polls, or other objective measures (I could find no such citation, in any case). Behar may be forgiven for his unfounded assertion. He was desperately looking for a way to explain the obvious to me, that people have always found this sort of thing interesting, and they always will. Americans are by no means unique in their yearning to explore their roots: Icelanders, for example, make it a national pastime, often maintaining meticulous family records going back many centuries. What may distinguish Americans from the world's other potential genealogists, however, is their frequently woeful ignorance of their roots, which goes hand in hand with the potential for surprises to be found along the family tree. Americans constitute the perfect market for genetic ancestry tests, which became commercially available in the early 2000s. Within the first three or four years of their existence, these companies drew a couple hundred thousand clients.

One of the pioneers in this business was a company called DNAPrint, one of a slew of biotech start-ups that began big around 2000 and soon fell on hard times. What saved DNAPrint was that its researchers stumbled upon the idea, and the technology, for performing genetic ancestry tests. DNAPrint is in Sarasota, Florida, a city of low houses huddled close to the road, mailboxes specifying "back" and "front" units, of pickup trucks and grazing cows—many, many cows grazing everywhere—and otherwise a conspicuous absence of human activity outside the islands that are the city's tourist centers. At first glance, the nonfarm economy is limited to restaurants and proud white resort hotels. The building that houses DNAPrint is a modest low office structure along a wide dusty road, with nothing to the left and nothing to the right of the parking lot. But I saw a Jaguar in that parking lot as I drove up. Biotech money, I thought.

"I don't know," chief research scientist Matt Thomas told me. "It certainly does not belong to anyone who works here." In the first quarter of 2006 DNAPrint Genomics would post losses of nearly a million dollars. At one point during our conversation Matt called the company "Tony's DNA shop in Florida"—Tony being Anthony Frudakis, a molecular biologist who had found some venture money in Florida and moved there to start a company in 1999. The idea was to locate genes that influence or determine the way people respond to various medications, which would have made the company invaluable to the health-care system and to humankind in general. In 2000—the year DNAPrint went public—by the Food and Drug Administration's estimate, 2.4 million patients had adverse reactions to medication and more than 125,000 people had died as a result, making prescription-drug reaction a leading cause of death in the United States. Common chemotherapy drugs can lead to heart failure; anesthesia can kill unexpectedly. Florida, with its aging, relatively affluent, and consequently much-medicated population, seemed like a good place to do this sort of research.

Frudakis and his team decided to begin their pharmacogenetic research with eye color. The idea was to conduct what is called a proof-of-principle study: Eye color, they thought, was like drug response in that it was a trait affected by more than one factor, but all of these factors were genetic. After he explained this to me, Matt Thomas said something that sounded almost impossibly obvious: "And what we quickly saw with the eye color stuff is, we had a number of samples from different populations around the world. And some of them—genetic locations and markers we were looking at, we saw strong segregation among the populations. So basically a particular marker, an allele or a version of the marker, was very prevalent among, let's say, Europeans, and less so among Africans, or vice versa." The scientists started talking with an associate professor at Penn State University, who was just getting into molecular anthropology, which is essentially measuring the differences between populations at the molecular level. "And this kind of opened up some new avenues for us, some new interests of 'maybe the best way to understand drug response and the eye color project is to understand the world's populations better.' And that led to some of the developments in our ancestry testing. The idea behind the ancestry testing is, we have another way of stratifying a population."

In other words, by looking at gene sequences on their computer screens, Frudakis's team arrived at the same conclusion as many physicians across the United States who had started asking patients about their ethnic backgrounds. The physicians did this because, for example, an Ashkenazi Jew who had lost a mother to premenopausal breast cancer—someone like me—had become a clear candidate for genetic testing. The Florida scientists thought that maybe figuring out that a particular drug response was more characteristic of one ethnic group or another would ease the search for genes that determined the reaction. In both cases, this is what Matt Thomas called "stratifying a population." If that sounds perfectly obvious, it was not: It would be another three years before the African American Heart Failure Trial made headlines across the United States. The result of that trial, which was halted early because of excellent preliminary results, was medication certified for use in African Americans at risk for heart failure.

The African American Heart Failure Trial enrolled over a thousand people who self-identified as African Americans. The only term less specific might be "human." People of African descent come from the most varied gene pool on the planet: This is what makes geneticists think the human race originated on that continent. In African Americans, that gene pool was altered further by the admixture with the "host population," which generally includes people from everywhere else. What is most striking is that, despite the imprecision of this particular race classification, statistics on the effect of, say, heart disease show a striking difference between African Americans and Caucasian Americans.

Presumably, the different effect of both heart disease and medication on people of different races is the work of a set of gene variations that, when identified, will eliminate the need to classify patients by race. That will happen, no doubt, but it will take a while. In the meantime, researchers could at least attempt to break people up into smaller groups and try to answer such questions as whether people of North African extraction should be classified as African Americans for the purposes of these studies and these treatments. Someone from Morocco, Tunisia, or Libya would probably look like a Caucasian to most Americans and might, if living in the United States, self-identify as anything at all. But how should such a patient be treated by medical doctors? If DNAPrint could tell a drug manufacturer that much, it would make itself indispensable.

Somewhere along this thought process—probably when DNAPrint ran out of cash—someone had the idea of using the results of this research to create a test, commercially offered to people curious about their ethnic background. DNAPrint teamed up with Mark Shriver, the molecular anthropologist at Penn State, who was starting to use gene sequencing to try to develop molecular measurements of the things anthropologists measure: things that make us who we are. I visited Shriver's lab and talked to a group of young female graduate students who showed me how they did their research based on the material at hand: several thousand eighteen-to-twenty-year-olds who were willing to have themselves studied for ethnic markers. The technology was apparently simple: Shriver's lab had collected DNA samples from various parts of the world. "Parts of the world" were fairly broadly defined: Sub-Saharan Africa, for example, was represented by Nigeria and the Congo, while East Asians were a combination of Han Chinese and Japanese, a few Koreans and a few Vietnamese. They compared the samples, isolating bits of gene sequence that seemed to differentiate one group from others, and designated these parts of the genome markers of a particular population. Then they started testing people, giving them results that would inform them that they were, say, 80 percent European, 10 percent African, 9 percent East Asian, and 1 percent Native American. The 1 percent would most likely be what researchers call "noise in the test," meaning either a mistake or the result of imprecise classification, while the percentages of African and East Asian might well surprise the subject.

Amy Bigham, the senior graduate student in the lab, showed me around. DNA-testing equipment tends to fall into two categories: things that look like printers and things that look like toasters. There were also some unusual pieces of machinery, like a three-dimensional camera, a contraption of many lenses and lightbulbs on movable metal arms that allowed the researchers to photograph a subject's head from all sides at once. The digital image could then be used to take all sorts of measurements of facial and other features. Amy cheerfully told me that in addition to taking subjects' DNA samples and personal histories—their families' ethnic backgrounds and their self-identification—they also used to take measurements of their skulls. Then she showed me poster boards hung all along the hallway. Subjects' pictures were mounted on the boards, next to sheets with their self-reported data. Flipping up the sheets exposed the subjects' test results, which were presented in the form of a pyramid, with the bulk of the subject's ethnic makeup at the bottom and what might be statistical noise at the top. The boards posted in the hallway featured subjects whose results could be curious: someone who looked white but was largely of African descent, or someone who turned out to carry markers from most of the world. It all struck me as borderline distasteful, like the mention of the skull measurements. An image from a Nazi propaganda film showing the cranial measurements of a Jew immediately flashed through my mind.

"You stopped that because it was kind of creepy?" I asked.

"No." Amy smiled, clearly confused. "Because our craniometer broke."

DNAPrint drafted Shriver and used his samples and some of its own to create a test called AncestrybyDNA. In the first couple of years of its commercial availability, about twenty thousand people paid two hundred dollars apiece to learn which ethnic groups were represented in their genome. People who tested as Europeans could pay roughly the same amount again for EuroDNA, which would tell them in which Indo-European region their DNA likely originated. Another upgrade was offered for those who wanted to explore their partial Native American–ness.

DNAPrint was probably the first of several companies that started offering DNA-ancestry tests at around the same time (though this was apparently the only publicly traded company among them). Prices ranged up to nine hundred dollars, the price of the "Male Genealogy Package" from a company called Genelex. The difference in their approaches was that most of the companies used Y-chromosome and mitochondrial DNA tests to determine haplotype while DNAPrint used markers on autosomal chromosomes. Both approaches had proponents who argued that their way was more accurate or informative. Some companies had an apparent area of specialization: Family Tree DNA focused on Jews, offering a test for the Cohen marker, among others; DNA Tribes concentrated on Native Americans. DNAPrint also seemed to attract a disproportionate number of people seeking to confirm a family legend of having a Native American ancestor. Many were disappointed, even enraged, by a negative result, a reaction Tony Frudakis diagnosed as American Indian Princess Syndrome.

Aside from informing people that they did not have a Cherokee (for some reason, Matt told me, it always seemed to be Cherokee) grandmother, DNAPrint's technology proved useful in forensics. If the scientists had indeed been able to find genes responsible for eye color or other features, crime-scene DNA samples could be used to create portrait descriptions of suspects. As it was, DNAPrint could offer only very rough suppositions about appearance, based on the suspect's apparent ethnic roots—but even rough suppositions could prove very useful when the police knew nothing.

The Baton Rouge Killer used a different murder method every time, but he had a signature: removing the phone from the victim's belongings. That made him a serial killer, and all American police officers are taught that serial killers are white. The task force formed to catch the killer was looking for a white man, and as part of its search it sent samples of his DNA—tissue left at the crime scene following apparent struggles with the victims—to DNAPrint. The AncestrybyDNA test said the killer was of African or African American provenance. The task force changed the focus of its search and, in 2003, captured Derrick Todd Lee, who was black. He was later convicted of five murders, based largely on DNA evidence.

In another case, in a rural community in California, the fairly decayed body of a female murder victim was found. The forensic pathologist judged the remains "probably Vietnamese," which suggested to the investigators she might have been a mail-order bride. DNAPrint's test, however, showed she was Native American. That, in turn, changed the way the pathologists reconstructed the woman's skull to obtain a picture of what she might have looked like. That murder still had not been solved, though, at the time I visited Sarasota.

~

Matt Thomas was thirty-five, stocky, with light hair and a round face of the sort that always looks boyish. He looked like the kind of person whose genetic ancestry test could yield no surprises.

"I'm 99 percent European and 1 percent African," he confirmed. "The 1 percent African is probably noise in the test."

European: That is one broad category. "Being Dutch and Finnish are pretty different things," I said, choosing my examples deliberately. Both populations are North European, but the Finns are a closed population that is quite genetically distinct.

"Right," said Matt. "But part of that Dutch and Finnish definition is a social construct."

This was silliness. Here, wearing a white lab coat, sat a biochemist turned geneticist who made his living by telling people what their genes told him about their identity—and he was telling me that ethnic, national, and geographic identities were a social construct. In another minute he went even further, telling me that closed groups based on ethnoreligious identity—such as Ashkenazi Jews—were a social construct. And in another few minutes I learned why.

"Let me say this," said Matt. "My mother converted to Judaism. I practice. But there is no way I'm going to have any genetic markers. So it goes back to the definition of groups. And because you do not share genetic heritage, does that make you any less a member of that group?"

Most people convert to Judaism because they are planning to marry a Jewish person. Matt's mother converted for a different reason: She liked Judaism best of all the religions around. Matt's father was Catholic, and Matt and his brother were raised with both faiths and allowed to choose once they became adults. Matt chose Judaism. He later married a Lutheran, and now, if they had children, they planned to give them their choice of religions, as his own parents had done. The only problem was, right before their eyes—and, really, with Matt's active participation—Jewishness was reverting to being an ethnic rather than a religious category.

Five years earlier, when I adopted my son, Vova, my partner and I devoted all of maybe an hour to discussing whether to perform a conversion ceremony for him. He was officially declared a Jew by a court of three Reform rabbis gathered around a *mikva*—a ritual pool—in Providence, Rhode Island, in October 2001. Now, if he was asked who he was—as he would be, for his gorgeous swarthy looks were a relative rarity in Russia, where we lived—he would tell people he was Jewish, I thought. I had no idea that it was a matter of months before the first companies would offer him the opportunity of asking *them* who he was, for a small fee. At this point these tests were too imprecise for our purposes—he would probably have found out only that he was of mostly European and perhaps partly Middle Eastern descent—but by the time he becomes an adult he will likely be able to find out just what ethnic groups his biological parents represented.

~

"So I can tell you that your mother was Jewish, or of Jewish ancestry," Bennett Greenspan told me when I called him for comment on my DNA. Bennett Greenspan was president and chief executive officer of Family Tree DNA, a company that grew out of his lifelong interest in genealogy. Greenspan, who was of Ashkenazi extraction, started the company in 1999, drafting Mike Hammer and Doron Behar, among others, to serve on his scientific advisory board. Now the company offered a variety of ethnicity tests, including specific mitochondrial DNA and Y-chromosome DNA tests for people who believed themselves to be Ashkenazi Jews. I did, so I asked my brother Keith, who, unlike me, possessed a Y chromosome, to do a cheek swab and send it to Family Tree DNA.

A few weeks later, I could log on to a page that said, "Keith Gessen, Kit Number 65289," and featured a little square with two hands, palms out, thumbs touching, and the words "Cohen match." That is right: My brother had the Cohen haplotype. "Do you have an oral tradition of that?" asked Greenspan.

"No," I answered honestly. "But we are Soviet-raised, so we don't have much of a Jewish oral tradition of anything." This was especially true of our—that is, my brother's—Y chromosome. The earliest known source of that, my great-great-grandfather Ilya Gessen, was drafted into the czar's army in the mid-nineteenth century.

"Oh, I know!" exclaimed Greenspan, exhibiting a knowledge of history characteristic of true genealogy enthusiasts. "The twenty-five-year death sentence." The standard term of service in the czar's army was indeed twenty-five years, and for Jewish males, it officially kicked in at the age of twelve. Many were drafted even earlier, some as young as eight, and served even longer. Few survived. But those who did, and were ultimately released from service, received the right to live outside the Pale of Settlement. This was a reasonable and per-haps necessary measure: Over the decades of service, the soldiers had lost touch with their families, their culture, their language, and, usu-ally, their religion.

My great-great-grandfather was one of the very, very few who not only survived but also managed to resist conversion to Chris-tianity. He settled in a village in the south of Russia, married a Jew-ish woman, and had nine children. It seems he did not remember much about the Jewish tradition—just that he was Jewish. Was he a Cohen? If he was, was this something he remembered? No such in-formation reached my generation. At the same time, he named his firstborn Aaron.

In 1899 Aaron finished high school and took off for St. Peters-burg, where he would go to university. This was a period of strict quotas and state-mandated anti-Semitic discrimination in Russia, and it was the year of the Dreyfus affair. My great-grandfather would never have been able to hide his Jewishness, but he did what he could to make it stand out less: He changed his name to Arnold, thereby perhaps obliterating the last remnant of the Cohen oral tradition in my family—until, that is, my brother was tested.

Strictly speaking, the fact that Keith was a match for the haplo-type did not prove that we came from a Cohen lineage. First, this was the philosophy behind Skorecki's original study: The DNA testing was there to illustrate the oral tradition, not to be used to prove or disprove anything. And there was another consideration. "Remember," said Greenspan, "Aaron and Moses's first cousins would have left Egypt alongside them—I mean, you've got these famous first cousins. Their sons would have had the same Y-chromosome markers but would not have been Cohanim."

Greenspan talked about Moses and Aaron, and especially their cousins, as though they lived practically next door, or at least yesterday. He was only slightly more distant when talking about my foremother: She turned out to be one of the four. K1a1b1a was her name. Family Tree DNA had 181 of my presumed relatives on file: A number of their e-mail addresses were supplied along with my brother's test results.

~

The Family Tree results reached too far back to tell me anything useful—or anything at all, really, about the origins of my mutation. That information came while I was in Israel doing research for this chapter. I sat drinking tea with my cousin Natasha and an elderly friend of the family. Conversation turned to one of my maternal great-grandmothers, Bat-Sheva, an extraordinary woman by all accounts. She had been a rebellious shtetl girl who forced her way into a cheder—a religious school that never admitted girls. When she was still young, but already married and no longer studying, her hobby was translating Pushkin into Hebrew. I remembered her well: She died in her late eighties, when I was twenty years old.

The family friend asked about Bat-Sheva's parents. I did not remember any family conversation that had reached so far back.

"Her mother's surname was Magidina," said Natasha. "She married a much older man and had several children, Bat-Sheva being the oldest. When Bat-Sheva was pregnant with her first child, her mother was already very ill and she knew it. But she traveled to where

Bat-Sheva was living, put in place everything for the baby, then went home, had surgery, and died either during the operation or soon after."

"What did she die of?" I asked.

"Some sort of cancer."

Here was the story, then. My great-great-grandmother had a mutation that led to cancer that killed her when she was still in her thirties. She had passed the mutation on to her daughter Bat-Sheva. As my great-great-grandmother lay dying, Bat-Sheva had a son who also inherited the mutation. The son, my grandfather, passed it on to my mother, who died at forty-nine. She passed it on to me. And there was no way to tell what it was that allowed Bat-Sheva to live into her late eighties without developing cancer, and whether, whatever it was, I had it, too.

Chapter 3

THE POST-NAZI ERA

At a breakfast table at a Viennese hotel, I was telling my companions about this book.

"But you cannot test for race!" one of them objected, horrified. She was German.

True, I said, unless the person tested wants to interpret the test that way. And many do, because today's racial characteristics are still the most useful shorthand for the information genetic testing gives us about ourselves. The third person at the table, an Ashkenazi Jewish woman from the United States, pointed out that Ashkenazi Jews carry certain mutations for which genetic testing is available and useful.

The first woman cringed. "You are constructing biological identity out of anxiety!" she shouted. "Have you even read Foucault?"

We had read Foucault, but we both felt, at the risk of sounding old-fashioned, that the facts were too bad for Foucault. The German woman tried to withdraw from the conversation altogether, noting resignedly that she was a postmodernist.

"This is the post-Nazi era," the American Jewish woman quipped, eliciting a joyous laugh from me and a frightened one from our German colleague.

But here we were. The German woman felt she had no right to move into the post-Nazi era. The other woman and I, both being Jewish, could use the privilege of historical victimhood to make that decision for ourselves. There were two of us and only one of her, and we could laugh and tell her that this was the reason Israel would lead the world in the science and application of medical genetics.

We were in Vienna because each of us had at one point received a journalism fellowship from the Institut für die Wissenschaften vom Menschen—literally, "the institute for the study of man," located here in Vienna, where man had historically been studied to excess. This was the geographic heart of Europe and the home of the twentieth century: the city where Hitler spent his formative years; home of Sigmund Freud and birthplace of psychoanalysis; home of Theodor Herzl, the founder of Zionism; and, not coincidentally, a site of one of that century's creepiest experiments in the application of genetics.

After breakfast I took the tram to an outlying neighborhood, where I searched the plain 1960s apartment buildings for the home of an old man who had lived to tell the tale. He had lived long enough, that is, and he had also made a life out of telling his story.

～

Johann Gross was short and portly, dressed in black trousers and a white undershirt, and he could not breathe. An oxygen tank nearly as big as the man himself sat puffing in the middle of his tiny apartment, and a white plastic cord stretching from it delivered oxygen directly to his nose. His mouth was occupied alternately by a foul, raw-smelling cigarette and a salad-green plastic inhaler. Wheezing constantly and halting occasionally, he told me his story. He had lived in the United States for four years some forty years earlier, and he could tell his well-honed story in English, so I came without an interpreter. It seemed, though, that Johann Gross could not understand anything I said. So his speech was a monologue.

When Johann Gross was eight years old, he had been living with a foster family for four or five years. He had no memory of his biological family: His mother had abandoned them when he was a year old, his father was a drunk, and little Johann lived in an orphanage before he was placed with his foster parents. As he remembered now, it was a good foster family: He was fed, schooled, and perhaps even loved there. But then Austria was annexed by Nazi Germany, and this meant that Johann Gross had to go home to a father he did not remember. His father was a very poor man: He was born missing an arm and had never really worked. State care for the children of the poor and the infirm contradicted some basic tenets of Nazi public health policies. Most important, using state funds to raise someone like Johann Gross meant engaging in what eugenicists called "counterselection": increasing the chances for survival, success, and procreation for the progeny of those less fit—and, in Gross's case, apparently genetically damaged.

"We had not much to eat. I slept in a little bed with my sister. I was hungry all the time." Johann Gross was inducted into Hitler-Jugend, which obligated him to spend his days going door-to-door with a tin cup, collecting funds for the Nazi cause. After about six months, eight-year-old Johann Gross ran away, tin cup in hand, back to his foster home. His foster mother called the child welfare service, which sent not a social worker but a Nazi Party officer, who expelled Johann from the children's organization on the spot and assigned him to an orphanage. After three or four escape attempts from the orphanage, the little boy was deemed "asocial" and placed in a psychiatric institution, on a children's ward called Spiegelgrund, or "mirror reason." The old man opened his book for me: He had published a memoir called *Spiegelgrund*. It was illustrated with his own drawings. He pointed at one of a wheelbarrow loaded with children's naked bodies, a mess of arms and legs with a little girl's body lying awkwardly on top. "I saw this," he said.

Most of the children at Spiegelgrund suffered from psychiatric conditions or mental retardation. From 1940 to 1945 they were studied, then starved or medicated to death and studied some more in the process. After they died, their brains were excised and placed in formaldehyde jars for further study.

Johann Gross watched the murders of the sick children from a short distance. He himself was considered problematic but not sick, so he was not killed. "I had a normal body. I think what saved my life was I was always very good in school," said this old man, who had had perhaps four years total of formal education. "They were not so quick to kill. When people were psychopathic, they were killed." Johann was not sufficiently disturbed to warrant a researcher's interest, so his brain was allowed to remain, functioning, in his body. He was, however, medicated systematically, excessively, and painfully after each of his ten escape attempts. "They gave me injections on my arms and legs," he remembered, visibly cringing sixty-five years later. "The injections on my arms—I would be on the toilet for twenty-four hours. The injections on my legs, I would be walking on the floor on my hands, I couldn't get up, and then it hurt for three weeks, and then I would run away again." While most of the children at Spiegelgrund received injections as part of a study, the treatment of Johann Gross may have been simple punishment: pure torture.

"The man from whom I got the most injections was a man with my name, this Heinrich Gross," Johann Gross told me. Heinrich Gross, a psychiatrist, administered what the Nazis called "euthanasia" to the disabled inmates of Spiegelgrund, and preserved their brains, which he kept studying after the war. He continued publishing on the Spiegelgrund brains into the 1970s.

Heinrich Gross was tried and convicted by a Viennese court in the late 1940s, and sentenced to two years in prison for his part in killing the children at Spiegelgrund. The Austrian Supreme Court threw out the verdict on a technicality, though, leaving Heinrich Gross free to embark on a second career as a neurological psychiatrist and court expert. In 1968 he became head of a research institute in Vienna. In 1976 he was called on to testify in the case of a man who had been interned at Spiegelgrund, like Johann Gross, as a "hardly reformable" youth. Heinrich Gross quoted from the man's Nazi file in court—and the man confronted his former doctor. The media exposed Heinrich Gross as a Nazi criminal, but it took another twenty years for the Viennese prosecutor to bring a criminal suit against him. The court ruled that Gross was demented and unfit to stand trial. "They didn't do him because they didn't want to do him," Johann Gross told me.

The brains of Spiegelgrund were finally buried in the year 2002.

I took a bus to the scenic outskirts of Vienna to see the hospital named for Otto Wagner, the Austrian Art Nouveau architect, another of the city's great twentieth-century figures. I entered and climbed up a very steep path in the middle of the vast grounds, punctuated with all manner of signs and arrows, including red plywood temporary-looking signs that pointed to something having to do with "NS Medizin." The warm autumn sun forced me to squint. The place seemed full of young people, some apparently medical students. Two teenage boys ran down the gravel path toward me, screaming, unable to stop.

I entered Building V, near the top of the hill, through a lovely sunporch. The room where the Spiegelgrund museum was located was Otto Wagner–stylish: all the whitest of white, sun-drenched, with double-height ceilings and a checkered tile floor. A wispy-haired young man, not much older than the teenagers I had just seen, sat at a little desk in the corner. As it turned out, he knew next to nothing about the exhibit he was overseeing: He was just performing his alternative civilian service to avoid being drafted into the army. This was a shoestring operation that had been recently mounted by a group of enthusiasts, led by a city councillor who failed to get reelected.

A circular semienclosed wall formed a room within a room. Inside, photographs of children killed at Spiegelgrund were tacked to the wall, captioned with first names, surname initials, and ages, from two to seventeen. Most looked obviously mentally impaired. Some looked terrified. One, Friedel F., looked exactly like my daughter: long thin eyebrows, huge eyes, ears that stuck out, and a bad haircut. Friedel F. had been nine when she died.

Displays lined the rest of the room, aiming to educate the casual visitor about Spiegelgrund and Nazi medicine.

"In the late nineteenth century a new discipline developed in Great Britain somewhere between anthropology, medicine, and biology, for which its founder Francis Galton coined the term 'eugenics,'" explained a printout pinned to one of the stands. "While 'valuable' individuals were to be promoted ('positive eugenics'), carriers of an allegedly inferior genotype should be systematically excluded from reproduction ('negative eugenics'), which would lead to a genetic improvement of humankind. The idea met with public approval in many European countries and North America." The rest of the page listed Nazi eugenic efforts—forced sterilization, "euthanasia," and the fight against intermarriage—while attempting to walk the middle line between two traditional interpretations of the relationship between the science founded by Francis Galton and Nazi eugenic policies. One tradition interprets the National Socialists as enemies of science who perverted the spirit of eugenics. The other tradition, implicitly accepted by all who treat *eugenics* as a suspect term, has it that Hitler's Germany was a land where science prospered, and in this fertile environment Galton's ideas yielded the only fruit they possibly could: crime against humanity.

～

Some of the definitions of racial hygiene, a concept that predates Hitler's rise to power by several decades, sound conspicuously like possible definitions of contemporary medical genetics. One of the founders of the field, Hugo Ribbert, wrote that a goal of racial hygiene was "the prevention and conquest of diseases afflicting the entire human race, diseases from which each of the various races might suffer in similar manner." His colleagues objected to the contamination of the value-free science of racial hygiene with "vulgar race propaganda." These concerns notwithstanding, racial hygiene, or eugenics, lent itself extraordinarily well to use by the Nazis, who claimed that "National Socialism [is] the political expression of our biological knowledge." At the same time, eugenics was perceived as the cutting edge of science not only in Germany but in the wealthy countries of Europe and, perhaps most of all, in the United States. In the period between the two world wars, Germany was far from unique in adopting laws based on eugenic ideas. In 1924 President Calvin Coolidge signed into law a bill that restricted the inflow of immigrants to the United States in accordance with then-dominant eugenic ideas. Coolidge himself had made his views known earlier. "America must be kept American," he had said. "Biological laws show...that Nordics deteriorate when mixed with other races."

In Nazi Germany and annexed Austria, the practical application of eugenics proceeded in rapid stages. First, all forms of social deviance and difference, from Jewishness to Marxism and from homosexuality to criminality, were framed in biomedical terms. (Even the concentration camps and, some years later, the liquidations of the ghettos in Poland, were framed in terms of "quarantine.") Second, public health policies changed to accommodate this understanding. Medical care for "the weak" endangered the race because it interfered with natural selection by allowing the unfit to survive. Public health had to concern itself with the good of the race as a whole, not just the individual: This was the view of Alfred Ploetz, the man who coined the term *racial hygiene*.

The first of the racial-hygiene legislative acts was passed by the Nazi government in July 1933. It was called the Law for the Prevention of Genetically Diseased Offspring, which mandated the forcible sterilization of individuals whom a genetic health court found to suffer from any of a number of diseases then believed to be genetic, including numerous psychiatric conditions, hereditary epilepsy, Huntington's chorea, genetic blindness or deafness, and severe alcoholism. (By this time, similar sterilization laws were on the books in twenty-eight American states and one Canadian province.)

Unlike sterilization, Hitler's euthanasia operations were mandated not by public law but by the Führer's secret directives, illegal even in Nazi Germany. The body appointed to oversee the child-murder operation, which began in 1939, was called the Committee for the Scientific Treatment of Severe, Genetically Determined Illness. This was a cover name, but a telling one. The doctors who served on the committee, as well as those who carried out its directives, believed that the proper scientific treatment for severely disabled children was death—a mercy death, in their opinion, for their lives were not worth living. The committee did not issue orders; doctors simply received expanded powers, which included the right to commit murder, and acted in accordance with their beliefs. Four years later, the "euthanasia" program expanded to include healthy children of undesirable races, such as the Jews. The adult "euthanasia" program, initiated shortly after the child-murder campaign, claimed the lives of roughly four hundred thousand psychiatric patients.

After the annexation of Austria, Vienna became the second-largest city of the Third Reich. Its officials raced to catch up with the implementation of racial-hygiene policies. The ground was fertile enough: As elsewhere in Europe, eugenic ideas were popular. Vienna had a city office of marriage counseling, created to stem the tide of anti-Darwinian selection. Attendance, however, was voluntary, and apparently modest. The new regime acted quickly to streamline Vienna's health-care system, adding new offices and entire new fields, such as "hereditary and racial health care." Public health offices were charged with inventorying the city's population, consolidating records culled from psychiatric hospitals, the police, and the office of youth welfare, to create a list of the genetically suspect. All those defined as inferior would be denied welfare benefits and health care. Public health officers were to pay particular attention to an individual's potential productivity. This was why Johann Gross, the son of a disabled man, lost his place at the orphanage but ultimately was allowed to live: He was right to tell me that it was his good school grades that saved him.

~

The easiest way to think of medicine under the Nazis is to believe that science was abused, and so, in some way or another, were the doctors who applied it. The problem is, that does not seem to be true. "One could well argue that the Nazis were not, properly speaking, abusing the results of science but rather were merely putting into practice what doctors and scientists had themselves already initiated," writes Robert Proctor, a scholar of Nazi medicine. "Nazi racial science in this sense was not an abuse of eugenics but rather an attempt to bring to practical fruition trends already implicit in the structure of this branch of science."

The line of thinking launched by Francis Galton in the nineteenth century produced the desire to improve the human race, and particular races among humans. As knowledge accumulated, so did the drive to act on it, leading to the policies of National Socialism. That is why my German colleague cringed as I described my research. That is also why the very word *eugenics* sounds, if not exactly obscene, then at least accusatory. The problem with this view, though, is that it really does make it impossible to mount any credible defense of the contemporary science of medical genetics.

Where does one draw the line between Nazi eugenics and contemporary genetics? For one thing, many of the racial hygienists went on to second careers as human geneticists, drawing on the experience gained while serving the Nazi regime—just as Heinrich Gross continued his study of disabled children's brains even after he was no longer allowed to kill them for his research. The hardest place to try to mark a division between the Nazis and us is at the science itself. Many of the enduring obsessions of racial hygienists have either remained germane research topics for geneticists or have been revisited in the last few years. These include the studies of twins to estimate the genetic components of everything; the heritability of behavior, such as a propensity for violence, crime, or alcoholism; and Jewish intelligence. Even the once-ridiculed claim that Jews have a higher incidence of certain genetic diseases has now yet again been affirmed by scientists.

What makes matters even more complicated is that some of the Nazi public health measures live on: For example, the restructuring of the Viennese health-care system, designed to prioritize public health over individual care, is still in effect. Indeed, it is difficult to object to the basic argument advanced by Alfred Ploetz, who claimed that on a policy level public health trumps individual medical care. In 1936 Ploetz was nominated for the Nobel Peace Prize for his public health work.

Some researchers have focused on the economic arguments inherent in Nazi public health policies, which calculated the value—and care-worthiness—of human life according to the person's potential productivity. But economic arguments are extremely important in contemporary public health policy as well. Professional journals regularly publish articles arguing, for example, that widespread testing for certain conditions is economically viable because care for full-blown disease is very expensive. In the mainstream media, discussions on, say, the cost and economic defensibility of life support for hopelessly brain-damaged patients are commonplace.

A term that used to sound so damning as to provide a potential demarcation line was "lives not worth living." But in recent years, American legal practice has virtually rehabilitated the concept by allowing so-called wrongful-birth lawsuits, in which plaintiffs argue that they, or their offspring, were born as the result of medical errors or negligence—for example, because a disability was not diagnosed prenatally. The basic premise of such lawsuits is, certainly, that some lives are not worth living. The conditions at the heart of "wrongful-birth" suits—such as severe mental disability, for example—tend to be the same conditions that the Nazis believed rendered lives unworthy.

Of course, the easiest place to draw a line is at forced medical interventions. Nazi "euthanasia" was actually murder. Nazi sterilizations were involuntary. A basic tenet of contemporary medical genetics, to a degree even greater than in Western medicine in general, is that any tests or resulting care must be chosen by the patient voluntarily. But starting in the early twentieth century, a number of European countries and more than half of all American states enacted forcible sterilization laws. The language of the German law was, in fact, quite scrupulous: It allowed sterilization only in cases where genetically defective offspring was a likelihood and criminalized the sterilization of heterozygous carriers. It seems obvious now that forcible sterilization, no matter how finely defined, is indefensible. But the question remains: Were such laws the first step down a preordained slippery slope, which Germany simply slid down much faster than other countries would have, or was that catastrophic path unique to Germany?

I wandered the city of Vienna, thinking about this. As a historian, I was less than a dilettante. But I had chosen to write a book on a topic over which a giant historical shadow is cast. I had no other option now than to try to draw my own dividing lines.

Of the industrialized powers of the twentieth century that could lay claim to serious scientific achievement, the first one conspicuously to reject eugenics was the Soviet Union. Indeed, the Soviet Union banished genetics altogether. That did not serve to prevent the institution of racist policies, including wholesale deportations of entire ethnic groups—it was just that the Soviets, instead of claiming that members of the group were inherently inferior, ascribed to the groups behaviors, such as cooperating with the Nazis, and then laid collective blame. Nor did it prevent nationalism: Even several years after the collapse of the Soviet Union, in the early twenty-first century, more than half of all Russians owned up to supporting "Russia for the Russians." Indeed, the only thing the rejection of genetics succeeded in doing was holding back Russian medical science: Russia did produce a crop of ambitious young geneticists once the ban was lifted, but in the absence of any institutional support, most of them emigrated. (In fact, about a hundred of them settled in Chicago, where they were doing most pioneering work on improving the human race—see chapter 12.) It serves as a good reminder that even in the twentieth century the rifts and crimes of nationalism existed independently of the science of either genetics or eugenics, perhaps because they resulted from differences that are real. The advances in genetics have served to remind us of these divides, which American culture in various ways and with varying success has tried to obliterate, especially when it came to the Jews, subsuming them into the general Caucasian population and deeming them a religious rather than ethnic or racial group. This time genetics came with a warning label, like cigarettes.

Cigarettes, doctors will reluctantly tell you, are dangerous only when consumed in large doses: Very light smokers can have the pleasure without the risk of lung cancer. But only a very few people (see chapter 11) are inured to the temptations of cigarette addiction and can go on smoking five or fewer cigarettes a day for years. I was one of those smokers: two or three a day, for nineteen years. Genetic research, like most knowledge, can be like a drug: It leaves most of us wanting more. To separate the knowledge effectively from collective behavior, we must take it in small doses: apply it to individuals and not groups. Once it is used to generalize and create population-wide policies, the risks skyrocket.

My own task now was to apply my new genetic knowledge to myself.

PART 2

The Present

Chapter 4

INDECISION

The technician asked me if I was claustrophobic. I am not. I climbed onto the table and lay facedown, lowering one of my breasts into a specially made hole, and the technician disappeared behind a glass door, from there to pull me into the MRI tube and begin. I lay very still, as instructed. There was a series of loud, rapid-fire bursts, like shots from a machine gun. Then a whistling sound—sniper fire. Then something in the machine churned heavily, signaling the approach of a new series of shots, the way a low-cruising plane announces an imminent bombing with its engine noise.

I had spent years working as a war correspondent. I had lived this. You lie still as you can, usually surrounded by people you cannot see or feel, and you wait, listening. You name every noise. It is not frightening exactly: There is no adrenaline rush like you get when you run from a threat or stare it down. There is an eerie calm, and it is all in the waiting.

When I went in for an MRI of my other breast a week or so later, I told the technician the machine reminded me of a bomb shelter.

"Oh my God," she exclaimed. "We had an older woman in here who lived through World War II somewhere in Europe—she said the same thing!"

Then the technician seemed to breathe a sigh of relief. "Oh, you've been to wars," she said. "Then this is nothing! You'll be fine."

Right. What I have always loved about war reporting—what, I think, generally gets people hooked on that absurd line on work—is that the danger is so clear-cut. You either catch a bullet or you do not. Most journalists, in fact, do not. Then you go home, and you are safe.

The life of a genetic mutant could not be more different. First, the nature of the risk is fuzzy. The mechanisms by which the vast majority of known deleterious mutations work are unclear to scientists. Mutations are most often identified by comparing epidemiological data with the results of genetic testing. In a 2002 study, for example, researchers set out to find out what role genetic mutations played in the rates of breast cancer in Pakistan—the highest in Asia (with the exception of Jews in Israel)—and ovarian cancer, one of the highest in the world. The study looked at 661 women: 341 had breast cancer, 120 had ovarian cancer, and 200 made up the healthy control group. Forty-two of the women with cancer turned out to have one of six mutations in either the BRCA1 or the BRCA2 genes (most of the mutations seem unique to Pakistan). The researchers concluded that genetic mutations were an important factor in the high rates of cancer, and, in addition, that consanguineous marriages (marriages between relatives) were associated with some of the cancers.

But the key word here is *associated.* Most of the knowledge that drives the decisions of mutants like me concerns associations, not causes. What caused the cancer in the other 419 women in the study, the overwhelming majority? And why would consanguineous marriages be important? That would suggest that a recessive gene—not a dominant one like BRCA1 or BRCA2—was implicated. Could it be that there were other causes, other mutations? Could it be that even the known BRCA mutations required a co-agent, possibly another mutation or a combination of mutations? Could a causal relationship be proven at all?

The study was conceivably a huge step forward for some Pakistani women, who may now be able to take advantage of genetic testing, but the field is so young that every step forward serves best to illuminate the vastness of the unknown.

Many of the advances in genetics have become possible as a result of sheer computing power available to scientists today. Our outposts at the genetic frontier are giant boxes of data—hangar upon hangar filled with machinery that works around the clock to slice and recombine. Most of the studies look like the Pakistani one: A wealth of epidemiological information on one side is matched up with sheaths of genetic code on the other, with little regard for the missing logical links.

By the logic of this science, a human being would best be sliced up into microscopic samples, each one to be checked for flaws and irregularities. In essence, the state-of-the-art approach to genetic-cancer prevention is based on the same carpet-bombing philosophy: A mutant like me is generally advised to undergo two mammograms a year, two MRIs (actually two procedures each time, owing to the number of breasts on the human body), get felt up by a breast oncologist four times a year, and get as many ultrasounds as are necessary to clarify the results of any of these procedures. Each of my own mammograms occasioned follow-ups and enlargements, all of which were largely inconclusive. My first MRI showed a mass in my left breast. I had an ultrasound-guided biopsy. I waited for the results for two weeks, in accordance with the usual rules of such things: calm at first but reaching a near-desperate level of helpless anxiety late on a Friday afternoon, when I still had not heard from the oncologist. Then she called: The mass was benign. I sat on a bench, hugging my cell phone, looking up at the warm late-April sun, taking deep breaths.

Almost exactly five years earlier I had sat on another bench, under a different sun. I was in Budapest, having just arrived there from Belgrade after spending six weeks in Yugoslavia during the NATO bombing campaign. I remember looking up at the blue sky, taking in the warmth, breathing like I imagine a diver must breathe coming up from a great depth. I was remembering what safety was like. It was disorienting and exhilarating.

Now, in Cambridge, Massachusetts, I felt none of that. I had been given a reprieve. But it was strictly temporary: I was now a professional patient. I would always be ill until proven healthy, and then I would have to prove it all over again in another month or two.

In the much-discussed and much-amended lists of breast-cancer risk factors there is a fascinating indicator: Previous breast biopsies increase the risk that a woman will be diagnosed with breast cancer. Is this because biopsies, by invading the tissue, increase the risk of cancer? Is it because any atypical mass, even if it is benign, means something has gone wrong? Or is it because women who are at risk are more likely to have biopsies? Any of these? All? No one knows. In my case, it did not matter: It was clear that with every passing year and every test, my risk would get greater. For my doctor, a sweet woman around my age, with two small kids and an Ashkenazi Jewish background—a woman very much like me—it was a foregone conclusion. When I suggested she stop treating me like a cancer patient because I had no cancer yet, she responded: "You have no *detectable* cancer."

And then there was the question of the ovaries. Eleven years before my mother died of breast cancer, her beloved aunt had died of ovarian cancer. It was apparently linked to the same mutation—my mutation, which is associated with a vastly increased risk of ovarian cancer.

But to say "vastly increased" is to say almost nothing. Ovarian cancer is a rare disease. I had met gynecologists who had never seen a woman with ovarian cancer.

One of my newfound mutant friends referred to her "down doctor" and her "up doctor," the physicians who helped her manage the risk of each cancer. I, too, came to feel like I had something like a bipolar disorder. In some important ways these two kinds of cancer are the opposite of each other. Breast cancer in young women is difficult to treat, while ovarian cancer is considered highly treatable. At the same time, while death rates from breast cancer have been dropping, death rates from ovarian cancer have stayed mortifyingly high: 71 percent will die within five years of diagnosis. This is because ovarian cancer is almost never found early enough to be treated effectively—while early detection of breast cancer is becoming increasingly common. Doctors, I found, tend to be terrified of ovarian cancer: Those who have seen it know that it makes them feel helpless.

Ovarian cancer is like breast cancer's poor neglected cousin. Breast cancer is everywhere. Actresses get it, singers get it, famous writers get it, and in August 2005 mobster daughter and television personality Victoria Gotti was accused of pretending to get it. In America it has been so glamorized and lionized that it sometimes seems like a rite of passage. Breast cancer is a modern disease— primarily because it is imagined as a disease that can be overcome. Ovarian cancer is a cancer like Susan Sontag described cancer: intractable, unimaginable, unspeakable.

At a lovely dinner in New York City one summer evening people told me about a heartbreaking book by a woman who, they said, wrote her own death from breast cancer. One of my companions, an editor at the publishing house that had reissued the book, sent me a copy. It turned out to be *The Furies*, by Janet Hobhouse, who died in 1991, at the age of forty-three, of ovarian cancer. Both of my companions had substituted the cancer that has come to feel familiar enough to be less than completely frightening.

In the book, Hobhouse describes discovering her diagnosis and being told of her treatment options: "I was told about surgery and chemotherapy and it all sounded like a course of beauty treatments and everyone was quite jolly about it. And yet I had seen what I had seen on my way up to the doctor's rooms and I knew I didn't want to be in a cancer hospital. It was a place that would take you out of your garden-party clothes, hide your lipstick and turn you into a gray, rumpled bedding. Once they had you, they took your colors away, put you in a world like early TV, black-and-white, reduced, fuzzy imagery on a tiny screen. I didn't know what all the merriment had been about in that office; it felt like a trick, even my part in it, like Pleasure Island in *Pinocchio* before everyone grows ears and gets shipped off for the slaughter."

This was very much what I felt like. I had been energetically welcomed to a club that used the confident language of science and progress to usher me down a corridor that led, I knew, to a lonely, unglamorous, and even rather untechnological death from cancer. I did, I was told, have the option of getting out, or at least of sticking one foot permanently out the door, by getting all the potentially offending parts cut off before they went bad. Then my breasts and my ovaries would be sliced up and examined in every microscopic detail, and only then, if they were found clean, would I be allowed to rejoin the world of the living, to fear plane crashes or stray bullets or nothing at all.

In the two or three months after my test results I lived through the MRIs and the biopsy, tried to get warm under a sun that did not comfort me, and spent hours upon hours staring blankly at the computer screen, unable to read the studies and articles that I knew should inform my decision. The data was not sinking in. I was not in denial, and I was not trying to escape from the responsibility I had laid upon myself. Fragmented thoughts of the mutation, cancer, and my surgical options spun incessantly around in my brain, but I was spinning my wheels. The decision was not coming; in fact, my mind seemed incapable of generating a single thought whenever I tuned it to the mutation frequency. Finally one night I had the idea of writing a series of articles about trying to make my decision. I pitched it to the online magazine *Slate*, where the editors liked the idea, and, positioned gloriously outside myself, if only for a few weeks, I set about trying to understand how people can make decisions in the age of medical genetics.

Chapter 5

A DECISION AT
ANY COST

I was getting my first lessons in using the medical system. Some of the experience would be familiar to me—and to most Americans—from, say, dealing with public school systems or with cell phone service providers. It was marked by that inimitable sense of smashing your head against a glass wall that arbitrarily separates you from a clear and obviously beneficial goal. What makes the experience of trying to procure medical help special, of course, is that you are trying to manage not a piece of technical equipment or even your child's essential education—areas where missteps can generally be remedied—but your own singular and mostly irreplaceable body. Over the next couple of years I would live through several moments that overwhelmed my imagination with their absurdity. There was the time my insurance company refused to cover one of my breast MRIs but paid for the other—because only one of them revealed a lump. There was the time I called to get a necessary medical document, wormed my way through the labyrinth of extensions and keywords to talk to a real live person, who, upon hearing my request, transferred me, quite purposefully, back to the voice mail system. But in December 2003, after my first iffy mammogram, I was just getting started. If I wanted to get the genetic test, my doctor explained, I could circumvent the long line by signing up to see not the famous head of the testing program but one of her fellows. If I tested positive, my doctor continued, I could jump the line and go directly to the head herself.

Judy Garber, director of the Cancer Risk and Prevention Program at the Dana-Farber Cancer Institute in Boston, was a kindly woman in her midforties. She wanted to know whether I was "finished" with my childbearing. I was thirty-seven. I had a six-year-old and a two-year-old who was still nursing. For most of the preceding two years my partner and I had been out of sync on the subject of another baby: When I agitated for one, Svenya argued in favor of waiting, and vice versa. But women in my family had had children well into their forties—in fact, my mother had had a miscarriage not long before her diagnosis of breast cancer—so I felt we could safely stay on this seesaw for a while.

But now Dr. Garber and her associate Katherine Schneider, a former president of the National Society of Genetic Counselors, were demanding an answer, in their soft-spoken way. I balked. Svenya sat next to me on the couch, stone-faced, her hands stuffed into the pockets of her coat. It so happened that the night before, we had cooked and hosted a dinner for sixty people. We were both exhausted and hungover. Svenya had spilled coffee on herself in the car on the way over. Both of us felt like frivolous intruders in this world of grave and clear-cut decisions.

This session was called posttest counseling. But it did not feel like counseling as I had imagined or experienced it: a slightly fuzzy conversation flowing warmly toward a conclusion that begins to feel obvious and, therefore, comfortable. Rather, it felt like an exam I was quite possibly failing.

I later understood why. For the past three-quarters of a century the key word in genetic counseling has been *nondirective*. The philosophy of dispensing information without guidance dates back to the first scientists who began providing pregnant women with an assessment of possible genetic risk to their future babies. Decades before DNA was discovered, they drew on basic knowledge of the few diseases that were known to follow Mendelian laws (Huntington's chorea and phenylketonuria were among the first such diseases; both are debilitating neurologic disorders, but in the case of the former symptoms do not usually set in until adulthood, while symptoms of the latter begin soon after infancy). Eager to distance themselves from the eugenics movement, which had dominated American genetics in the first quarter of the twentieth century, these early counselors withdrew from making value judgments about a woman's decision to terminate or continue a pregnancy. Surely it helped that the first genetic counselors were academics—Ph.D.s more often than MDs—who could establish the sort of distance with the possible mutant in their care that a family doctor could never have managed. In addition, they had no definitive tests at their disposal, so their suppositions about potential risk to an unborn baby were just that: inferences based on a woman's and/or her husband's family history. Finally, they were counseling pregnant women in an era when abortion was performed at the discretion of medical practitioners, and while this discretion seems to have been wide enough to accommodate the fledgling field of medical genetics, neither the counselor nor the counselee was actually in a position to make the abortion decision.

As time went on, things shifted, but the result, weirdly, remained the same: Abortion was legalized, but genetic counselors, who by this point were no longer men with Ph.D.s in science but women with master's degrees in social work (the first specialized program for genetic counseling was created at Sarah Lawrence College in 1969), chose to underscore their neutrality with regard to abortion—now because it had become paramount to acknowledge the woman's autonomy in making the decision.

The term adopted by genetic counselors—nondirective counseling—was actually coined by the psychologist Carl Rogers in the 1940s. Rogers himself later changed the term to *client-centered therapy*, and these days his technique is known simply as Rogerian therapy. The crux of the approach is the belief that a therapist best serves as facilitator rather than guide, helping the client to unleash his own potential for positive change largely by reflecting his reality back to him. Rogers's key belief was that the client, not the counselor, should set the eventual goal of the therapy. The underlying assumption, however, remained: A person went into therapy seeking change. In the field of genetic counseling, on the other hand, change was often impossible.

The first predictive genetic test for adults became available in the mid-1980s. This was the test for Huntington's disease (which used to be known as Huntington's chorea). The test was somewhat unreliable—at that point it identified markers rather than the gene itself—and terrifying for the potential test subjects, since there was (and is) no treatment for the debilitating and ultimately fatal condition. The field of genetic testing did not really take off until the mid-1990s, when predictive testing for familial cancer syndromes—most notably for the BRCA mutations, but also for mutations that lead to colorectal and other cancers—became available. The rhetoric of genetic counseling remained largely unchanged, despite some counselors' doubts: *Nondirective* remained the key word. And when it came to familial breast cancer, it was reinforced by the circumstance of largely liberal educated urban women being counseled by other liberal educated urban women, whose lingua franca was the rhetoric of a woman's right to choose what to do with her body, even though they were now talking about breasts and ovaries, not pregnancy.

But unlike prenatal genetic counselors, the genetic counselors at high-risk-cancer centers have a goal. The goal is cancer prevention: not to extend a woman's life or improve its quality but, quite singularly, to prevent her death from cancer. And it is impossible for a medical professional to conceive of different approaches to cancer prevention as being equally valuable—the way a Rogerian counselor might imagine different paths to self-improvement. Genetic counselors know perfectly well how to prevent cancer. In their universe, the idea of pretending not to know the "right" answer becomes as elusive and misguided as the ideal of objectivity in journalism. In fact, it leads to the same results: Just as a journalist pretending to be objective merely hides his bias behind what passes for neutral language (but is really just stilted), all the while trying to steer the reader to the conclusion he believes, so the genetic counselor creates the illusion of neutrality by avoiding engaged conversation but still tries to get the counselee to do what is best for cancer prevention. So it was that the two nice women cocked their heads to the side and pushed me to decide whether I was having any more babies.

I stalled. The women moved inexorably toward their conclusion. Once I made up my mind, they said, and either had a baby or did not, I should have my ovaries removed. I balked. In the weeks between having my blood drawn and hearing the results, I had researched early surgical menopause. Its unpleasant effects, I had found out, included increased risk of heart disease, high blood pressure, osteoporosis, cognitive problems, and depression—as well as inelastic skin and weight gain, which in this context seemed downright frivolous to mention.

I mentioned the big ones. Judy Garber said softly, "The payoff is keeping you here." This was certainly not nondirective. It was also quite possibly not true. I had looked at many studies, most of which indicated that an oophorectomy (surgical removal of the ovaries) was indeed effective in preventing ovarian cancer and lowering the risk of breast cancer. But I had also found a single study that provided some statistical analysis and concluded that a prophylactic oophorectomy would extend the average life expectancy of a mutant like me by between 0.03 and 1.7 years. The authors of the study, who included one of the women sitting in front of me, concluded that the increase was significant. I understood that the numbers looked so small because they took into account the roughly 50 percent of women who would not have gone on to develop ovarian cancer in any case. I also understood that looking at statistics when attempting to make a decision about my lone life was not particularly useful—but still it looked like an absurdly small gain in exchange for drastically lowering my quality of life and giving up the chance of more babies. The payoff, it seemed, was not so much keeping me here as increasing the chances that I would die of something other than cancer—whenever that happened. I politely suggested I could just shoot myself tomorrow: That would prevent my death from cancer with a 100 percent probability. The joke remained suspended in the thin air between us and the counselors, and with it, our disengagement from one another was complete.

Before our awkward conversation ended, Dr. Garber did mention the possibility of a preventive mastectomy. She said it quickly, as though in passing, but I tried to grab this conversational straw by making an immediate objection. "I'm still using them to feed my daughter," I said, smiling. The counselors seemed to take this comment as a categorical refusal to discuss the subject, and moved on, leaving me confused. After all, that same study had credited prophylactic mastectomies with increasing life expectancy by as much as 5.3 years.

Svenya and I walked silently to the parking garage, where I flashed my distinctive small blue Dana-Farber card: I was a cancer-center patient now, and this qualified me for free parking. The air smelled of February cold and spilled coffee. I felt nauseated and lonely, as one does on a gray day after a party.

～

What I did over the following couple of months was, I imagine, normal under the circumstances. Sometimes I was obsessed with the subject of cancer and my decision. Sometimes I forgot all about it. Sometimes I lay down with my daughter on a futon on the floor in her room and, as she nursed and drifted off to sleep, I thought that sacrificing physical parts of myself and even my youthfulness was a small price to pay for continuing this happiness. "If you are anything like me," my physician had said, "you are looking at your kids and thinking you just want to be around for their college graduation." I had had much bigger plans, but I was learning to think small.

I had the MRIs. I got the phone call from my doctor telling me there was a lump. I waited for my biopsy date. When it came, I spent a day walking around Cambridge with a chemical icepack under my arm; it ended up leaking and damaging a good silk shirt. I waited for my results. I got used to the idea that I had cancer. I imagined surgery, chemotherapy, and radiation. I considered having an affair to give my body a proper send-off. Then none of it happened. I had to return to trying to make my impossible choice: live and wait for the cancer to come or start carving up a body that felt utterly healthy.

We were in Cambridge that year because I had a fellowship at Harvard University. The Nieman program is designed to give so-called midcareer journalists a chance to spend a year at the university in the pursuit of intellectual enrichment. I had spent nearly twenty years writing articles, and I had recently had a baby, so I viewed the year as a chance to jump-start a brain that had effectively ground to a halt. It worked: Everyone around me was actively, energetically, and even expressively thinking, and very soon I felt the wheels start to churn inside my head. Now that I felt myself paralyzed by an impossible choice, I decided to try to harness some of the best brainpower in the world to make my decision for me. It seemed clear enough that the genetic counselors had catastrophic tunnel vision. It seemed I was not up to the task. What would the economists say?

Economics, as a field, had been one of my discoveries as a remedial college student. Like most journalists, I had at times been compelled to write about financial matters, and, like most journalists, I labored under both the fear that my ignorance would be discovered and the misapprehension that economics was the boring and confusing science of money mass and aggregate values. As it turns out— handily—the dismal science is in fact a study of the way people make decisions. Also handily for me, economics was at that time undergoing a revolution. The 2002 Nobel laureate in economics, the psychologist Daniel Kahneman, and his coauthor Amos Tversky had shown that for nearly three centuries economics had operated under the blatantly incorrect assumption that people made decisions rationally and in their own best interest. Rather, they argued, people relied on intuition and immediate perceptions, distorted by built-in bias—and only sometimes, if they were intelligent and aware of their mistakes, could they correct their decisions rationally. Kahneman's Nobel lecture was devoted to the concept of "bounded rationality."

Naturally, just over a year later the Harvard course called Psychology and Economics was oversubscribed. I was auditing it, struggling with the calculus but reveling in reading about beautifully designed experiments. The two instructors talked about things like lotteries, the Milgram obedience experiment, and an experiment run on eBay, the online auction site, showing that people were more likely to bid—and to bid faster—on a lower-priced CD with a higher shipping and handling fee than a higher-priced CD with a lower fee, even though the total amount of money spent would be the same. Most of the studies they examined exposed some of the mechanisms through which people regularly make decisions that are demonstrably wrong. It was the best kind of course. Twice a week, it made you feel smarter than other people, and, because it concerned itself with things like lotteries, purchases, and fateful decisions, it carried the promise of helping you become a winner. So now I went to see one of the instructors. It seemed like it would be a very good idea to sit down with him and do some math.

David Laibson was sitting, slightly hunched, in front of his computer monitor, scrolling down a spreadsheet and talking on the phone. He was saying things like, "This one is a point-four chance" and "She is no more than a point-five chance." It turned out he was scrolling down a list of prospective graduate students, each of whom had a rating quantifying the chances of his or her accepting an offer from Harvard. I found this extraordinarily appealing. I felt that if David and I could find a way to frame my own predicament in similar terms—a 0.4 chance of cancer weighed against, say, a 0.5 chance of salvation—my own rationality would know no bounds.

I quickly explained that I was working on an article about the way people go about making decisions based on genetic information—using the breast cancer genes as an example. David immediately confounded my expectations by doing what anyone would have done: try to get out of answering the question. He told me about two New York University economists who have demonstrated that people do not really want to know what will happen to them, especially when it is not clear what to do with the knowledge. I countered with two other studies, specifically concerning the BRCA mutations, which showed that women do want to know. "They just say they want to know," David responded. So I told him, instantly eliciting the cocked-head look. It was an awkward, even ridiculous moment. I had met David as a fellow auditing his class. I had come to his office as a journalist interviewing an expert. Now these seemed like false pretenses: I was a woman feeling very lost, asking a man I barely knew what I should do with my life and my health.

Later I read the studies David had mentioned. Two New York University economists, Andrew Caplin and John Leahy, looked at the usefulness of information given to patients by doctors: in essence, whether the patients benefited from knowing more or less. The ultimate answers were intuitively predictable: It depends on the patient, and on the news. Some people prefer to know more, while others like to know less: A much earlier study classified them as "monitors" and "blunters," concluding that the former benefited from information about an upcoming stressful medical procedure while the latter suffered from it. Along the way, the economists concluded that doctors are not very good at handling bad news, or perhaps even information in general, and made the titillating suggestion that things like test results may better be conveyed by a machine.

A third economist, Botond Köszegi at the University of California at Berkeley, reclassified "monitors" as "information lovers" but made an important new observation: Regardless of whether patients are inclined to seek information or avoid it, they prefer the same sorts of tests—ones that, in the economist's lingo, have "an upside potential surprise more than a downside potential surprise." It may sound obvious that good news is preferable to bad news, but it was actually an important insight into the motivation of people who go for medical tests. Köszegi pointed out that one is highly unlikely simply to go "check for cancer" but will seek a second opinion if diagnosed with cancer. In the first instance, there would be either no news or bad news; in the second, there was always the hope, however slim, for a pleasant surprise. So it made sense that women like me—those who had watched their mothers die of cancer—would opt for genetic testing. Not so deep down, we all believe we will develop it, so we go in hoping for a negative test result, which, in our minds, would be akin to winning the lottery. I remember being almost elated when, before my test, the genetic counselor pointed out that my chances of having the mutation were fifty-fifty—compared to near certainty, these had seemed like terrific odds.

I also reviewed the studies with which I had countered David's argument. I had unintentionally overstated my case. One of the studies, conducted in Israel, looked at people's attitudes toward being tested for Huntington's disease or for a nameless disease very much like it: rare, incurable, debilitating, diagnosed by genetic testing with absolute certainty (everyone who has the mutation will go on to develop the symptoms), and generally developing in midlife. About half the subjects said they would want to know they had the mutation. In the second stage of the study, people were asked whether they would want to know if they carried a mutation correlated with a disease very much like breast cancer: common, treatable, and sometimes curable, less than certainly diagnosed by genetic testing (survey subjects were given theoretical odds of 60 percent), and also generally diagnosed in midlife. The respondents did not appear to care that this disease was far more common or even that in this case the presence of the mutation was an imperfect predictor; all that mattered to them was that this hypothetical disease could be cured. Now the contingent of those who would want to know grew to 80–93 percent.

The other study, conducted in Boston, was a telephone survey of two hundred Jewish women, who were asked whether they would want to be tested for one of the BRCA mutations. The results were less dramatic: Only 40 percent said they would want to know—as many as said they would not. This seemed to make sense, though: The Israeli study was a little more recent, and Israel has applied genetic testing more widely, so it had penetrated the culture. Still, even 40 percent is a very high rate.

Did that mean the economists were wrong? Not necessarily. In fact, probably not. The Israeli paper also reviewed older studies of willingness to be tested for the Huntington's gene. Before the gene was actually discovered, about half of those who had Huntington's in their families said they would want to be tested. Once the test became available, only about 15 percent of potential subjects chose to take it. It is much harder to calculate how many of the potential carriers of a BRCA mutation are choosing to be tested: There are too many variables to be able to estimate the pool of subjects. But when I went to see one of the authors of the Boston study a couple of days later, she answered my question: "We don't have women stomping down the door to get tested."

Finally, I even found a study that justified my own choice. A group of researchers at Georgetown Medical Center asked relatives of carriers of the BRCA mutations whether they wanted to know their own status. A majority agreed—but among the minority who did not, depression frequently took hold. The findings were so striking that the researchers called their paper "What You Don't Know Can Hurt You." Remarkably, the New York University economists had used this exact study to bolster their point that some people really do not want to know. I got the point that even if they did not, they perhaps should.

But then, I am an information lover. Now, though, sitting across from the young economics professor and waiting for him to tell me what to do, I wondered whether I'd been too passionate, even rash, in my pursuit of knowledge.

He sat back down at his computer and opened a new Excel file. In the left column he entered ages, year by year, starting at thirty-seven (which happened to be his age as well as mine). Across the top, he placed the options: "Oophorectomy," "Mastectomy," "Oophorectomy and Mastectomy," and "Do Nothing." Now we had to devise the formulas for figuring out the value of life under each possible decision for each possible year. "Let's normalize not being alive as having a utility of 0," suggested David, reasonably enough. "Further, let's normalize a year of healthy life to have a utility of 100. Should we assume that life without breasts is equally good?" That was unexpected, but it sounded right. Sure, I said, provided the surgery went well.

Now what would be the value of life with cancer? I could recognize that question as a classic problem: The value would get higher as probability got lower. The longer a woman lives after the initial diagnosis, the more likely it is that her life will once again approach normalcy. David asked if we should simply equate a diagnosis of cancer with death, which had a value of 0. Now this was a clear cop-out: If we did that, the correct choice would instantly become obvious—do everything in order not to die—but we would intentionally have ignored all complexity. So then David decided to calculate the values more precisely—that is, to perform the act of alchemy for which I hoped: to express life in numbers.

He switched from his computer screen to the huge dry-erase board opposite his desk. After a couple of false starts, he wrote:

$$V(G) = 100 + ß\,[(.98)V(G)+(.02)V(illness)]$$

Translated, that said, the value of a year of life in what we consider a "good state"—apparent health with none of the possible disadvantages of surgery—is equal to 100 (this is the utility we had assigned to a year of normal healthy life) corrected by "the discount factor" B, which consists of the normal mortality risk for young people, which is about a quarter of 1 percent, and the 2 percent annual risk of getting breast cancer. In brackets on the one side is the 98 percent likelihood of a good life and on the other the 2 percent likelihood of a diagnosis of cancer.

Now we had to decide what the value of life with cancer would be. David asked me some of the sorts of lottery questions I had concluded he most enjoyed: Would you take this gamble if, say, your chances of developing cancer were 80 percent? 60 percent? 40 percent? Any percent? He wrote my answers on the board and, stepping ever so lightly over some logical steps, we came up with an answer that was perhaps random but also reasonable: The utility of life with cancer, upon first diagnosis, would be 70. This seemed generous—certainly more generous than our original attempt to pose a diagnosis of cancer as equal in value to death—but also somber enough: You do not die right away, but life is never quite the same either.

$$V(illness) = 70 + ß\,[(.8)V(illness)+(.2)X0] = \text{(skipping the algebra here)} = 347$$

What that means is that while life—actual life—with cancer might have a utility flow of 70 utils per year, which is not all that far from normal healthy life, the likelihood of death makes the value of one's entire life with cancer just 347. Plugging this value into the first equation, we find out that life in the "good state" has a value of 4,752, even allowing for the 2 percent annual risk of breast cancer.

I already knew that if I got a mastectomy, the utility of one year of life would be essentially the same. The value of my entire life, however, would actually increase because the risk of breast cancer would be cut by 90 percent. Still, after the age of forty the risk of ovarian cancer would kick in, also to the tune of about 2 percent a year, bringing the value of my now breastless life right back down to 1,500. What if I got an oophorectomy? Sure, that would cut my ovarian cancer risk by 95 percent. But it might make me instantly old, susceptible to heart disease, high blood pressure, osteoporosis, and so forth. I suggested David look at my collection of all the age-specific risk statistics of these diseases. David's solution was far more elegant: He just changed our "discount factor" to reflect the mortality risk of an older person—someone over sixty-five. The assumption here was that the mortality risk is a good enough expression of all the disadvantages of life after menopause. That mortality risk is about 1.5 percent a year, rather than a quarter of a percent, as it is for young people. The utility of life in a state we termed "Good/old" turned out to be 95 (as compared to 100 for good/young). And the value of my entire life?

$$V(G/old) = 95 + \text{ß}'[V(G/old)] = 9,500$$

Huh? Did that mean it was better to be old and cancer-risk-free than young and at risk?

"Hey," said David, "being old isn't so bad. I can't do many of the things I could do when I was twenty, and I don't miss them."

I thought about this. Physically I did not feel that different from when I was twenty. True, to achieve this state of well-being I had to drink a lot less, smoke not at all, and work out a lot, but these did not count as costs. (On the other hand, when I was twenty, I could actually follow this kind of math with pleasure.)

"You know what?" continued David. "Having a 1 percent chance of dying from being old just doesn't compare to having a 2 percent chance of getting cancer." Actually, it seemed to me that it did: It was only half as bad.

"It seems the doctors are right," David concluded. Surgery was the solution. We stared at each other for a minute; this was an outcome neither of us expected. I had been almost sure we would prove that the absurd-sounding recommendation of a preventive oophorectomy was wrong. The intimacy of the interview grew uncomfortable again.

"This is probably what your brother would tell you," David suddenly said. "If your brother were an economist."

It was already dark outside, and pouring rain, when we left the building. My babysitter had stayed longer than planned. David had had two guilty phone conversations with his wife. He opened his umbrella and ran for his car, and I jumped on my bicycle and sped home, making currents in the puddles, getting soaked, feeling strong and a little silly and generally like my life had a utility of 100 a year, possibly even more, now that I also felt that much more competent for being able to put a number on the value of riding in the rain.

~

Nancy Etcoff studies the important things in life. She wrote a book called *Survival of the Prettiest: The Science of Beauty,* and when I met her she was working on a book on happiness. She was also teaching at Harvard Medical School and at the college itself, where I was auditing her class on the psychology of happiness. It was a wonderful class, one that made me very happy—not a surprising result of an hour of talking about nothing but what makes people happy. In my state, the distraction of discussing happiness with a roomful of very bright twenty-year-olds was nearly a lifesaver.

In her book Nancy wrote that breasts are one of the things that distinguish humans from other animals: We are the only mammals who, starting at puberty, have rounded breasts all the time, whether or not we are breast-feeding. We are also the only species that views breasts as sexual. American culture, she argues, is as obsessed with breast size as it is with penis size.

No wonder the genetic counselors mentioned ovarian surgery so much more easily than the preventive-mastectomy option. Plus, having one's breasts removed is a much more difficult surgery than an oophorectomy. The ovaries are small and invisible from the outside, and the surgery can be performed as a laparoscopy—a tube is lowered into the abdomen through a tiny incision and the ovaries are snipped and pulled out through the tube, all in under an hour, leaving only a miniature scar. In another light, choosing whether to lose breasts or ovaries was choosing between the visible expression of femaleness and its invisible essence. "The visible always trumps the invisible," said Nancy—meaning, the visible is harder to sacrifice.

I asked her what makes a woman—or, rather, a beautiful woman. Breasts, she said. Also, the shape of the torso (narrow waist, wider hips), skin tone (women's skin is less ruddy than men's), lips (fuller than men's), hair (no male-pattern baldness). In other words, it is the ovaries—or more specifically, the estrogen they produce. Without her ovaries, a woman will often gain weight, especially around the midsection (there goes the torso), her skin will lose elasticity, her lips will lose their fullness and color, and her hair will thin (though not necessarily the same way a man's hair does). Hair and skin, Nancy writes in her book, are essential to our idea of beauty and attractiveness: They are "polymorphically arousing and primal in their appeal." There follows a breathtaking sentence: "Within each inch of skin are sweat glands, oil glands, hairs, blood vessels and nerve endings through which we shiver, shudder, sweat, blush and quiver." An organ like that could not be jeopardized lightly.

Another sexual attribute that goes out with the hormones is the libido. It occurred to me that I was facing a choice between feeling desire and being desired. I would need my ovaries for the former and my breasts for the latter. I said I would rather be desirous: I figured I could always find one person to desire me—my actual partner, with any luck—but that would bring me nothing but frustration if I could not feel desire myself. Nancy seemed skeptical of this idea: She said it could be devastating to feel that there was no one in the world who wanted you. And worse, to see the reason in the mirror every time you took off your shirt.

In her book on beauty Nancy writes a lot about what makes a woman attractive to a man. As its title suggests, the book is an attempt to apply Darwinian interpretations to today's realities—an extraordinarily convincing attempt. In pondering why female medical students state a preference for men who will make even more money than they will, while male medical students seek the opposite in a mate, Nancy suggests it had to do with the hunter-gatherer division, which mating heterosexuals seek to maintain in some form. When she writes about women's attractiveness, she focuses on the signs of her fertility. In fact, she notes, men are most drawn to women who look like they have never been pregnant, an attraction that can be seen as a form of intelligent investment, from the point of view of procreation: Women who have not yet hit their peak fertility (women under twenty, in other words) have their most fertile years ahead of them. Of course, we live in a time when the visual signs of the most attractive age are easily and frequently faked—and when fertility itself is manipulated more and more successfully—but human instincts, Nancy argues, trail well behind progress.

Books like Nancy's—and it is a wonderfully written book that sold well—serve a dual purpose. Those of us who read them get to feel superior to those who are described within: We can always cite our alternative priorities, unorthodox family structures, and advanced fertility techniques as proof that we are not governed entirely by Darwinian instincts. At the same time, they justify our less refined feelings—in my case, my reawakened vanity and deepening concern with sexual desire. I was not single; I was not particularly young; I was not even a practicing heterosexual. Nancy cites studies showing that lesbians pay youth and other traditional features of female attractiveness no heed; I was pretty sure the respondents were lying. Many of the women I had interviewed played the wise adults. They said things like, "I already have a mate." Then they also sometimes said that they were happy with having had the surgeries—despite the fact that they had no libido and no longer liked looking at themselves in the mirror, ever. This struck me as counterintuitive, which is to say, utterly insane.

But Nancy also writes about something else: Beauty does not guarantee or even significantly increase the probability of being happy. "Beauty gets you a mate," she said, "but it doesn't make you happy." That is very strange, because beauty yields any number of everyday advantages, of the romantic and the purely social sort, and beautiful people do have higher self-esteem—and both self-esteem and small daily pleasures and satisfactions are known to matter a lot in happiness. Still, beautiful people, on the whole, are not happier than those who do not look as good. There are two possible, complementary explanations. First, one can always be more beautiful, so good-looking people are not immune to feeling dissatisfied with their looks. Second, happiness—like beauty, but quite separately from it—may be genetic.

In 1996 two University of Minnesota psychologists, David Lykken and Auke Tellegen, published a new study of twins in a series of studies of twins that they had undertaken. The state of Minnesota maintained a twins registry—including twins that had been reared apart—that allowed these researchers to compare everything and anything about them. Their method was classic for geneticists: They collected data on both monozygotic and dizygotic twins, issuing questionnaires to various pairs of twins twice, at intervals of either ten or four and a half years, and then compared it. Whenever the results for monozygotic twins indicated a markedly higher correlation than for dizygotic twins, the researchers theorized that the trait had a genetic component. In general, virtually anything they set out to study seemed to have a genetic component—with the notable exception of choice of romantic partner, which, Lykken and Tellegen concluded, was entirely a matter of chance.

Their conclusion that happiness is largely heritable did not mean that one was either happy throughout life, starting at birth, or unhappy forever. The people they studied got less or more happy over time. Only half of the people interviewed were equally happy at the age of twenty and the age of thirty—but monozygotic twins were much more likely to be as happy as their twin either was or had been. The psychologists concluded that happiness was roughly 80 percent heritable. In other words, they wrote, "It may be that trying to be happier is as futile as trying to be taller and therefore is counterproductive."

The idea of happiness as a genetic trait appealed to me. First, it affirmed my sense of reality: I generally felt as if I had been born to be happy. Second, it seemed to restore a kind of fairness: I might have inherited the cancer mutation from my mother, but I had lucked out of whatever unhappiness genes she seemed to carry. When I was much younger, our relationship veering between mutual torment and cold indifference, my undefeated ability to enjoy life had sometimes felt like revenge. Something of that old feeling resurfaced now.

But then the mathematics of chance in happiness genetics are even less solid than in cancer genetics. "If the transitory variations of well-being are largely due to fortune's favors, whereas the midpoint of these variations is determined by the great genetic lottery that occurs at conception, then we are led to conclude that individual differences in human happiness—how one feels at the moment and also how one feels on average over time—are primarily a matter of chance," Lykken and Tellegen write at the conclusion of their paper. In other words, a predisposition to being happy is like a predisposition to being healthy or thin: Lousy lifestyle choices and plain bad luck can spoil the best of chances.

So what do we know about happiness? Scholars of happiness have offered a number of definitions and ways of measuring happiness, but have largely settled on the concept of subjective well-being, which boils down to the simple premise that people generally know when they feel contented. Some people are more likely to feel happy than others. Religious people are a bit happier than nonreligious people. People who feel optimistic about their future are happier than those who do not, but though this was an important consideration for someone in my genetic boat, it posed a clear chicken-and-egg problem. Married people are happier than those who are not—though this, too, may confuse cause and effect: Happy people may be more likely to get and stay married.

Here I was, considering doing drastic things to myself in the hopes of making myself live a longer and healthier life. But could I do violence to myself in the interests of longevity and still be reasonably sure that the person who came out the other end would be someone worth having around?

Health is important to happiness—but not as important as one might think. People can be extraordinarily resilient in the face of shocking damage to their health. People rendered partially or completely paralyzed by accidents have a way of regaining their sense of well-being within eight weeks. Burn victims, who can suffer profound physical and psychological consequences, tend eventually to reach a good quality of life. In this sense extreme misfortune has as little lasting impact as great fortune: Winning the lottery, like getting a new job, being promoted, or even getting married, improves our sense of well-being only briefly. In the long run, people confined to wheelchairs are as happy as those who can walk unaided, wealthy people and good-looking people are as happy as those less fortunate, and even elderly healthy people are only marginally happier than their peers whose health is failing.

Did that mean I had nothing to fear? Could I count on the genetics of happiness to keep me level as I battled the genetics of cancer? Nancy pointed out that cutting off apparently healthy breasts and ovaries would be strange. She might have been trying to avoid calling it insane. After all, people generally aim to fake youth and beauty, not purposefully to annihilate them.

There are only a couple of things that have been found to make people profoundly less happy for a long time. One is being widowed—and this was certainly an argument for doing anything I could to prolong my life. The other is developing a degenerative disabling condition. Surgical menopause could well turn out to be just such a condition, at least for a few years.

Nancy pointed out that for some women, menopause actually brings a sense of liberation. No more mood swings. No more unwanted sexual attention. "As women get older, they broaden their definition of beauty," she said, "to emphasize aspects of strength, confidence—intangibles."

"What do those look like?" I asked.

Nancy was momentarily stumped. "I think it's more about self-presentation and style—projecting more than meeting signals." She then suggested that genetic counseling for a woman in my position ought to include a computer model of how she would change with age—or as a result of premature aging brought on by surgical menopause. I liked the idea of giving shape to the suggestion of surgical tinkering—creating a visual expression in place of a statistical model.

Much harder to grasp was what it would feel like. The few available studies of women who had undergone preventive mastectomies seemed to show that they were happy with their choice. There were a couple of problems with these studies, however. First, women who were likely to have been most conflicted might simply have refused to participate. Second, once the results were unpacked, they looked far less positive. Take a British study that concluded that "for a majority of women there is no evidence of significant mental health or body image problems." In fact, only 21 percent of the women in the study said their body image had not changed for the worse as a result of the surgery. More than half said they felt less physically and sexually attractive, and a third said they felt less feminine.

On the other hand, the women said these were slight changes, not profound ones. And it did seem that they paled in comparison with the psychological benefits of the surgery. Several studies concluded that women were much less anxious about their health and their future, and much less likely to become depressed than mutation carriers who chose not to have surgery. In other words, it seemed, cutting off her breasts could make a woman happy.

~

When it came to menopause—early surgical menopause, to be precise, caused by the removal of the ovaries—my efforts to pin down the probability of specific outcomes appeared very nearly doomed. Most doctors seemed reluctant even to discuss some of the possible side effects that worried me most, such as cognitive difficulties and memory loss.

I interviewed a number of women who had undergone surgical menopause. My sample was unrepresentative and almost certainly skewed toward those who had handled the surgery well enough to be willing to discuss it, for publication, with a complete stranger. As it was, I heard everything from "It's no big deal" to accounts of debilitating effects ranging from depression to utter physical collapse. My own doctor assured me that depression, in most cases, stems from sleep deprivation, which is treated successfully with hormone-replacement therapy. "But word recall is never recovered," she added matter-of-factly. Word recall, as it happens, is essential to my craft. Would my professional competency and, to some extent, my identity be too much to sacrifice for a couple of years' increase in life expectancy? Would my mind be too much to give up in exchange for peace of mind?

I may have been exaggerating. Most specialists dismiss any discussion of the cognitive consequences of surgical menopause as "anecdotal"—which, generally, is medicalspeak for, "We haven't studied it, so it can't be considered proved."

It is remarkable how little the effects of surgical menopause have been studied. Hysterectomy is the second most frequently performed surgery in the United States (following cesarean section). For the last two decades of the twentieth century, about 60 percent of American women had their uterus removed by the age of sixty-five—and the average age at hysterectomy was a mere 42.7 years. In more than half of the cases, the ovaries came along for the ride. A groundbreaking study published in 2005 found that these unnecessary oophorectomies caused women's deaths from heart disease and hip fracture. But what actually happens to the lives—or, rather, to the minds—of the roughly four hundred thousand American women who have their ovaries removed each year? No one really knows.

A study of the effect of surgical menopause on women who did not have cancer found that they suffered a significant decline in all the mental functions that were measured: digit span (the number of digits a person is capable of memorizing at one time), visual memory, logical memory, and mental control (essentially, the ability to concentrate). This study was conducted at the University of Egypt and involved just thirty-five women who had an oophorectomy and another eighteen who served as a control group. Another study, carried out in California, failed to find any clear negative consequences of oophorectomies—but it compared women's test scores to those of other women rather than to their own presurgery test results, as the Egyptian study did. A third study, in Italy, looked specifically at the effect of oophorectomy on memory and concluded that surgical menopause was "a critical negative event within the female brain, in particular when it occurs prematurely." I loved the wording—"a critical negative event"—like an earthquake or a hurricane in the brain, coming in an instant and leaving devastation in its wake. This certainly seemed to add up to more than anecdotal evidence—except that three studies, one of which was inconclusive, is pitifully little proof in medicine.

It took a while to find someone willing to discuss the effects of surgical menopause with me. I finally located Susan Bauer-Wu, a nurse scientist at the Dana-Farber Cancer Institute. She stressed that she was not an MD but rather a doctor of nursing, "which is important, because I have a whole-body perspective." The jab at MDs, who view bodies as the sum of spare parts, may have been unintended, for Bauer-Wu did not seem the type of woman who would jab anyone on purpose. Bauer-Wu turned out to be a thoughtful woman who spoke very carefully and softly, but in well-formed, nonspecific phrases that gave me little to latch onto.

We spoke in a fake exam room: Dana-Farber, which attracts constant media coverage, has one set up for taping television interviews with doctors. Once again I heard everything I feared I would. Bauer-Wu enumerated the certainties of surgical menopause: depression, memory loss, difficulty with word recall, and that feeling I remembered so well from the postpartum year, the feeling of having all your senses padded with cotton. She also left me no doubt about the loss of libido: "It just happens to everyone." There would not be an exception in my case.

Women who undergo surgical menopause can have far more extreme symptoms than women who go through "the change" naturally, sometime in their fifties. This is probably because the levels of hormones, estrogen and progesterone, drop suddenly rather than diminishing over time, and also because the hormones disappear completely—while the ovaries of postmenopausal women continue to produce a little bit of estrogen for years. Being the only one among your peers to go through the process certainly doesn't help.

"Over time it gets better," said Bauer-Wu. Did she mean it actually gets better? No. "I can't say that women are living without symptoms. But it gets better to live with."

So, what if it were you? I asked. Or someone you love?

Bauer-Wu struck me as the kind of person who has opinions but doesn't believe it is often socially appropriate to share them. This is also what made her answer, if she was going to give me one, especially valuable.

"If it was myself or a family member," she said, "I'd help them look at the statistics. But they are just statistics. It all depends on where women are in their lives." I'd heard such platitudes dozens of times, and my heart sank at enduring more polite nonanswers. But then she said something that surprised me. She said maybe some women should not get oophorectomies, that maybe they should get regular surveillance instead. She was the first medical professional I had met who suggested that surveillance for ovarian cancer might be meaningful and effective. She said she knew women who had ovarian cancer that was caught early and cured.

A few hours after the interview I received an e-mail message from her: "I've continued to think about this issue since we met this morning. I still think that close surveillance (i.e. CA-125 with vaginal and abdominal ultrasound every 3 months) is a reasonable option for some women, especially those concerned about quality of life issues after surgery. Of note, I doubt most insurance companies will pay for such monitoring." Setting aside the implication that there are women who are not concerned about quality-of-life issues, as well as yet another indication of the absurdity of insurance practices, this was the brightest ray of hope I had seen yet.

Bauer-Wu suggested I contact Dr. Ross Berkowitz, professor of gynecology at the Harvard Medical School and director of gynecology and gynecologic oncology at Brigham and Women's Hospital. Berkowitz had a deep, lulling voice and a professor's measured pace of speech, which made note-taking easy. I interviewed him over the phone, because he was a very busy man, and also partly because I hated going to Brigham and Women's, where my mother had her mastectomy and where her downward spiral began.

Berkowitz studies early detection of ovarian cancer. The symptoms of ovarian cancer are infuriatingly vague: an enlarged abdomen, diarrhea, and/or constipation. The available screening tools are ultrasounds and the blood test for CA-125, which stands for "cancer antigen 125," and which is often elevated in women with ovarian cancer. The problem is, by the time the level of CA-125 becomes clearly abnormal, the cancer has usually spread. Ultrasound is generally even less sensitive. Berkowitz's lab had been working on finding additional markers, substances for which tests could be run alongside the test for CA-125. They had identified six potential candidates. I expected Berkowitz to say that, with effective surveillance methods apparently coming soon, women might not be foolish to keep their ovaries. In fact, I assumed he would advocate surveillance over surgery.

Berkowitz said that women at risk for ovarian cancer should get oophorectomies. Even women who are not yet forty? I asked. (To the extent that the statistics can be trusted, the risk of ovarian cancer shows up around forty or later, depending on the specific BRCA mutation.) "For a thirty-seven-year-old woman with the mutation," said Berkowitz, unwittingly describing me, "as long as they've completed their childbearing, it's not an unreasonable option. It's not so rare for these cancers to be diagnosed before the age of forty."

But surveillance might be available so soon, I countered, and the costs of surgery were so great. "You are talking about a disease that's 80 percent lethal," he responded. "And yes, hot flashes are unpleasant. And one needs to prevent osteoporosis, and you need to treat vaginal dryness in various ways—but I think death is worse."

American medicine takes essentially an instrumental view of the body: the body as a car. These things need regular checkups. Doctors are mechanics whose primary goal is to ensure that the functions are performed properly. The primary function of the body is reproduction—hence the ease with which a woman's reproductive organs are discarded once the function has been performed.

I heard this attitude distinctly in Berkowitz's comment, and I described my feelings in *Slate* magazine, thereby infuriating several other doctors, who wrote me letters about the ungrateful nature of patients like me and the disrespectful ways of journalists like me. All of which made me think that I had stumbled upon an important disjunction between the medical profession and the lives it is called upon to heal.

One particularly vivid example of this thinking is the treatment of intersex newborns—babies who, as a result of any one of a dozen genetic conditions, are born with ambiguous-appearing genitals. Historically, doctors have "assigned" these babies to the male sex if they could grow up to impregnate a woman and to the female sex if they could not. Surgery was—and often still is, despite patient advocates' growing objections—performed very early in life to bring the child's appearance in line with his or her sex assignment.

The most famous case of this kind of surgery occurred in a baby who was born not intersex but a normal male—or, as the medical literature puts it, "an XY individual." David Reimer's penis was burned off during a botched circumcision when he was eight months old. Doctors ultimately convinced his parents to have the boy's testicles removed and a vagina constructed and to raise him as a female. One of their key arguments was that, as a man, he would never be able to have normal heterosexual relations, marry, or have children.

David Reimer's transformation into a female was a disaster. When he finally learned the truth, as a teenager, he refashioned himself as a male, had a mastectomy (breasts had grown because he was taking estrogen) and a phalloplasty—a penis was surgically created for him. He eventually married. At the age of thirty-eight he committed suicide.

David Reimer's story, discussed extensively in the late 1990s, surely contributed to a fine-tuning of the mainstream thinking on intersexuality: In the day of the genome, genetics is edging reproductive anatomy out as a deciding factor in gender assignment for intersex babies. In other words, being "an XY individual" may now hold more sway than being an individual with a penis. What remains unchanged, though, is the surgical habit: Even as advocates for intersex babies—most of whom were born intersex—argue that surgery should be delayed until the person can make up his or her own mind about it, doctors tend to want to use their scalpels to complete "sex assignment."

This, of course, is nothing new. For thousands of years people have been going under the knife in the name of normalization and disease prevention. Consider a list of surgical procedures that became routine before they were abandoned because they proved either useless or harmful: bloodletting, routine tonsillectomy, routine circumcision (the United States remains the only country where most male babies are circumcised whether or not their parents' religion requires this), repeated cesarean delivery, internal-thoracic-artery ligation (a once-touted method for improving blood flow in angina patients), gastric freezing (a procedure used in about fifteen thousand duodenal ulcer patients in the 1960s, before a placebo-control study exposed it as ineffective and dangerous), prophylactic portacaval shunting (major surgery that relieves blood pressure in the liver). The effectiveness of surgery is much harder to evaluate than the effectiveness of medication: A placebo-controlled study would necessarily involve sham surgeries, which are dangerous and generally considered ethically unacceptable. So irreversible surgical procedures become a part of standard medical practice faster and more easily than do chemicals.

It came to make sense to me that the frontier of genetic medicine was, in fact, surgical. The simple and decisive nature of surgery is seductive. It is a one-shot deal. Being able mentally to reduce one's own body to a collection of parts creates a powerful sense of control. Cancer, particularly hereditary cancer, makes this feeling of control especially desirable. Cancer is one's own cells gone awry. Cutting out the potentially offending organ before it has a chance to betray you shows the body who is boss. The greater one's fear of cancer, the greater the temptation to cut.

Perhaps the only force that can moderate this response is another fear. In the year before my genetic diagnosis I had battled postpartum depression. The sense that my mind was a traitorous, unreliable partner had only recently left me. My fear of experiencing that again was, I was starting to think, greater even than the fear of cancer. After my own fashion, I was coming around to Susan Bauer-Wu's "whole-body perspective."

∼

I spent the period from the mid-1980s to the early '90s writing about AIDS. It was rare luck as a beginning reporter to be able to sink my teeth into one of the big stories of the century. I learned everything important back then. I learned that how authoritative someone sounds is a measure only of how firmly this person believes in the information, not of how accurate the information is. I remember interviewing people who said there was no way the Food and Drug Administration's drug-approval process could be shorter than ten years (at this writing it is down to 25 percent of that, in some cases), people who said there would never be an AIDS vaccine (there still is not, but there likely will be), people who said HIV does not cause AIDS (it does), and people who said there would be no effective therapies for generations (there already are).

Now I relearned the fine art of listening to medicalspeak as I went from expert to expert. I spent a couple of hours with an oncologist going over the fine points of cancers that mutation carriers might develop and how they differ from cancers that occur spontaneously. It seemed that breast cancer in mutation carriers might or might not be more aggressive and harder to treat than in noncarriers. Ultimate survival rates were not significantly different. On the other hand, mutation carriers seemed to have a significantly higher risk of recurrence of breast cancer. The story with ovarian cancer was different: It seemed it might be more treatable in mutation carriers than in noncarriers. But all of these statistics quickly started to run together: Compared to my overall risk, these were tiny, insignificant amendments.

As I shifted the piles of data around in my head, I found myself thinking about my days as an AIDS reporter more and more often. I remembered two men with AIDS coming to fisticuffs in the office of my magazine when one accused the other of not being an "empowered person with AIDS" because he was not getting a then-state-of-the-art treatment called aerosolized pentamidine. I remember a robust, asymptomatic thirty-year-old HIV-positive man dropping dead from a heart attack after procuring some Compound Q, a Chinese concoction that was briefly hyped as the wonder drug. I remembered the zest with which doctors and patients rushed to administer antiretroviral drug "cocktails" to anyone who had tested positive for HIV, and the slow disappointment that set in as the drugs' side effects, ranging from the aesthetically unappealing to the potentially life-threatening, emerged—until generally accepted practice switched to starting treatment later in the progression of the disease. I also remembered the great controversies—still raging in some impoverished parts of the world—about whether and why anyone should get tested if patients were helpless to do anything about the results.

Most important, though, I remembered the peculiar sense of belonging to a community that lived by a separate set of values. Its heroes were long-term survivors, whose special attitude, knowledge, and medical decisions were believed to account for their better-than-average fortune. Its currency was information: People educated themselves in the language of medicine and traded information through newsletters and endless conversations (this was before the Internet). Its members paid their dues by subjecting themselves to experimental treatments ranging from taking unapproved medication to drinking their own urine to traveling to Mexico to have their blood put through a heating machine. The guerrilla researchers would report back, passing around their notes and parading their bodies—apparently intact, which was itself a triumph—and, often enough, a following would immediately form.

This was something else I recognized: a peculiar sort of crowd mentality. In a sense, my time as an AIDS reporter now seemed like it had been a rehearsal for my current medical trip. Then I had been an engaged observer: I lost many friends to AIDS, but I was never even at risk myself. Now I found myself becoming a member of another community that was inventing new rules for living under the threat of death. There turned out to be an organization called FORCE (Facing Our Risk of Cancer Empowered), which united women with various BRCA mutations. On its online forum, women traded scientific information, surgery experiences, and insurance-handling tips, vented, and sometimes cried in the virtual comfort of a community populated by clever and sometimes hokey usernames.

The virtual identities belonged, for the most part, to well-educated urban women, who tried to be as open-minded and accepting of one another as one can be with strangers. But, as in any closed community, conventional wisdoms took shape and took hold. By the standards of FORCE, it was right to choose to have both of the surgeries—the mastectomies and the oophorectomies. It was even right to have a particular sort of reconstruction—and at a particular medical center at that. The women of FORCE may or may not have been a representative cross-section of American BRCA mutation carriers: The single existing study of the behavior of BRCA mutation carriers showed that none of them opted for a preventive mastectomy within two years following the test, while a whopping 46 percent (and 78 percent of those older than forty) had an oophorectomy.

The FORCE women lived by the laws of progress: They made their decisions in accordance with the latest medical evidence (even when it was not complete or when it ran counter to previous knowledge), and they applied the latest medical techniques to themselves. This was where I learned about the cutting-edge breast-reconstruction techniques, which involved using a woman's own fat tissue, from the belly, the back, or the buttocks, to build new breasts from natural tissue that would be connected to a woman's own blood vessels.

Sue Friedman, the founder of FORCE, was the sort of woman who believed in controlling her fate and her health. She was a conscientious exerciser, a vegetarian, and generally a bit of a health freak, and at thirty-three, in 1996, she was diagnosed with breast cancer. The cancer was particularly intractable: She had a lumpectomy, which left her doctors uncertain they had gotten all the cancer, then a mastectomy a month later, but it came back in another eight months. Around that time she read about the BRCA mutations in a Jewish newspaper. No one had told her about the possibility that her cancer was genetically determined—possibly because she had no obvious history of breast cancer in her family. She was tested, turned out to be positive for the same mutation I have, started to look for information on her condition—and found next to nothing. So she started a Web site, which turned into a nonprofit organization, which turned into a full-time job, which helped spin a community, which I now found myself warily joining.

She was a natural leader for this kind of community. A veterinarian, she had no difficulty with the medical lingo, and she inspired confidence in the newcomers, all of whom she took the time to welcome personally, albeit virtually. She set an obvious example of proactive mutation management: She had all the surgeries. When I called her for an interview, she was happy to share her personal story but reluctant to talk about her experience following the oophorectomy. The effects of her own surgery had been disastrous. Things were all right at first, but a couple of years after the operation she found herself unable to get up off the couch. She termed it "severe life-altering fatigue." Antidepressants barely took the edge off it. The medication she took for her severe joint pain also did little to help her. It took her and her doctors a year to figure out she needed hormone-replacement therapy—a risky choice for a breast cancer survivor, because it may contribute to the risk of a recurrence, but then in Sue's case it was no choice at all. Within two weeks of starting the hormones, she was off the antidepressants and the pain medication, and off the couch as well. "Mine is not the most common experience," she cautioned me after telling the story. But neither was it very uncommon.

What Sue Friedman was doing was what Daniel Kahneman, the psychologist who won the Nobel Prize in economics, termed "framing." There is nothing wrong with it—indeed, it is inevitable that our options are always framed one way or another. A classic example of framing, used in universities all over the world, goes something like this.

Your city is facing an outbreak of a deadly exotic disease, which is expected to kill six thousand people. You must choose between two different programs proposed for combating the epidemic. If Program A is adopted, two thousand people will be saved. If Program B is adopted, there is a one-third probability that all the six thousand people will be saved and a two-thirds probability that no one will be saved.

People generally choose Program A, which offers the comforting certainty that at least some lives will be saved. But the problem can be phrased differently.

If Program A is adopted, four thousand people will die. If Program B is adopted, there is a one-third probability that no one will die and a two-thirds probability that six thousand people will die.

Now people will tend to choose Program B.

More than twenty years ago, Kahneman's collaborator Amos Tversky conducted a study of the framing effect on medical decision making. He asked groups of patients, doctors, and graduate students to choose between surgery and radiation as a treatment for lung cancer and found that the results varied significantly depending on how the decision was framed: It was important whether the treatments were named, whether information was presented in terms of life expectancy rather than cumulative probability, and whether it was framed in terms of the probability of living rather than the probability of dying.

As framed by the genetic counselors and the BRCA community, the oophorectomy problem went something like this: If you do not choose surgery, there is a roughly one-in-two probability that you will develop ovarian cancer, which is likely to kill you; if you choose surgery, there is a near certainty that you will not develop the cancer. The doubters among us reframed the problem in terms of the risk of other diseases and decreased quality of life—issues that simply were not as salient to the proponents of surgery.

Kahneman and Tversky identified a range of heuristics—shortcuts in decision making that people have a habit of using. One of them is the "availability heuristic," which means simply that a familiar experience is recalled and imagined more easily than, say, an abstract example, and influences decision making to a far greater extent as a result. For most of the women living with BRCA mutations, the familiar experience was watching their mothers, aunts, cousins, sisters, or grandmothers die of cancer. I interviewed a twenty-two-year-old woman who had been beating down the doors of surgeons in her town in Michigan asking them for a preventive mastectomy and oophorectomy. Two refused out of hand, citing her age, and one told her to come back in a year and tell him whether she still felt she wanted the surgery. From her perspective, it was an outrageous, and outrageously unfair, example of medical paternalism. After all, if she had come asking for a breast enlargement, she would not have been sent away; in fact, even if she had come asking for a sex reassignment, she probably would have been able to find someone to remove her breasts and ovaries and chase that with shots of testosterone. At the same time, even as I interviewed her by instant messenger, I probed for ways to tell her to wait, to enjoy her sex life while she had her ovaries and breasts, to have children and nurse them, and to rejoice in being so young that by the time her risks really kicked in, there might be a cure. I clearly was not the first person to reach for these arguments, and, just as clearly, it would be no use. She had seen her grandmother die after struggling with both ovarian and breast cancers for nineteen years. She had seen her mother live through a debilitating course of treatment at the age of thirty-four and again at forty-two. What drove her was not even so much a fear of death as a fear of cancer as a way of life—her family's way of life.

She so wanted to avoid continuing the family legacy that she was certain she would not want to have children who were biologically her own if they risked inheriting the mutation. The joys of sex and motherhood were vivid and easily available experiences for me but for her, they were things she would not miss if she never experienced them—and she might have been unable to enjoy them anyway, if she was consigned to living with the constant sense of doom.

I spent hours talking on the phone with a fifty-two-year-old public-health educator in Portland, Oregon. Nancy Prouser was quite certain that she decided to have a preventive mastectomy because she had watched her mother die of breast cancer at fifty-eight. The fact was, though, that Nancy had inherited her BRCA2 mutation from her father, who had successfully battled breast cancer at seventy-six and was still alive eleven years later (about 1 to 2 percent of breast cancers are diagnosed in men, and it seems that many of these may be caused by the BRCA2 mutation). Nancy had the mastectomy with implant reconstruction, and an oophorectomy, and she became a walking consciousness-raising program on genetic mutations. In addition to running a program training clinicians to do breast exams, she lectured on genetic mutations and met with women who were referred to her by genetic counselors. She told me she had "gone into the bathroom at Starbucks to let a woman see and touch my reconstruction." She was a convincing advertisement for preventive surgery—at her age she would otherwise be more likely than not to have developed one of the cancers—and she was disarmingly frank about the huge blow her body image had sustained. When I asked her about the loss of libido, she laughed: "I'm glad you reminded me! I have none. I forgot." She had not been trying to hide her loss: It was simply outside the frame.

My own availability heuristic, clearly, stemmed from the vividness of my mother's death from breast cancer and the remote memory of my great-aunt Eugenia's death from ovarian cancer. I had been fourteen when she died. I was aware of her diagnosis, and I was even involved in some of the caretaking, making runs from one end of the wintery city to the other to deliver a jar of beluga caviar that, I think, was supposed to help with the depleting aftereffects of chemo. I accompanied her to the hospital after weekend visits at home a couple of times, and the sight of the cancer center in southern Moscow, a huge gray concrete eyesore that seemed to swallow people whole, still fills me with dread. But I had been shielded from watching her deterioration and death; I recalled her illness as a taxing routine rather than a disaster. It simply was not as scary. And twenty-two years later, as I struggled with my decision, I learned that a woman I knew and respected a great deal had been diagnosed with ovarian cancer at twenty-one. She was now seventy. My own unrepresentative, skewed, utterly unscientific sample seemed fated to calm me and make me hesitate even as virtually everyone else told me to proceed to get my ovaries removed—whether or not I was asking for advice.

An old and dear friend, who was diagnosed with breast cancer at thirty-six and soon opted for both a mastectomy and an oophorectomy, wrote to tell me I should do the same thing. A friend's sister, whom I had never met, also wrote, telling me of her experience battling advanced breast cancer in her early forties and suggesting I consider having an oophorectomy, as she had. And, incredibly, a New York City breast surgeon I called to schedule an interview for this book began to advocate for an oophorectomy for me during our first phone conversation.

In his Nobel lecture on "bounded rationality," Kahneman explained his theory that people have two systems for making decisions:

System 1, which is intuitive and immediate, and System 2, which corrects it.

System 1 involves what he called "natural assessments," which come easily to mind. They include similarity: People immediately think of situations that remind them of the current one—as women with the mutation tend to recall their mothers' suffering and, often, death. It helped me to realize that I was talking not only to women who came from cancer families but also to doctors who came from cancer practices. My own doctor, for example, told me that doctors believed breast cancer hit the best and the healthiest. This was when I was fishing for reassurance: Surely I had improved my chances by being fit—a veritable gym freak—and eating well. She shot me down with the story of a woman who had had it all: good health and a good mind, great friends and a wonderful family, and had broken her doctors' hearts by wasting away from breast cancer in a few short months. In fact, there is some evidence for the idea that exercise and a low-fat diet help keep genetic cancer at bay—but my doctor was exposing her own heuristic: The tragic story of that one patient simply stuck out in her memory more than other, more predictable breast cancer stories.

This went to another of Kahneman's "natural assessments": what he called "affective valence." Researchers of risk perceptions have long noted that people tend to fear a horrible death from, say, cancer, far more than a death from heart disease, which kills more Americans than does any other condition. Cancer has a greater "affective valence" than heart disease: It is scarier.

Kahneman also observed that people are not very good at estimating probability. They try, but they have a clear tendency to overestimate the frequency of events that are easy to recall. He compared the way people deal with probabilities to the way they assess physical attributes such as distance. He used the example of a mountain in the distance. The human mind knows that the blurrier the mountain's contours, the farther the mountain is. This is the heuristic. But on a foggy day, the contours will appear even blurrier, and on a very clear day they may be unusually sharp. A person will tend to overestimate the distance to the mountain on a foggy day and underestimate it on a clear day. This is the bias.

For women terrified of breast cancer, all days have a way of being foggy. Studies have consistently shown that American women think their risk of breast cancer is higher than it actually is—even when, as in cases of familial cancer, it is in fact very high. I knew I was doing this, too: Being told that I had as high as an 87 percent chance of developing breast cancer in my lifetime sounded an awful lot like being told that I would certainly develop breast cancer. Ovarian cancer, once again, was a different matter: Here the lifetime risk was in the 40-to-50 percent range.

What people do with fifty-fifty risk is a question that has long fascinated scholars of decision making. Kahneman and Tversky based their famous Prospect Theory on a study of fifty-fifty gambles. They concluded that most people will reject a gamble with even odds of winning and losing—unless the possible win was twice as big as the possible loss. This is exactly what my newfound community of mutants was doing: They were engaging in what economists call "risk-averse" behavior by cutting out their ovaries, thereby opting out of the fifty-fifty lottery.

Meanwhile, what I had done was overanalyze the decision-making processes of other women in order to delay making my own. But I had promised my editor and readers at *Slate* that I would come to a decision at the end of my research. The editor might even have said I should not feel obligated to keep the promise—but prolonging the limbo would have defeated the purpose of going public. I wrote:

> But I have to make the decisions. So, here they are. I plan to get a bilateral preventive mastectomy with immediate DIEP flap reconstruction as soon as my daughter is weaned, possibly as early as this

coming autumn. I am not going to get a preventive oophorectomy, at least until I am 40, and will aim to avoid getting one at all, which means staying up-to-date on early-detection research and forcing my physician to do so as well. This means that for the rest of my life I will bear the physical marks of my mutation and will have to stay obsessively on top of medical research and my own health. This is the sort of thing that eventually happens to most people—but for me, like so many other things, it has to happen earlier.

I wrote that in May 2004. Then, secretly, I decided to give myself some time to get used to my decision—or not.

Chapter 6

THE FATHER OF
HEREDITARY CANCERS

Back in the 1960s and 1970s, members of a particular extended family in Green Bay, Wisconsin, used to spend virtually every weekend together. There were eleven siblings—five sisters and six brothers—and they, like their parents, tended to have a lot of children and not a lot of money. So gathering at one or another sibling's house to hang out and drink was what they did for entertainment. The kids came along and were generally left to their own devices. This turned out to be a recipe for creating community: The kids formed a group that would hold together for decades, perhaps for the rest of their lives. As it happened, the cousins of that generation were mostly girls—about fifteen of them around the same age, all born in the late 1950s or early 1960s: Sheila, Karen, Donna, and a dozen more.

In the late 1960s three of the five sisters were diagnosed with breast cancer. All three survived their first bouts, but all succumbed to a recurrence by the early 1990s. Sheila's mother died first. A fourth sister died of pancreatic cancer at the age of fifty-two, a few years later. "They were all in their thirties, and they were all having cancer together," Sheila remembered years later. She was twelve when her mother died.

Sometime in the early 1970s—no one remembers quite when—there was a family meeting at one of the remaining siblings' houses. A doctor from Madison, Wisconsin, wanted all the women and girls of menstruating age to give blood. He explained that there seemed to be something going on in the family, something about everyone having cancer. That somehow made it official: Three sisters having breast cancer at the same time was not a freak accident but a fact of medicine, perhaps of fate. The cousins, who were all in high school by then, started talking about cancer obsessively. They dubbed themselves "cancerphobes." They assumed they would die of cancer before they reached their early forties, which seemed a long, long way away but still a freaky prospect. Some of them even joked they would have their breasts cut off to keep the cancer at bay. This, though, was morbid humor, not a suggestion anyone would have considered seriously.

Whoever came to draw the family's blood in the 1970s seems to have left no traces at the University of Wisconsin at Madison, where he may have been a graduate student. His idea of studying only the females in the family was not particularly sophisticated—in fact, it was plain wrong from the point of view of genetics—but the very concept of identifying a family that seemed to have cancer running through it was forward-thinking at the time. It was only in 1966 that another young Midwestern doctor, Henry T. Lynch, had, to the derision of his colleagues, suggested that cancers might be heritable.

Lynch had taken a bit of a circuitous route in his career, serving first in the U.S. Navy, then completing his course work toward a Ph.D. in genetics before finally going to medical school. In 1961, his first year as a resident, at St. Mary's Hospital in Evansville, Indiana, Lynch met a patient coming out of D.T.s and conscientiously tried to interview him about his reasons for drinking. The patient's story proved unexpectedly coherent and compelling: Everyone in his family had died of cancer, he said, and he expected the same fate. The dread drove him to drink. As a former geneticist and future oncologist, Lynch was fascinated. He took down the patient's history. Then he took down a number of other family histories. Gradually, the idea that cancer could run in families went from a crazy intuition to a scientific theory. Somewhere along the way, Lynch discovered that back at the turn of the century a Michigan pathologist named Aldred Scott Warthin had described a family that had cancer running through it—"cancer excess" was the term used in the literature. Warthin had stumbled upon this knowledge in much the same way as Lynch: The pathologist's seamstress, who had grown severely depressed, divulged her family's devastating medical history at the doctor's prodding. The seamstress later succumbed to the family cancer, which in retrospect appears to have been a particularly virulent form of colorectal cancer now known as hereditary nonpolyposis colorectal cancer, or, simply, Lynch syndrome—so named for Dr. Henry Lynch, who in 1966 first described two Midwestern clans as having "cancer family syndrome." As it happened, he and Warthin had stumbled not just upon similar families but upon families whose members suffered from the same syndrome, which caused them to develop colorectal cancer at an early age—forty-four, on average.

There was more than a seven-decade gap between Warthin's original research and Lynch's pioneering paper. Another thirty years would pass before the genetic—and therefore sometimes familial— nature of cancer would become not only accepted fact but a cornerstone of research in oncology and genetics. In the intervening decades, Lynch would become so accustomed to staying ahead of the curve that the fact of his own scholarly eminence seemed to evade him. Not only was he the first person to describe familial cancer, Lynch also seems to have been one of the first doctors to float the idea of preventive surgery—prophylactic mastectomies and oophorectomies—in the 1980s. "They thought I was crazy," he told me when I asked him about his colleagues' reaction. Later he also suggested prophylactic colectomies—the removal of the stomach—for patients who hailed from families riddled with stomach cancer. What did his colleagues think this time? "They thought I was crazy," he repeated, without so much as a chuckle.

By the time I interviewed him, in the fall of 2005, Lynch had written more than a hundred papers on "cancer family syndrome." I once read the abstracts in one sitting, which yielded a wildly speeded-up version of the history of the science, and of Lynch's own evolution. In the 1960s and 1970s he had doggedly chased genetic clues, without much success. He first brought up the idea of prophylactic surgery in a 1977 paper—it seems, out of an exasperation bordering on desperation: There were people dying of cancer—*young* people *predictably* dying of cancer—and there was nothing he or his colleagues could do about it. More to the point, he seemed to grow convinced that there was nothing his colleagues wanted to do about it.

In 1988, when he had been writing about hereditary cancer for over twenty years, Lynch used an article in a medical journal to attack his colleagues for failing to diagnose the syndrome Warthin had first described nearly a hundred years earlier. The problem was that physicians failed to ask patients about their families, wrote Lynch. Two years later Lynch complained in the journal *Cancer* that doctors were failing to educate themselves about hereditary cancers. In the early 1990s he seemed to use every opportunity, in any journal, to call on his colleagues finally to learn to take down patients' family histories. Gradually, he grew belligerent, albeit in a mild, Midwestern, doctorly sort of way. "Despite more than two decades of documentation in the literature, many physicians fail to recognize the clinical features of these syndromes," he wrote in *Diseases of the Colon and Rectum* in 1993. Lynch's campaign to educate, cajole, and occasionally browbeat physicians into being aware of hereditary cancer culminated in 1998, when he agreed to serve as an expert witness, testifying against a doctor at a malpractice trial initiated by a family of a twenty-year-old woman who died of a Lynch syndrome cancer that had not been properly diagnosed despite the mother's attempts to draw doctors' attention to the family's history of cancer. He then described the medical case, the misdiagnosis, and the trial in an article in *Diseases of the Colon and Rectum*. Genetic mutations connected with many of the cancers Lynch had spent his life studying—colorectal, breast, and ovarian—had already been identified, and now the doctor was really mad. "The question, 'Is cancer hereditary?' has been answered beyond any doubt," he wrote in the journal *Seminars in Oncology* in 1999. "Have clinicians acted on the importance of hereditary factors in cancer so that this knowledge might be translated into patient benefit? Data showing that 59 percent of patients... still die of metastatic colorectal cancer suggest that the answer is no."

In 1994 scientists pinpointed the two chromosomes harboring the two genes that suffered the mutations that likely caused Lynch syndrome (three more mutations have been found since). Later that year scientists managed to clone two defective genes, making an early test available for potential carriers. In an article in *Anticancer Research*, Lynch and his wife, colleague, and coauthor, Jean (a nurse), proudly reported that more than twenty-five years after identifying the syndrome, they were now in a position to suggest that their patients do something about it: have the colon removed prophylactically rather than wait for the cancer to develop. The surgery remained controversial for years. After all, the Lynches were suggesting that young, apparently healthy people have major surgery that would certainly lower their quality of life, risking frequent defecation at best and incontinence and nutrition problems at worst. Lynch continued to dismiss the idea of nondirective genetic counseling, in his gentle, nonjudgmental, but very insistent way. He was known to call some of his patients regularly, just to remind them to consider the option of a preventive surgery.

Lynch's hobby of collecting information on what he called "cancer families" grew into a scientific preoccupation, and eventually into what may be the world's largest collection of such information. By the late 1980s his papers would make reference to families that had been under investigation for two decades or more. He found families with outstanding epidemiological features, like six siblings with cancer (five brothers with colonic cancer and one sister with uterine and laryngeal cancer, which the woman's son also developed). He described a Navajo family that he observed for thirteen years, before identifying the mutation that caused the family's colorectal cancers and drawing blood from fifty-one members of the family. Of the twenty-three who wished to learn their results, seven were positive for MHL1, one of the Lynch syndrome mutations. "Reactions ranged from full acceptance," wrote Lynch, "to traditional Navajo reasoning such as the family had been cursed." Lynch tracked more than 170 families with Lynch syndrome. Soon enough he was even able to generalize about patients' decision making. He reported in 1996 that more than half of those who discovered they carried an MSH2 mutation considered the option of a prophylactic subtotal colectomy (it was too early to tell how many would actually go through with the surgery).

In 2003 he published the results of a huge study involving over ten thousand members of seventy-five families with breast and ovarian cancer and forty-seven families with Lynch syndrome. In the course of the study, 1,408 people were tested for cancer-causing mutations, and Lynch and his team then analyzed the consequences for the families as a whole. As it turned out, the results of those 1,408 tests affected 2,906 people—all the people who were tested plus their direct descendants. Lynch's team came to a startling conclusion: 77 percent of those affected went from "at risk to non-carrier status." It was a bit of a statistical hat trick. Members of any cancer family are reasonably assumed to have a 50 percent chance of having inherited the mutation. If 1,408 people were tested, roughly half of them should have tested negative. All their children would then automatically shed their "at-risk" status. That's how the researchers got their 77 percent figure. But the study still made an important point. If you take a cancer family and study it, you will inevitably be the bearer of good news. Everyone in a cancer family is, as the Wisconsin teen-agers put it thirty years ago, a card-carrying "cancerphobe"—that is, everyone believes he or she will get cancer. Those who learn they are mutation carriers will only have their fears confirmed; the other half will be able to celebrate its liberation from the legacy.

Along the way, Lynch seems to have coined two terms, a poetic one and a technical-sounding one. A person's *cancer destiny* is what he believes the various mutations predict—in theory, at birth, or even earlier. And *family information services,* he believes, can help confront or even change that destiny. A family information service, a Lynch invention, is a surreal experience: generally a weekend when the many members of a family Lynch and his colleagues have studied and mapped meet to hear lectures on their particular family syn-drome—and to give blood. These are invariably family reunions of sorts: People who haven't seen each other in years, have never seen each other, and, sometimes, have never even suspected one another's existence, come together to exchange niceties and some of the most intimate information about one's health anyone is ever likely to di-vulge. Most of them learn something new about their family and about the cancer that makes it different from all others. Most of them will also learn something about their destiny. Lynch's hat-trick paper on "risk adjustment" is one of the best arguments in favor of these odd gatherings. For most participants, they will ultimately yield good news. For the rest, they will likely provide essential, possibly life-saving knowledge.

～

In November 2005 I lucked into the largest of these family infor-mation services Lynch had ever conducted. I arrived in Green Bay, Wisconsin, a bit late—the Packers were playing that weekend, and last-minute travel arrangements were tricky. On the Friday night be-fore the weekend's family reunion, Lynch was speaking at a dinner with a group of local physicians. He was well into his lecture when I got to the restaurant. He hovered just outside the door to the small dining room—most likely because this afforded his aging eyes enough distance to see his own PowerPoint presentation, but it gave the impression that he was too large to fit in the room. Extremely tall, with a broad frame rendered a bit shapeless by age, at seventy-seven he had enough uncertainty of movement that he occasionally looked like he might topple over, like a tower. His face was similarly obscured by signs of aging: One eye squinted a bit behind gold-rimmed aviator glasses, and while the eyes seemed to be smiling, the mouth was frozen in a sort of permanent frown. His speech was loud, certain, and clear, with that rare ring of something that has been repeated countless times yet has not grown automatic.

Lynch was talking about hereditary cancers and the importance of knowing patients' family histories. The eighteen doctors in the room seemed marginally familiar with the topics: Lynch's ideas were not revolutionary for them—indeed, it seemed a given that they were the new medical mainstream—but none of them seemed to have personal experience in applying them. They asked whether it was all right to put women on hormone-replacement therapy following a prophylactic oophorectomy. The jury was still out, explained Lynch and his assistant. The doctors asked more questions about prophylactic surgery: of the ovaries, the uterus, the breasts, the colon (not as big a deal as it may seem, Lynch made sure to add), and even the pancreas.

Then a shiny young man and young woman from Myriad, the company that holds several international patents on BRCA mutation testing, worked the room, giving a well-informed and informative unabashed sales pitch for testing and prophylactic surgery, both oophorectomies and mastectomies. They did their own PowerPoint presentation, handed out cards, chatted with individual doctors, and paid for dinner. Lynch's assistant stole a peek at the bill and made an impressed face; then someone sitting nearby pointed out that just a few full BRCA tests, at roughly five hundred dollars a pop, would cover the expense. As I found out the next day, the sly Dr. Lynch, who for years had pursued his family information services as an un-funded extracurricular activity, did not himself use Myriad, which had been waging a fight in essence to monopolize testing. He sent his blood samples to Ontario, where a research lab did the sequencing for free.

Early the next morning, Sumedha Ghate, a local genetic coun-selor, swung by my motel in the hospital's minivan, on her way to pick up the Lynch party: the doctor, his assistant, and Jean. Lynch veritably bounced out of the hotel. "This is so wonderful," he boomed, folding himself into the minivan. "I've never seen anything like this! How many physicians were there? Seventeen? Eighteen? And they were not doing it for money, and they were not doing it for credit—they just came because they were interested." This from an international medical star who for a while now had been collecting several awards a year—many of them with words like *lifetime achievement* in the title. But after decades of being the object of his colleagues' ridicule and suspicion—the crazy man who thought can-cer ran in families; the rogue oncologist who suggested cutting off parts of one's body *before* cancer struck—Lynch seemed sincerely surprised and excited to discover that the medical mainstream had fi-nally caught up to him. He had spent so long trying to convince his profession that genetic medicine was its future that he had missed the moment when the future actually arrived.

～

The halls of Green Bay's St. Vincent's Hospital were plastered with arrows pointing to the family information service, a token not only of Sumedha's enthusiastic diligence but also the relative scale of the event: It was not every day that as many as a hundred people would gather in a single room at St. Vincent's. At the end of the labyrinth, Karen, a woman in her forties who had once been one of the "cancerphobe" cousins, sat at a folding metal table covered with fly-ers, sign-up sheets, name badges, and colored dots to place on them to identify the wearer as belonging to one of four branches of the family. She greeted all comers, many of whom she had met over the phone, handing out the materials and making conversation.

"You know," she was telling someone, "Mamie died in childbirth, and she had a daughter, Annie, and I thought it ended there. But it turns out she had two children. The guy married and moved away. And I contacted them, and what do you know: The girl has breast cancer young, and the guy had prostate cancer young. And I called the guy's wife, just cold-called her, and she was so receptive. She said, 'I have kids, I'll be there.'" The listener was a woman who was one of the semiofficial family historians: Every branch had one.

One of the branches, this one distinguished by huge clear-blue eyes and red paper dots on their name badges, brought a family his-tory typed in caps by one of its matriarchs, and a photograph from the early 1900s of Frank and Nellie, the Dutch immigrants who were ancestors to everyone present. The family story was a fascinating document in the way of all family stories, with its casual unfolding of generations of a fast-growing dynasty: Couples generally had be-tween six and twelve children. But I also found it captivating for what was utterly absent from the story—the family curse, the mutation that had crowded more than eighty people into a single basement room in the hospital. Among the droughts and thunderstorms in the document, Pearl Harbor and the Depression, suicide and tuberculo-sis, there were only three references to cancer—certainly no more, and no more significant, than you would find in any other family narrative of similar scope.

Frank came to the United States from Holland in 1843 at the age of twelve. His future wife, Petronella (Nellie), was born in Holland that year and brought to America a few years later. They married in 1865. One of them carried a mutation that causes breast and ovar-ian cancer and possibly raises the risk of some other cancers. It is likely to have been Frank, who lived to eighty-seven: Being a man, he would probably have carried the mutation and not developed cancer. On the other hand, Nellie, who lived to seventy-five, may have had cancer: The family history, written by one of her granddaughters, says nothing about the cause of her death but indicates that she had been ill. Frank and Nellie had six children; two daughters died very young, and three sons and a daughter lived to adulthood. In a bit of bad genetic drift, all four of them inherited the cancer-causing mutation.

The family history I read happened to cover the family of Frank and Nellie's second son, Bill. He died at seventy-six of throat cancer, which may not have been related to the faulty gene. He and his wife had twelve children. Two daughters died in childhood; one son lived to eighty-three but never married. Another daughter, Frances, had epilepsy. In her twenties she was engaged to be married, but, wrote her younger sister, "the doctor told her she should not get married since there was a chance that the children would have seizures. So she gave up that plan." Frances stayed at her parents' house, up the hill from the sand-and-gravel pit that was her father's business, until her death at fifty-six. This bit of genetic guesswork from the 1930s is a remarkable detail. In fact, familial epilepsy is rare—the risk of a person with epilepsy having a child who develops epilepsy is 4 percent or less—and there is no indication that Frances's epilepsy was the result of a genetic mutation that she could pass on to her children. The potentially deadly mutation in Frances's family—the one that was passed on with a 50 percent probability—was at that point unknown, unnoticed, and consequently undescribed.

Eight of Bill's twelve children went on to have children of their own—forty-four of them in all. This was the oldest generation of the family represented at the meeting in Green Bay, and many of them by this point had great-grandchildren. Altogether Dr. Lynch's office had data on more than a thousand members of this family—including the descendants of Bill's siblings.

"Mother lived with us from 1951 to 1964," I read in the family story, eight single-spaced pages, typed in capital letters. "She had broken her hip in the early 1960s. This took the spunk right out of her. And then when Christy died in January 1962, she had a hard time dealing with it. They had been very close for a mother and daughter, even raising their children together. Mother felt it should have been her that died, not Christy."

Christy, Bill's oldest daughter, died at sixty-seven—young by the standards of this family, apparently made of enviable genetic stock. For a detective seeking traces of the cancer mutation, this mention of her death would be the only clue. Although her much younger sister does not write this—and may well have not know this—Christy died of ovarian cancer.

Christy had fourteen children—nine daughters and five sons. Starting in 1985, three of the daughters were, one after another, diagnosed with breast cancer. In 1993 Joyce, the second-youngest daughter, was diagnosed with advanced ovarian cancer. This was when a researcher in Madison, Wisconsin (not the graduate student from the 1970s), who had started to put together a pedigree of Frank and Nellie's children, suggested Joyce get tested for a BRCA mutation. She agreed—and tested negative. As it turns out, she was tested for the known "Jewish" mutations. A year later, when researchers in Holland had documented another mutation with apparent roots in Holland—a sizable chunk missing from the BRCA1 gene—Joyce was tested again, and this time turned up positive. Vonnie tested positive a year later; Rita tested in 2001. In the end, ten of Christy's eleven surviving children got tested for the mutation; six turned out positive. With their permission, the Madison researcher shared their status information and contact details with cousins they had never met (or, in one case, whom they had known as neighbors but had not suspected of being relatives).

Four of the sisters, one with her daughter, now sat in metal folding chairs in the hospital conference room. There was something that came close to being unique about these women, even by the standards of Lynch's remarkable "family information services." It was their ages. Here was eighty-two-year-old Rita, a former dairy farmer, who had survived breast cancer. Next to her sat seventy-two-year-old Vonnie, who battled breast cancer twice but had been cancer-free for nearly twenty years. Then there was sixty-five-year-old Joyce, who by this time had lived twelve years past her diagnosis of Stage III ovarian cancer—a virtually unheard-of feat. (Seventy-nine-year-old Bernice, who tested negative for the mutation, came just to meet new relatives.) "I think we have bad genetic genes," explained Joyce, "and good fighting genes."

"We know something that everybody doesn't know," added her older sister Vonnie. "We need to share the survivorship. I had really severe chemo, nineteen drugs a month, and I made it. That's because we don't just sit back and listen to the doctor." The doctor would probably beg to differ. Lynch suspected there was something about this blue-eyed branch of the family that improved their odds of survival. Their cancer seemed to develop later, and, with the long-ago exception of their matriarch, Christy, they did not seem to die of it. The later onset may have been the reason: The older the patient, the less aggressive and slower-growing her cancer is likely to be. Alternatively, there may have been something about these women's bodies that made them resistant to cancer in spite of their defective BRCA1 gene—and this may explain why the cancer came on late and, apparently, weak. It was Lynch's hope that the vials of blood his staff carried away from the family information service would shed some light on the difference between this branch of the family and the others.

~

Standing in front of this large gathering of people of different ages, professions, states of health and general well-being, holding a microphone his booming voice hardly required, Lynch looked like a preacher. He talked like one, too. He was here to teach. He covered the basics of Mendelian inheritance in less than five minutes and got into specifics. "What we call the natural history of the disease is simply when is it going to occur on average, and the answer is: early. Much earlier, as a rule, than the population expectations. The average age of onset of breast cancer in patients with BRCA1 and BRCA2 is about forty to forty-five years of age. We have patients as young as the age of twenty. Unusual, but we do have them. In the twenties and thirties it becomes more common, and then it gets to the forties and fifties, and then less common when we get to the sixties and seventies, but even there we will have individuals affected. And there is a branch in your family where there is a much later onset in several members of the family. If we followed the American Cancer Society or the National Cancer Institute recommendations for starting mammography, say, at the age of fifty, we'd miss all those individuals of the young age. We would recommend starting mammography at age twenty and doing it annually from then on.

"So, early onset." Lynch took a quick breath. The audience was riveted: They were having a consult with the biggest name in hereditary cancers, and he was pulling no punches. "Now, bilateral. That simply means both breasts. In this hereditary variety both breasts are at high risk. Now, what does that mean? Well, if I have a patient with a BRCA1 gene and she has cancer of, say, right breast, I'll talk to her about the opposite breast." That may have been a stock joke, but he laughed and the family reunion laughed along with him. "They need to know that the risk to the opposite breast is very high. They need to know the limitations of mammography. Mammography is not a perfect science like physics and mathematics. They miss lesions and then you go back and say, 'Here it was.' Because we are dealing with shadows, and maybe that shadow was a significant one in retrospect. So, given those factors, women may need to make a decision: Should both breasts with the whole nipple be sacrificed by prophylactic mastectomy?" Lynch was on to his main point now. He made the case for prophylactic mastectomies and oophorectomies.

He stopped just short of saying, "You need to do this." Indeed, what he said was, "I am not going to say, 'You need to do this.' But it's something that the woman should consider, knowing the risk, and the risk for breast cancer over your lifetime approaches 80 to 85 percent."

It would be silly for Lynch, who pioneered the idea of prophylactic surgery for hereditary cancer, to pretend not to have an opinion. He told me once he hounded a patient for a couple of years until the man finally decided to have his stomach out. The man had since become something of a proselytizer for the surgery, so he agreed to talk to me on the phone.

~

I called Larry in St. Cloud, Minnesota. Here was another rolling Midwestern voice. He told me about the family holidays of the years 1994 to 1995. First, there was Holy Thursday. That was the day Larry's oldest brother, then aged forty-four, was diagnosed with stomach cancer. He said the doctors were surprised, explaining to him that it was "an old man's disease."

"They were just going to remove a fifty-cent piece," said Larry. "When they opened him up, they took out a four-by-six-inch piece." He had a very difficult time recovering from the surgery, and just as he was getting back on his feet, he was diagnosed with a recurrence of the stomach cancer. He died six months later, in October 1995.

But before that, there had been Thanksgiving 1994, celebrated at Larry's sister's house. She cooked the dinner but was not eating any of it: She said she had a stomach flu—the latest in a series of stomach bugs over the past few months. Her diagnosis of stomach cancer came a few days later. "They knew right away that it had already progressed into other parts of her body and it was just a matter of time." She was thirty-nine years old, and her five kids ranged in age from ten to seventeen. She lasted just past her second daughter's high school graduation in June 1995.

Larry's other brother phoned on Halloween night, 1995. "He said, 'Guess what? I've just been diagnosed with stomach cancer!' I said, 'You are kidding!' He said, 'No, but I'm going to beat it.'" After watching his siblings' excruciating and ineffectual surgeries and chemotherapy, this brother decided to seek alternative treatment in Mexico. He went through what Larry described as "some food-cleansing program," came back feeling little better, and decided to have surgery. "They opened him up and closed him back up again." He was dead in March, at the age of forty-three.

In the space of a year and a half, Larry had lost three of his four siblings, all under the age of forty-five, to a disease he had never really heard of. He called the American Cancer Society and told his story. He got the phone number for Dr. Lynch's lab at Creighton University in Nebraska. In 1996 Larry, his parents, and the one remaining sister had their blood drawn. "We had pretty much ruled out environment," he explained. "One brother, when we were growing up, spent a lot of time with an aunt and uncle on their farm, because our parents were having a hard time financially. And then we all lived in different states." The odd thing about this particular hereditary cancer, however, was that Larry's parents, one of whom must have passed on the gene, were both alive and healthy. Lynch's lab identified the gene that caused the stomach cancer. Larry's mother carried it, as did Larry. What made them different from Larry's two brothers and sister, why they had not developed the cancer—and whether Larry's good luck would last as long as his mother's—the lab could not say.

In the spring of 1999 Lynch held his trademark family information service for more than fifty people from Larry's family. What doctors enticingly call "the pedigree" continued to expand, and in 2002 Lynch traveled to Minnesota again, for another of his sessions. "Some new, younger cousins showed up," said Larry. People got tested. Larry did not know how many of the cousins turned out to have inherited the gene. His sister tested negative, which left him alone with his dilemma.

When Lynch first brought up the possibility of preventive surgery, Larry "didn't give it a second thought." The famous doctor did not insist exactly: He persisted. He called Larry roughly every six months, to make sure the idea never quite left his mind. "He called me and e-mailed me in 2002 and told me he had just come back from Portugal, from a meeting of international doctors, and at that time a group of doctors had concluded that this surgery was very good at preventing the cancer. And he said, 'You are young, you've got children, I strongly recommend that you have the surgery.' And he went on and on, and I said, 'Doc, let me think about this.' And then the wife and I went on some long walks. And I would wake up and have a stomachache and then think." Larry paused. "Mentally, it was a heavy burden on me."

The surgeons took out Larry's stomach, an inch of his intestine, and an inch of his colon, to make sure no stomach tissue was left behind to develop cancer. He went back to work as a supervisor with the highway department three months later. Life changed, of course: "Well, I got to eat small meals. Today I went grocery shopping and I had a cup of coffee and a roll, and I'm full." Restaurant meals were out, as was drinking and eating at the same time: there was not room for both solids and liquids. To keep himself hydrated, Larry now had to sip water throughout the day. He lost about 40 pounds following the operation: going down from a not overly impressive 190 pounds to a very slim 150. "I put sandbags in my pockets on windy days," he joked. Hey, he was still around to joke about it. "I basically told my relatives that Dr. Lynch basically saved my life by finding the gene and talking me into surgery. I'd say in the next ten years I'll have a lot of nieces and nephews that will have this done."

They would not be the first to forgo their stomachs as a family. The largest family group to go for surgery counts eleven cousins—all of whom tested positive for the mutation in that generation of the family. Before scheduling his operation, Larry corresponded with a woman who had had the surgery done at the same time as her brother and several cousins, "and she said it was the best thing that ever happened to her." Larry showed the woman's message to his father when he told him of his decision: "He nearly fell off his chair, but the e-mail from her was good, knowing that my life wouldn't be turned upside down. And Dr. Lynch said chances of growing old weren't real good."

~

Back at the family reunion, the question-and-answer period, mixed with testimonials, had begun. A thin blond woman named Vicky rose and said she had had the surgeries—preventive mastectomies and a hysterectomy with an oophorectomy—four years ago and now wanted to know if soy products, known to contain certain estrogens, were safe to eat. Lynch said he did not know. His assistant, Carrie, said it did help with the symptoms. Karen, the family member who had taken the lead in organizing the event, asked the doctor to discuss the family's particular mutation. He explained that, in addition to breast and ovarian, the mutation seemed slightly to raise the risk of prostate and colon cancer. Sumedha added that one of the branches of the family had a lot of colon cancer, but it might not be related to the mutation under discussion. Then, quickly and without a noticeable transition, the conversation turned to the sort of reminiscences one would expect to hear at a family reunion—almost.

"I got the Creighton number from the American Cancer Society," said Sheila, a stocky woman with short dirty-blond hair. "I called Carrie and then I was gathering information for a while. I literally was at cemeteries looking up death dates so I could go look up the cause of death." Sheila was the driving force behind getting the mutation information together until she actually got tested—and turned up negative. Such was the strange logic of these things that her good luck destroyed her credibility as the local cancer organizer.

"Then we had more than five hundred people," said Carrie. "But we have now found more than two hundred here, through Sumedha.

I was referring people to her to get their results, and she said, 'I think I know another branch.'"

"My ovarian cancer started at Stage III," said Mary Lou, who had the characteristic chemo look: thinned silky hair and a swollen body, as though it were visibly pumped full of drugs. "I have had sixty-two chemos," she said. She seemed like she might be a little confused; her speech was slurred. "Well, there is no cure for my cancer. All they can do is chemo, shrink the tumor." Mary Lou had no question for Dr. Lynch. Mary Lou was the youngest of the five sisters in the generation that was "all having cancer together" in the 1970s. With her four sisters dead, she was the last representative of her generation, and when she developed ovarian cancer, she became the key to discovering the cause of all the family cancers. It was when she was tested that the researchers finally pinpointed the deletion in the BRCA1 gene.

Lynch ceded his place at the front of the room to a fortyish be-spectacled woman named Brenda, who now seemed intent on something of a pep rally. "My mother has lost her sisters, her mother. And has two daughters who are positive," she almost chanted. "But we also have a survivor gene, because there is great survivorship!"

Lynch stretched out his legs and leaned back in a plastic chair, looking very pleased. "This is the largest of all I've done, all over the world," he told me. He had held family information services in Uruguay, Colombia, Brazil, Argentina, Singapore, and elsewhere. "I can guarantee you that many of these people didn't really understand it before this and they still don't completely, but it's so valuable. And what does it cost? Almost nothing."

"Dr. Lynch, could you come to the front?" Brenda was not about to leave him alone. "Could you tell us why you do this?"

"I am interested in genetics," answered Lynch affably. "And as an oncologist, I really do not enjoy treating advanced cancer."

He latched onto a shy, beautiful blond who had been quiet all afternoon. Now it turned out she was a relative from Milwaukee whose father and grandfather had died of colon cancer. Lynch asked her for "any tissue that may have been preserved." He also told her to get a colonoscopy, quick.

In the other room, members of the family were having their blood drawn. "You know, Bob is positive," said Sheila proudly, speaking to no one in particular. "And he's got three kids and they are all here and they are all being tested. They are all in their twenties." A young man with five-day stubble and a baseball cap turned backward stood next to her, smiling smugly. I assumed he was one of Bob's kids.

~

By four o'clock the basement conference room had emptied out. Lynch and his team had gone back to Nebraska. The blue-eyed branch of the family—the long-term survivors—went out to early dinner. Others had gone home. Karen, the organizer, and I sat in plastic chairs and talked.

She told me about growing up in Green Bay, about her aunts getting cancer and her cousins developing the fear of it, about the second wave of cancer hitting more than twenty years later. "Amy, Nancy, Carla, Carolyn, and Cathy developed cancer in the mid- to late nineties. I started talking on the phone with Sheila, and we decided we had to find out about that Madison study." They remembered that graduate student or whoever he was coming to draw blood from the family when Karen was twelve. "Sheila called Madison; they found nothing but gave her the Creighton number."

The first family meeting included sixteen or seventeen female cousins. Sheila did a presentation, and everyone filled out the paperwork to get into Lynch's study. "They tested Mary Lou, Sheila, and a couple of others. They found the deletion. Letters went out to everyone saying, 'We have found the mutation, we want to test first-line relatives.' I asked my father to get tested. He didn't care but agreed to get tested anyway, though he was still of the thinking that it didn't affect the men. He was positive. I was tested in April 2001, got results in October, had surgery in November." An oophorectomy, hysterectomy, and mastectomy. Just like that.

The thing was, the trail had been blazed. In this family—no doubt in large part because of Lynch's influence—if people test positive for the mutation, they get surgery. "Misery loves company." Karen smiled. "Vicki's surgery was in October, mine in November, Terry's in December. When I found out I had the mutation, Vicki had just had the surgery. I went over to her house and said, 'I need to see what you look like.' She lifted up her shirt, and I said, 'Okay, I can do this.' And when I came out of surgery, there was Vicki waiting for me. And she said, 'I'm just here to say you can do this.' And when Terry had her surgery, we were both there waiting for her."

They were there for each other to talk about the first glimpse in the mirror ("Oh my god, what did I do?"), and menopause issues from memory loss ("I start reading a novel and toward the end I forget it all") to vaginal dryness to their new bodies and sex ("What did your husband say the first time you had intercourse?"). One of the cousins, Terry, had seen her mother diagnosed with breast cancer at the age of thirty-two. "Terry bathed her, when she was nine, after her radical mastectomy. Back then her chest just caved in." In those days surgeons took the chest muscle along with all the breast tissue. All the cousins had reconstruction, with saline implants. But, said Karen, the decision to have her breasts removed was perhaps hardest for Terry, who "remembered lying in her bed at night listening to her mother cry for hours."

As for Karen, who was now forty-seven, a stay-at-home mother of three boys, she had decided right away that she would never look back. And one other thing: "When we left Sumedha's office after getting my results, I said to my husband, 'I will never pine for that girl I didn't have.'"

~

I told Karen about my daughter. And my breasts. And my decisions. We compared notes. For an hour or so, I felt every bit like I was one of the "cancerphobe" cousins. Then I went back to my motel, ordered in, watched the rain fall outside the window as I chewed, and felt a vague sense of envy set it. Cancer, even in the so-called previvor stage, is the loneliest thing on earth. To think that some people go through it as families: Like much of what I had seen and heard that day, the thought boggled the mind.

Chapter 7

THE CRUELEST
DISEASE

Before my visit, Rob must have asked me half a dozen times whether I really wanted to make the two-hour drive east of Toronto to see him in the country. I assured him I did: I wanted to see him in his home and on his land, whose beauty, I could tell from our brief correspondence, was important to him.

The drive was dull: highway followed by a stretch of plain country road to the village of Hastings, which offered the trip's first visual respite with a couple of old red-brick buildings and a rapid river running through the center of the village. Another few miles down a side road, and I was in Rob's driveway, an unpaved path up a wooded hill. This is where my sense of my surroundings changed: Whoever chose this as a place to live had followed a very deliberate aesthetic.

Rob's house was exactly what it should have been: white walls, exposed wooden posts and beams, a glass wall to the porch overhanging the hillside, the smell of a wood-burning stove, a mix of vintage and new furniture arranged with a meticulousness that marked this as the residence of a comfortable urban gay man. Rob had prepared a lunch of soup and home-baked bread, with sliced California strawberries for dessert. This was a man who worked at comfort and harmony, and, from what I could tell, was very good at it.

We ate, drank a glass of white wine each, chased it with coffee (I was driving), and talked for a couple of hours. Rob was shy, thoughtful, and very eager to help with my project. When I transcribed the interview at home a couple of weeks later, I realized he had spoken almost exclusively in sentence fragments, often running on so that I had some trouble with the placement of punctuation marks. I had no way of telling if this had always been his way of speaking. A lot of people talk like that; it is perhaps less odd to the human ear than when a person consistently uses complete sentences. But Rob's impulsive and not quite grammatically correct speech may have been an early symptom.

Huntington's disease begins with signs that are best deciphered in retrospect. After the diagnosis, relatives usually remember that it was years ago that the affected person became sad or withdrawn, then depressed and difficult to get along with. Other common symptoms include lack of motivation, sexual problems, and suspiciousness or even paranoia. In other words, at the beginning the patient seems like any middle-aged person—the disease usually hits in midlife—having a bad day, week, or year. But the darkness never lifts. Instead, the symptoms become increasingly pronounced, and diagnosis usually comes after the patient develops characteristic involuntary movements, or chorea, which gave the disease the name by which it is still known in some places: Saint Vitus' dance. In affected patients, the gait becomes unsteady; they make endless jerky movements with their limbs and sometimes their head. Eventually, they will lose the ability to walk or even sit unaided. At the same time, their speech and memory will deteriorate, ultimately resulting in dementia. Their behavior, too, will grow more and more troublesome: They will be impulsive, disinhibited, socially inappropriate at best and aggressive at worst. Not long ago, Huntington's patients used to die on psychiatric wards, often misdiagnosed as having schizophrenia or another mental illness. Now most of them spend their final years in general long-term care facilities, where they die, bedridden, nonverbal. Many literally starve to death: Their loss of muscle control makes it impossible to swallow—they choke on every spoonful—and if a no-tube-feeding order is in place, death eventually follows. Other Huntington's patients die of the illnesses that always plague the immobile and the infirm: pneumonia, heart failure, or infection. Death usually comes fifteen to twenty-five years following diagnosis. The discoverer of the disease, a nineteenth-century Long Island doctor named George Huntington, described a patient in the end stages as "but a quivering wreck of his former self."

At the root of the symptoms is a mutation at the end of chromosome 4, where a pattern of three bases—CAG—is repeated an abnormally high number of times. In a normal gene, the number of repeats is between ten and thirty-five; a mutation carrier may have over a hundred repeats. The more repeats, it seems, the earlier the person will become ill. Still, geneticists cannot use the number of CAG repeats to predict when symptoms will set in: Even identical twins sometimes come down with the disease as much as seven years apart. People who have between thirty-six and forty repeats are said to fall into the "gray area," which means they may or may not eventually develop symptoms.

Rob had forty-one CAG repeats. Given the nature of heredity in Huntington's disease, his mother probably had the same number: outside the "gray area," but low as mutation carriers go. She was diagnosed at the age of sixty-three, very late by Huntington's standards. Her early symptoms were emotional. Her family grew concerned when she began having accidents. "Fender benders, left-hand turns, that sort of thing," Rob recalled. "Where you have to make a decision."

Difficulty with decision making seems to be one of the early typical symptoms of Huntington's. A person in the early stages of the disease will get hopelessly confused when faced with the need to choose between turning right and going straight—especially if, say, she needs to go to the bathroom at the same time. The options of turning either way or opening the car door and squatting to pee in the breakdown lane may all seem equally valid.

"They may go, 'I'm hot, I'll take my clothes off,'" Rozalia Andrejas, director of the Toronto and Area Resource Centre of the Huntington Society of Canada, described her clients' logic to me. "'Need to go to the bathroom? Good spot!' Trying to reason with the person is futile." Huntington Society supplies its ambulatory clients with business cards that read, "I have Huntington disease but I am very aware of what is going on around me. Sometimes I may be forgetful, have slurred speech, or have difficulty with my balance. It may take me some time to express my thoughts or answer questions. Thank you for your patience and understanding." These seem to be helpful in preventing Huntington's patients' arrests for public drunkenness or disorderly conduct.

The errant gene in chromosome 4 has been dubbed the huntingtin gene. It directs the cell to make the huntingtin protein, which performs an unknown function. What is known is that a mutant gene directs the cell to make the huntingtin protein with a higher-than-normal number of consecutive glutamines, which causes the protein to be processed differently and thus to accumulate in the neuron, which somehow leads to the death of brain cells. The exact mechanism remains murky to scientists, but the culprit of bizarre Huntington's behaviors and dreadful Huntington's symptoms, or at least most of them, seems to be the part of the brain called the caudate nucleus. This is a structure—two structures, in fact, since there is one in either hemisphere of the brain—that is responsible for transmitting information from different parts of the brain to the frontal lobes. The frontal lobes, in turn, are at least in part responsible for motor function, problem solving, spontaneity, memory, language, initiation, judgment, impulse control, and social and sexual behavior. In other words, for everything we do. A person in the very advanced stages of Huntington's disease—a decade or two after the initial diagnosis—has basically lost the ability to perform any of these functions. But a person in the early stages, whose caudate has deteriorated but remains largely intact, appears to be controlled by faulty circuitry: Signals flicker, producing the expected effect only some of the time, unreliably.

Decision making is the first ability to go. People start having trouble organizing their time, setting simple daily priorities. Those who know themselves to be at risk try to prepare by passing time management on to others. Rosie Andrejas described a client of hers, a high-powered lawyer who was still well enough to perform in court but had entrusted the making of his schedule entirely to his secretary. He was also pondering less-demanding fallback occupations—perhaps gardening. Another client, whom she described as "working in entertainment," was concerned with early signs of movement impairment. "We are not juggling fire and we are not swallowing swords anymore," Rosie told me quite seriously, looking at me through her round horn-rimmed glasses. "There is also a lot of travel involved, so this client tries to minimize the organizational component." In his professional activities, he had fallen back mostly on illusion: rope-cutting, card tricks, that sort of thing.

The early stage of Huntington's disease, when a patient is considered symptomatic but is still able to live independently and perhaps carry on at least some sort of employment, can last a long time—up to thirteen years. Eventually, though, a person drifts further and further away. "One of the saddest moments I remember is Thanksgiving dinner," Rob told me. "There were a number of us who were there. We always went out to my grandmother's. I think there were a couple of my cousins, and I had a couple of friends with me. A bunch of people there for dinner. Mom was still able, helping out with dinner and so on, and being there, but she was obviously affected. And she was still smoking. Her ashtray was a little ashtray with a picture of Bobby Burns, the Scottish poet. And she was stubbing her cigarette out. This little picture on the ashtray, and she was stubbing her cigarette out on his face. Somebody joked about it: 'Wonder how Bobby Burns feels about that.' So we all started to laugh. And Mother didn't get it. And she thought people were laughing at her, or she didn't understand why they were laughing. And it was a moment for her when she realized that she was out of it, behind a veil there, you know. And she just laughed and then she started to cry. And no one could explain—we tried to explain, but that was when I realized that communication had started to be a problem, because we couldn't really explain to her. That we were just laughing at Bobby Burns. Laugh with us, you know. Couldn't do it. Couldn't comfort her."

If a cruel deity set out to concoct the most punishing disease imaginable, it might have come up with Huntington's. The disease goes on for a long, long time. It gets worse slowly, inexorably, but unpredictably. And for much of its progression, the person trapped inside is aware of mounting losses, even if he or she cannot control them, grasp them, or, eventually, express emotions about them. It probably feels like a protracted bad dream, the kind where you try to run but cannot, try to scream but cannot, try to reach a target but cannot. What could be more horrible than that?

What could be more horrible than that would be to go down that long, darkening path after having watched someone you love go down it. Huntington's is hereditary: The vast majority of people who develop it have seen a parent succumb to it, and many are watching their siblings struggle as well. In this sense, Huntington's is like all other genetic conditions, including hereditary cancers: a person's fears and expectations stem from the parents' experience. But, perfect punishment that it is, Huntington's is also different. It is the only known human genetic disease that is what scientists call "truly dominant." A person who has two mutant copies of the huntingtin gene—one inherited from the mother and one from the father—will fare no differently from a person with just one bad copy. By contrast, people with two mutant copies of one of the BRCA genes are not born: Whatever the gene does wrong keeps the embryo from being viable. In a person who has a cancer-causing mutation, the healthy copy of the gene does its job for some time—possibly even well into old age—so that some mutation carriers never develop the cancer. The presence of one copy of a mutant huntingtin gene with forty or more CAG repeats is a 100 percent guarantee that the person will develop the disease: Such is the nature of "true dominance."

Fittingly, Huntington's was the first genetic disease for which a predictive test was developed. In 1983 a group of scientists located the huntingtin gene at the end of chromosome 4. In 1986 a test using linkage analysis became available. Starting in 1993 laboratories were able to offer a more precise and definitive direct gene test. The search for the gene was a family quest, initiated by a Hollywood psychoanalyst whose wife was succumbing to the disease. Their neuropsychologist daughter ultimately managed the search, and a second daughter, a history professor, movingly documented it in a book. Huntington's was the perfect test case for genetic testing: a disease that is fatal, incurable, and basically untreatable, but also "truly dominant." Each child of an affected person has a 50 percent chance of testing positive. A positive result tells a person he or she will definitely develop the disease and die from it, but it contains no information on when the symptoms will hit, or what they will be—whether it will begin with depression or aggression, with involuntary movements or with balance problems.

Recent research suggests that years before clinical symptoms become apparent enough for a diagnosis, the disease process has begun. One study examined 260 "at-risk persons"—meaning those who did not know their gene status, only that they had a 50 percent probability of having inherited the mutation. They were given various tests of cognitive abilities. When researchers went back to their test subjects two years later, 70 of them had been diagnosed with Huntington's; these were the people who had scored worse on the tests. In other words, they had been losing crucial brain cells and mental capacity long before a clinician would have diagnosed them with Huntington's.

Another study looked at the brain scans of mutation carriers who were considered presymptomatic, and concluded that visible changes were occurring, in the caudate as well as in other areas of the brain. Rob may have been in that study. "My neurologist got me into a couple of research programs," he told me. "They flew me down to Long Island three times. New York City, yes. North Shore Jewish General Hospital in Long Island. They were looking at whether they could see changes in PET scans before there were clinical symptoms. They wouldn't tell me, no, no. They gave me my scans. They were very colorful. I could see there were changes. Something is happening there, for sure."

I listened to Rob's clipped sentences on my recorder. His diction was also a little off—not disturbingly so, but as if he were sucking on a piece of candy. Had Rob always talked like this? Did he know he was talking like this now? Did he think he was showing symptoms?

"I think I have some little subtle things that only I know about," he told me. "Sometimes I lose my balance when I turn around. And I've always been a little bit disorganized, but, I mean, time management now. I have problems getting hold of my time, getting motivated. But that's always been my way." Here were the same questions again: Were Rob's symptoms actual symptoms, or just the normal expressions of speech habits, character, blood-sugar levels? And did the answers to these questions matter if he knew that sooner rather than later he would develop clinical symptoms of Huntington's, as everyone who carries an abnormal huntingtin gene does? Finally, did this knowledge itself matter if, even at the age of fifty-seven, having prepared himself for the onset of the disease that took his mother,

Rob would be unable to know when it actually came? Indeed, one of the particular cruelties of his predicament was that the certainty of his diagnosis—made on the basis of clinical manifestations rather than gene status—would come just at the point when his ability to understand the meaning of events began to erode. Virtually his entire adult life, from the moment he learned of his mother's diagnosis, when he was twenty-four years old, Rob's understanding of how Huntington's affected him personally was always and would always be slipping away from him.

"My parents told my brother and I together. It was quite a shock to everybody. It was a devastating moment for everybody, my mother's diagnosis, especially for her, poor duck. But for me—I was twenty-four, young enough that it seemed like something—I knew I was at risk, we knew about Huntington's disease, my parents, my mother was a social—well, both my parents were social workers, they'd encountered it in their own awareness, and they talked. I remember talking to them about it many, many years before my mother was diagnosed." A few minutes later, Rob picked up the lost thread: When he first learned of his mother's disease, "it was something I didn't have to worry about right now. I often thought about it over the years, especially after the test became available, and I was online discussing it with other people and families, and I spent a lot of time answering e-mail on the mailing list." Rob actually put up one of the first Web sites devoted to Huntington's disease, so he found himself in the position of amateur virtual counselor to many people at risk for the disease.

"Young people who were teenagers—a lot of them wanted to be tested, and I couldn't—at that time I was already forty-something, and I couldn't imagine a young person facing that test. I always counseled them carefully, because you didn't want to say—I wasn't an expert or anything, but my advice always was, 'No, don't. Wait. If you wait, then you've got the choice. If you take the test, then you don't have the choice anymore.' If the test had been available when I was twenty-four, I might have wanted to have it too. Immediately, you get the news that you are at risk, the test: Let's find out. What would it have done to my life? I think it might have destroyed me. Because I think in some ways I overreacted once I got the test, even though I was an adult." Just then lunch was ready, and Rob halfheartedly pleaded with me to take a break from the interview: "I make soup noises when I eat. I slurp." A lot of Huntington's patients in the early stages are embarrassed to find themselves unable to eat neatly any longer, because of decreasing muscle control. Then again, I forgot to ask Rob whether he had always slurped his soup.

Rob actually took the test twice. He pushed the idea of Huntington's to the back of his mind through his twenties and thirties, even as his mother slowly deteriorated. But after his fortieth birthday, Rob suddenly became obsessed with his risk. It seems to work like that with Huntington's: The fear kicks in on its own schedule. The French journalist Jean Baréma, who wrote a book about his own decision to take the test, described it this way: "That particular weekend, eighteen years after my mother's death, fear suddenly coursed through me, jolting me awake.... Quick—get the information!"

Rob's awakening to the fear coincided perfectly with the advent of the first predictive test for Huntington's. The early linkage test required blood samples from both parents, to attempt to determine whether the person had inherited the mutant gene from the affected parent. Even then, the results were not always conclusive: The test looked not for the mutant gene itself but for markers, signposts in the neighborhood that would indicate that the bad huntingtin gene was there. Rob's father was reluctant to give blood because, said Rob, "he was just in denial; he always would say, 'Oh, Robert, don't be stupid, you are not going to get Huntington's disease.'" Obtaining a blood sample from Rob's mother presented a different sort of problem: "She was fairly advanced in the symptoms, and it was kind of like an intrusion—you were never really sure if she understood what this blood sample was for. The trip to the doctor was a little bit sort of intruded on her space in ways that were not respectful necessarily." Ultimately, Rob got both samples and got the test. The results were inconclusive.

The direct gene test became available a couple of years later, in 1994. "I phoned them right back and started on a program to get that," he recalled.

If that makes it sound like a long-term, comprehensive undertaking, that is because it was. The first laboratories that offered genetic testing for Huntington's approached the task with grave responsibility—as though they were the first humans to attempt foretelling the future. The rules required that the person getting tested come with a partner, notify all at-risk relatives of the intention to get tested, and continue with counseling even after blood was drawn. The lab held on to the blood until the counselor gave the go-ahead for testing—usually about three months after the person's first genetic counseling session. The work itself took time too: The early linkage test occupied two lab technicians for two weeks, testing family samples and the sample from the potentially affected person, and then duplicating the study to confirm the results. It was not long before clients began to protest the process as excruciating and patronizing—and eventually guidelines began to relax—but Rob found the process reassuring. "I thought, *Hey, they are taking care of me!*"

Rob went to the counseling sessions with Bill, who had been his partner for about three years by that point. Rob showed me a picture from the early 1990s: two smiling bearded gay men, one a bit older, the other a bit heavier, sitting on a stone ledge, their legs hanging in the air. "He is fifteen years younger than I am." Rob smiled. "Poor baby. A lot to cope with."

It was Bill who burst into tears when the genetic counselor gave Rob the news. Rob was taken aback—not so much because it should have been his moment in the tragic limelight as because he felt compelled to take care of his distraught partner before he could deal with the knowledge he had just gained. "I don't think he ever recovered,"

Rob said. "He just pulled away afterwards." They stayed together for another ten years, and there is, of course, no telling whether they may have lasted longer if it had not been for the genetic test. But as with Huntington's, the signs of a relationship's disintegration seem clear in hindsight, and seem to reach far into the past.

"We wandered around, the two of us, for the next couple of days, in a haze of..." Rob trailed off. "After we left the hospital, we went to a park for a walk, and I can remember being there but I can't remember anything about what we talked about, and I didn't know how to be supportive except just to say that things will be okay and everything else, and I think what I needed really from Bill was a direct statement like, 'Don't worry, Rob, I will look after you.' Or something like that. Some kind of leadership, or at least a willingness to take part in the whole process and say, 'We'll figure it out.' But Bill, he was floundering even worse than I was."

Rob's father had died by this point, relieving Rob of the temptation to say, "I told you so!" Rob's mother was in a nursing home, bedridden and essentially nonverbal. Rob went to the nursing home every day, to sit by her bed and read to her. Sometimes he thought he saw "a flicker of a smile" on her face; that was the extent of her ability for interpersonal contact. She would live for another two years, but her son was all alone with their common illness.

Rob got his test results on a Friday. The weekend passed in a haze. On his way out the door Monday morning, he found a Post-it note from Bill: "Don't forget your lunch!" He was not in the habit of leaving his lunch at home any more than other people, but he was a marked man now. Although nothing in his physical or mental health had actually changed, Rob felt like he now stood at the threshold of his life as a man sick with Huntington's. "If I had known that in twelve years' time I would be talking to you, living in the country alone, and not too greatly changed from—you know—then I might have made different decisions. I started to sort of rush into doing things."

Rob quit his job. He did all the things genetic counselors had warned him against: He quickly made life-changing decisions, and he got ready to get sick tomorrow. At the same time, of course, he did exactly what the genetic counselors advised him to do: He prepared for the rest of his life, and he made plans and decisions in accordance with his new knowledge. The line dividing these two approaches is as obvious and as elusive as the line dividing the behavior of a man reevaluating his priorities in his forties from that of a man who has been told he faces a certain future of torment and death. At root it is the same thing, just much, much starker. One man, upon reaching midlife, has realized he will eventually die; the other has learned he will die a protracted, horrible death.

Rob had been working as an assistant at a Toronto reference library. He had taken the job intending to stay there "only long enough to figure out what I was going to do with my life"—that had been twenty years earlier, when he dropped out of college. He had grown to like his job, but had for years thought of photography as his true vocation. So he quit—only to discover the curse of all freelancers and artists: Time expanded and shrank at the same time, and he never seemed to get much done. He went back to work part-time for a while. Then an opportunity to house-sit for friends in the countryside came up. Rob moved, the arrangement dragged on, Rob got used to country living, and finally bought a place of his own: the house where I visited him.

He had some regrets. He would be better off financially if he had stayed in his job longer. Drifting into the country life might not have been the best idea either: "It stretched out without really knowing where it was going, it allowed me to get into that sort of— I didn't have any plan. And my relationship with Bill was falling apart, partially because I was spending so much time out here on my own. Anyway. Where was this going?"

I reminded Rob that we had started out talking about regrets. He said perhaps he should have stayed in Toronto. Later in the conversation, though, he said that going through with the genetic test "worked out okay. I wouldn't have done anything else." There was a slight possibility that Rob was exhibiting a behavior I had been told was characteristic of Huntington's patients: answering the same question differently each time, depending on how it was phrased or how the neurons happened to fire. But more likely, he was doing what we all do: continuously reevaluating his less-than-perfect predicament, now thinking he took a wrong turn somewhere, now thinking he was just where he should be anyway. "I'm prepared to stay here as long as I— obviously, I can't survive here if I can't drive. As soon as I can't drive, I'll have to do something else. And I should start making the sort of end-of-life things I want to do before I—while I still can, like go down an Arctic river in a rubber raft."

"It's a little cold, a rubber raft," I observed.

"I want to see the North."

"How will you know when you can no longer drive?"

"Oh, yeah. I'm very careful with driving. But so far, um, yeah, I'm very self-observant. I guess there will be a close call someday. Something will knock me for a loop, and then I'll have to sort of say, 'Well, it's time to give up.' I don't know."

"But you might just have an accident."

"Some jerk could have caused an accident. Oh, is today Thursday?"

It was Tuesday. This was not, however, a Huntington's moment. It was just that a garbage truck showed up on the wrong day for some reason. But it gave Rob a momentary scare—forgetting to put his garbage out might also be an early sign of the disease. He fretted for a minute, then observed that someone could walk through the door at that moment and he might forget how to introduce me. My saying that this sort of thing happened to me, too, was cold comfort. "The thing about Huntington's that's so horrible," he said, "is the mental stuff and the emotional stuff. I could accept any amount of physical disability and pain, but not being able to trust your own emotions and perceptions and your own awareness of reality—that terrifies me more than anything else." Imagine living with that fear every day. Now imagine living with the certainty.

~

In the early 1980s, when the possibility of developing a predictive test for Huntington's disease began to seem real, a number of studies attempted to predict how many people at risk for the disease would actually take advantage of the test. Every study showed that at least half and perhaps as many as two-thirds of those who were aware of their 50 percent risk of Huntington's would seek genetic testing once it became available.

"I believed all those people," said Barbara Handelin, a geneticist who had worked at the Massachusetts Institute of Technology lab that ultimately found the gene and who had gone on to run the first commercial laboratory to offer testing for Huntington's. "We all believed them. And so we got busy dedicating ourselves to coming up with these very—we hoped—very thoughtful protocols for how you go through testing." These were the protocols that required talking to family, coming with a partner, and sitting out a waiting period before the lab work ensued.

Nobody came. By early 1992—more than eight years after testing became available—only three hundred people in the United States and fourteen hundred people worldwide had completed the process. Even the Wexler sisters—neuropsychologist Nancy, who managed the campaign to find the gene, and Alice, an English professor who wrote a book about the quest, which began soon after their mother was diagnosed—never had their blood drawn. In April 1986 Nancy Wexler appeared on *60 Minutes* to talk about Huntington's. "I've always believed in knowledge for its own sake," she said. "And it is ironic that after working for precisely that, I'm now finding it much more complex than I ever thought it would be." Diane

Sawyer asked her if she had believed that she would take the test once it was discovered. "Absolutely," Nancy Wexler responded. "Yes. I never doubted it. And now I'm not so sure."

As of this writing, Nancy Wexler, now a professor at Columbia University and still active in Huntington's research, still had not been tested and had even become something of a campaigner for moderation in genetic testing. Speaking once again on CBS, in 2004, she said, "I think there's a huge amount of social pressure on people to get tested. I know that with me, if I were to go to bed every night thinking, I'm going to die of Huntington's, you know, why should I bother getting up?"

By the early 1990s researchers started looking at why people were *not* choosing to be tested. The following list of reasons emerged: increased risk to children if one was found to be a gene carrier, absence of an effective cure, potential loss of health insurance, financial costs of testing, and the inability to "undo" the knowledge. In other words, some people did not get tested because a positive result would raise their children's theoretical risk from 25 percent to 50 percent. Conversely, of course, a negative result would drop the children's risk to zero. Some people did not get tested because the result was pure knowledge: There was nothing one could do to postpone the onset of symptoms or to treat the disease itself. Some people did not get tested because they feared losing their health insurance or because the test itself cost too much, which is why I did the research for this chapter in Canada, which has universal health insurance. And, finally, some people were too afraid of what their own reaction might be, whatever the result.

There was another possible reason, the most universal human decision-making factor of all: habit. "Huntington's families grow up knowing, 'I've got a fifty-fifty chance of having gotten that trait,'" said Barbara Handelin, who by the time we met had spent twenty years thinking about this. "It's a sense of having lived with uncertainty. And it turns out that most people feel a lot more comfortable

with a state of uncertainty than certainty." I later found a study showing that people who chose to be tested were more likely to have
learned of their risk status as adults, rather than as adolescents. So it
may not be that most people prefer uncertainty—some just have had
to make lifelong partners with it.

~

In Sudbury (population 155,000), the largest town in northern Ontario, I did my interviews at a long-term care facility. I went there
with Julie Denomme, the social worker who ran the Northern Ontario Huntington Disease Resource Centre chapter and who had
arranged for the interviews. The long-term care facility was simply
a convenient, quiet location, but the choice of it was obviously ominous. Like most long-term care facilities in the area, this one housed
a couple of Julie's clients.

Julie wanted to introduce me to Abbee, a woman in her late thirties who was something of a success story. She had grown incapable
of taking care of herself out in the world. Julie's description was stark
and unappetizing: spoiled food in the refrigerator, a bathtub full of
dirty dishes, life-threatening weight loss. Placing Abbee in a long-
term care facility returned her to the world of the living, but the fact
that this world was populated with frail octogenarians served to amplify the sense of unfairness and insult that went along with Huntington's. Her disease, together with her situation, made her prone to
tantrums, sulks, and even hunger strikes. Eating is one of those
nightmarish Huntington's conundrums: It becomes more and more
difficult as the symptoms progress, but also more and more necessary
because the involuntary movements that are among those symptoms
burn up far more calories than a normal adult generally uses.

One of the nurses told Julie that Abbee had not been at supper.
Julie frowned: "This means she is upset." We walked—nearly
jogged—along long gray corridors, past doors through which I could
glimpse beds and televisions, two to a room, through heavy waves of

smells, institutional food mixing with medication, in search of Abbee. We finally found her in one of the smoking areas—not the one Abbee liked, explained Julie, and this might well be the thing that upset her. Abbee preferred to assert her separation from the other residents of the facility by using the other, more remote smoking area. Now that area had been closed.

Another of Julie's clients, Abbee's aunt Colette, was also in the smoking area, sitting in a chair, her knees propped against her walker. Colette's fingers were covered with cigarette burns, red and blistered and raw. Julie asked to see Colette's fingers. Colette, simultaneously surprised and embarrassed, asked for time to finish her cigarette. She had smoked it almost down to its filter, and now she tried to take one last drag, which proved impossible: She kept missing her mouth. Her hand slapped against her face, her mouth opening out of sync as her cigarette burned into her fingers, which had lost all sensation. Julie calmly asked why Colette did not use her "smoking robot," a mouthpiece with a long rubber tube that attached to a cigarette that was supposed to sit in an ashtray affixed to Colette's walker. Julie did not get an answer. Colette had large cigarette burns on her purple fleece sweatshirt.

Abbee, an average-build woman with long, not particularly clean, dark hair, watched the scene with evident disgust. As we walked back down the hall, Abbee explained that she was upset with Colette over something Colette had done—apparently hit someone during an aggressive outburst, but ages ago—and this was the source of her current state and her refusal to eat.

"It's not good, eh?" asked Julie.

"I don't know. I'd rather go to the other place. I've been getting a walker to sit on so I could smoke in the other place. But I can't go there now because they say they have too many siblings about smoking, so I'm angry at Theresa now."

"Oh, shnucks. Shnucks, shnucks, shnucks. So you are angry at something they did, and you are punishing yourself."

"Yeah, I know, but she is the one that's making me go here. I gotta go into my room."

"Because you are pissed off now. I'll come and talk to you tomorrow again, okay?"

"Mmmmmm."

"Is that okay? Because I don't want you to stop eating for a long time."

"I've only been eating one meal a day. Because in the other place I was so happy."

"Don't you hate change?"

"Yes, I know. That way I don't have to deal with Colette."

And so on. It was like talking to a four-year-old trapped in the body of a forty-year-old trapped in a place for eighty-year-olds. I got a chance to ask Abbee why she was there.

"I got diagnosed for Huntington's there and I had somebody come to my home but I had my slippers on so I slipped on my butt there so she said I can't be alone."

"Is that why you got diagnosed? Because you slipped?"

"I don't know."

"How old were you when you got tested?"

"I never was tested. No. I was going there, to do it, with my aunt, when she was going, but I chickened out of the testing."

"You didn't want to know?"

"I guess I was moving like it so I knew. When I was being tested with her, the doctor ask me, 'How do you move?' I say, 'I move fine, I feel fine.' Diane [Abbee's sister] said, 'I find Abbee moves a bit like my mother.' I said, 'Oh boy. I didn't want to know if I was being tested or not.'"

"Your mom had Huntington's?"

"Oh yes. Died at fifty. She was in a nursing home because she kept running away from home. Dad had to call the police to go find her. My cousin Susie didn't know about it, because we didn't talk to anybody about anything."

"Does your son want to know?"

"Oh no."

"How old is he now?"

"Ah, he is twenty-one. He right away wanted to know about Huntington's. But they are saying you can't do it for kids, eh? My brother went and got tested too, he lives with him."

"And what's his result?"

"Mmm, negative, yeah. He wanted to get tested, he says so."

"Your son wants to get tested?"

"Oh, he does. But they say he is too young, remember, I said?"

"But now that he is twenty-one, he can."

"Yes, I know. So he is going to college. Scholarship he got, you know what that is? They given him money for school, so it's free for him."

In the end we seemed to leave Abbee in a better mood. She even said that she liked Julie's pink leather jacket, and Julie promised to give it to Abbee once she stopped wearing it. I found Julie's way of handling the conversation extraordinary: She was caring and insistent without being either condescending or disrespectful of Abbee's privacy. She spoke the way a very good teacher speaks to an adolescent, that human creature who combines an adult appearance with childish fears and a sort of innocent self-destructiveness. Julie had been on the job for about four years. She had seen forty-five of her clients die. She had withstood battles none of us would want to witness—like when a nursing-home staffer started to feed a Huntington's patient who could no longer swallow but had signed a no-tube-feeding order. She had had her most firmly held beliefs turned upside down: She had even had second thoughts about her opposition to euthanasia. She expected this to be another death-heavy year: These things tend to go in cycles. She said she was "finding it extremely difficult to remain as good as I can be."

~

Wendy and I sat down for our interview in a conference room of the long-term care facility. Wendy was used to these sorts of places. Her mother, who had Huntington's, was in one. Her father, an alcoholic who had lost the ability to live independently, was in another one. Wendy was the assistant director of care at a third one. She spent all her time in them, working during the week and visiting her parents on the weekends. And she had made a very conscious decision not to find out whether she would also one day live in one.

Right around the time Wendy was learning about Huntington's disease in nursing school, she found out that it was the disease that had killed her grandfather. He had died at fifty-five. "That it was hereditary—that was the big one. And that it was the dominant gene. At that point my risk status was 25 percent, because my grandfather had had it and my mom had not been diagnosed. Plus, I was in a serious relationship that I'd just started, and I thought, What is he going to think? Are we going to have kids? And blahblahblah." It does get predictable after a while: Wendy's mother had certainly gone through all the same thought processes. She was twenty-five and pregnant with Wendy when her father died. Her sister had already had her children and attempted to talk Wendy's mother out of starting her own family, but Wendy's mother said, "You can't live forty years up the road" and went on to have two children. Out of five siblings, Wendy's mother turned out to be the only one who inherited the faulty huntingtin gene.

Medical training did not equip Wendy to recognize Huntington's at home. When her mother began to withdraw socially, it looked like depression, unsurprising in the wife of an alcoholic who was drinking more and more. Wendy's mother was a teacher, but even after her penmanship began to change perceptibly, the family pretended nothing out of the ordinary was going on. Wendy's mother was fifty when she finally tested positive for the mutation—even though she had pretended she was being tested to allow her children to learn of their risk. In her case the positive result amounted to a diagnosis.

When her risk was bumped up to 50 percent, Wendy already had a son. "I went through genetic counseling to determine whether or not I should be tested and whether or not I should have another child, and I didn't know what to do. And I was really cheated at that point because I wanted a big family and suddenly I felt like I couldn't have a big family, because of the risk: If I was gene-positive, what would it mean? I couldn't live with myself if I had given this to my children. So I went through the counseling, and it was the same geneticist that had tested my mom, and he was very familiar with our family line, and he and I together determined that I wasn't the type: There was nothing that could be done in terms of learning that I was positive. There was no miracle cure. It was a matter of a waiting game, and could I live with that? And the answer to that is, No. It would affect me and my life and my outlook, and I'd be sitting there waiting for the other shoe to drop. But I also did a lot of soul-searching, and I made the decision to have one more child, and I did, and I had a tubal when I was twenty-seven, right after he was born. And I have two wonderful boys. One that I carried when I didn't know and one that I carried when I did. So I also have to live with explaining to him someday, if in fact I am gene-positive. I wonder if he is going to be angry with me, if he is going to understand why I made the decision that I did."

Wendy planned to tell him pretty much what her mother had told her: that "you cannot live forty years up the road." Wendy was thirty-six when we spoke and her son was nine, and the fact was that, if she indeed carried the defective gene, she would be very lucky to be lucid enough to have this conversation with her son when he was old enough to ask a question that hard. Her own mother, at sixty-one, was in the advanced stages of the disease. She had been in a wheelchair for three years, because she had become incapable of standing or walking without falling down. The many falls she had sustained, several with impact to the head, might have been responsible for some types of mental confusion that were not typical of Huntington's patients, such as sometimes thinking she had seen relatives who had not actually visited.

Her mother's Huntington's symptoms, Wendy told me, included utter disinhibition, which in her case luckily served to amplify her cheery, outgoing personality. Where another Huntington's patient might have an aggressive outburst, Wendy's mom was just likely to be excessively loud in praising someone's blouse. Wendy insisted that she enjoyed her mother, with her newly unleashed sense of humor, and it was an hour into our conversation before she mentioned that during a recent visit her mother had grown angry with a nurse and attempted to bite her. She was not dissembling: It was just that at this point, Huntington's subjected her mother to more uncontrolled cheer than uncontrollable aggression—thanks, in all likelihood, to something that was fundamental to the woman's personality, to her children's unconditional devotion, and to some sort of neuronal luck. That this luck would sooner or later run out was nearly as certain as Huntington's continuing to progress in Wendy's mother.

Wendy tried to imagine herself living with the symptoms. "I could live up to the point where my mom is now. I could do that. So I slur a little bit, and I sit in a wheelchair and I don't stand. I can interact with my family, that type of thing, and have some sort of a meaningful relationship and make some sort of a contribution. But I don't know if I could do anything beyond that. I don't know if I'd want to be here to do anything beyond that." Considering assisted suicide, even if it concerned the consequences of a disease she had a 50 percent chance of avoiding, made Wendy feel like a hypocrite. "Religiously, I don't agree with it. I think we are all here for a reason and there is something greater than us and that's what keeps me going. I have faith. Without that, I don't know where I'd be." This was perhaps as close as Wendy came to explaining to me why she could not get tested. She had reached an uncertain equilibrium by relying on her faith. A positive test for the mutant huntingtin gene might have destroyed her ability to believe in a guiding power. No God could be that cruel.

~

Genetic counselors and social workers who handle Huntington's cases get very good at telling the ones who will go through with the test from the rest (although several of them told me they were famously bad at pinpointing the ones who would test negative or positive—unless clinical symptoms were already pronounced). They are the ones who always start Huntington's conversations with the social worker who comes to the house to check on the affected person. They are the ones who use the key phrase, "I've just got to know."

Genetic counselors try to weed out those who are too young or too ill-equipped to handle the information. Sometimes a twenty-two-year-old will agree to wait a few years, until he can form a contingency plan—and buy a life insurance policy. Sometimes a person who worries counselors will agree to see a psychiatrist. Ultimately, though, these days no one who really wants to get tested for Huntington's is turned away. Presymptomatic testing rates are higher than in the early 1990s but nowhere near the rates those early surveys predicted. In the different Western countries where predictive testing for Huntington's is available, roughly one in five at-risk people choose to find out if they carry the mutant gene.

Scott was one of those who needed to know. Handsome, tan—his job had him outside most of the time—he fit the profile of the Huntington's information lover. He was forty-eight, and it was less than a year since he had found out he was at risk. His father's disease set in very late—at seventy-eight—and he had hidden it from his children. He and his wife attributed the involuntary movements—including "rabbit-chewing," constant, uncontrollable, and intensely embarrassing, with the tongue always finding its way out of the old man's mouth—to old age. It was not until a series of catastrophic events at his parents' home pulled Scott into their household that he learned of the diagnosis.

His mother had to be hospitalized for knee surgery, so Scott was helping out his father. Just then the older man fell down in his driveway, hitting his face. Scott took his father to the hospital, where they were referred to a neurologist who said, "I think your dad knows something he is not sharing with you." Outside the doctor's office, Scott's father confessed he had been diagnosed with Huntington's.

That was the first Scott heard of the disease in his family. It is possible that this is one family where people have to live a long time to see symptoms set in, so many of its members were spared a Huntington's death. On the other hand, one of Scott's paternal aunts—his father's sister—died relatively young, some years after being diagnosed with paranoid schizophrenia. She might have been one of those misdiagnosed Huntington's patients.

Scott surfed the Net and went to see a geneticist. "It was funny because the doctor, the genetic counselor, kept saying, 'Are you sure?' And they had the form for me and I was just, 'Give it to me, I'm going upstairs.' 'Are you sure? Do you want to take some time to think about this?' 'No, I'm doing this!'"

When I spoke to Scott, he had already given blood but was still two months away from getting his results: In rural Ontario, genetic information was handed out strictly when the specialist came to town, which was not often. Scott assured me he was fine with waiting—he did seem perfectly calm—and explained the plan. "If I have it, I will tell my boys. If I don't have it, I will still tell them what Grandpa has and I will tell them that I'm okay and they are okay." His sons were twelve and twenty, and before going to have his blood drawn, Scott had taken out life insurance policies on them. He also planned a weekend fishing expedition he had been promising his sons for a decade. He had resolved to do the things he had been putting off, to work less and spend more time with his family, and perhaps to travel finally outside Ontario.

The thing was, Scott clearly did not believe he carried the gene. Occasionally, very occasionally, he casually diagnosed himself, the way all Huntington's family members do. "Every time you stumble, every time you forget something, automatically: Is it starting? If I get out of bed in the morning and I'm tired and I kind of do a little wave or something, and then I see my wife do it, I think, *oof, it's both of us!*" But he also did something else all people at genetic risk do: He applied the odds to his own family. I had spoken to people who said they could not get tested because a negative result would mean their siblings were positive—odds being fifty-fifty, after all. Wendy had something of a no-testing pact with her lone brother. And Scott, it was clear, believed his only brother was the one who inherited the disease. Scott's brother was also awaiting his test results when I interviewed Scott; he refused to meet with me, which, judging from what Scott told me, was unsurprising: Unlike Scott, he found the testing process excruciatingly difficult.

"You know, I said something to him before I knew my dad had Huntington's. I saw him at home, and he couldn't sit still, and his movements were just all over the place. And I said, 'You need to go see a doctor.' And he said, 'What for?' I said, 'You are just different.'" A short time later, when the brothers found out their father had Huntington's, the difference was named.

That sort of certainty can certainly be wrong. In Toronto I interviewed a young woman named Liz, who had learned of her father's Huntington's at sixteen: He was diagnosed in his early forties. She was tested at twenty-one but could never bring herself to return for the results. There was no point: She was so forgetful and experienced such frequent uncontrollable movements in her extremities that she needed no confirmation of her diagnosis. She resolved never to marry or have children. But at twenty-six she became serious about a man and decided to go back for her results. Liz was so surprised she was negative that she did not tell anyone about her results for a few days. Her memory problems turned out to be attributable to, well, not a great memory. The involuntary movements were symptoms of rest-less legs syndrome, a neurological condition that can be intensely un-comfortable but seems almost pleasant compared with Huntington's.

Scott and I agreed I would call him in a couple of months to ask about his results. I called and called, and he never picked up the phone. I submitted the manuscript to the publisher with a place-holder paragraph at the end of this chapter, saying that the self-confident handsome man I had met in Sudbury had tested negative while his brother had canceled his appointment with the genetic counselor a couple of times before finally going in to find out that he carried the mutant gene.

Finally, when the book was about to go to press, I asked Julie De-nomme to find out what really happened. It had now been more than a year since Scott was tested.

Scott's brother was indeed positive, and symptomatic. Scott him-self had also tested positive for the mutation. His number of CAG repeats, however, fell into the "gray area": thirty-eight, with thirty-six considered the upper limit of the norm and forty spelling certain Huntington's. The low number of repeats might have been the rea-son Scott's father's symptoms set in so late in life—and it might even mean that Scott himself would never develop the disease. But while this was good news, of a sort, it sounded to me like another cruel joke of Huntington's: Scott the information lover could get no in-formation. He still had not told his sons. The man who had planned so well, who knew exactly what he would do if he tested negative and what he would do if he tested positive, was now apparently stuck not knowing what to do.

Chapter 8

THE SCIENCE
OF MATCHMAKING

Miriam kept her five-by-seven index cards in a plastic box. She used paper clips to join the cards and photographs of the young women they described.

"I had to throw out her picture because we didn't need it anymore." She smiled. "If you had seen her picture, you would have said, 'Not pretty!'" There was no telling, of course, what I would have said, but Miriam's son Yehuda had said, "No way!" (or so Miriam told me).

But first Miriam had collected all the information she could on the prospective bride. She called her friend Bella, who ran a camp where the girl had worked as a counselor, and took notes. "Very attractive," Miriam scribbled on an index card. "Not a Harry." Harry is derogatory slang for someone whose ways have become too American. In addition, Bella told Miriam, the girl's mother was "not like an American Henrietta." She also said "personality 1,000 percent," "perfect for Yehuda," and "not European yeshivish, but definitely yeshivish." Miriam was not sure what "European yeshivish" meant, but she knew Yehuda, who was a Talmudic scholar, wanted a girl he could talk to, so it was good she was "yeshivish."

Miriam also called the principal of the young woman's high school, and scribbled down more superlatives on the other side of the index card. "Petite, thin, pretty, adorable, sparkling eyes. I love spending time with her. She is light, she is radiant." The principal said she singled out the girl back in eighth grade—the principal's first year at the school. "I went to her and I said, 'When you finish high school and you finish seminary, I'm going to give you a job. Come back here and become a teacher in high school,'" said the index card.

The principal also commented on the young woman's mother: "Very special lady. Not pretentious. She is down to earth. She is not intimidating." This was reassuring to Miriam, because the prospective in-laws were a rabbi's family, and a rabbi's family can seem a world apart, out of reach. Miriam's husband was a real estate developer, and Miriam herself was an educated woman and a published author. Her family lived in an elegant apartment on Manhattan's Upper West Side, and by all measures were a successful, worldly family, but even so they could be intimidated by a rabbi's family should the family choose to intimidate them.

Choosing a mate for her son was more than choosing a wife: It was looking at a potential partnership for the entire family. Miriam had seven children, and it would be reasonable to expect that each of them would marry into families that had as many or more, and would in turn have between five and twelve children of their own. Orthodox Jewish families tend to expand quickly to the size of a large village, and planning such a village takes concerted, reasoned effort. Miriam's eldest son and eldest daughter had already married. It was now Yehuda's turn. Next would come Rachel, who was in Israel for the year. A couple of calls from matchmakers had already come in for her, but Miriam had not yet started another box of index cards: She did not expect Rachel to start dating until she returned to New York in the summer. With any luck, Yehuda would be married by then. It was now November.

There was another notation on the index card: The principal said that one of the rabbi's children had had two children who died, and another had had a child who was born premature and had "complications." So Miriam called up one of the family's references, a man named Yan who was clearly in a position to advise her: First, one of his children had married into the rabbi's family; second, he was himself a professional matchmaker. According to the index card, Yan recommended the prospective bride as "easy-going, warm, affectionate." He mentioned she had been valedictorian of her high school class. He also highly recommended the family. As for the mysterious illnesses that worried Miriam, he reassured her. He said both the rabbi and his wife came from large families unaffected by World War II, meaning their genetic histories could be plainly viewed. Neither branch seemed to carry any genetic illnesses. The two dead children were, he said, "a freak accident." The index card said: "No other indications, no markers to check for this, same odds as if you married anyone else." The premature birth in the other family, he said, was an unrelated event.

"He had done his research," said Miriam. She was reassured. She set up a date. Yehuda found the girl nerdy-looking. Her clothes were too large for her. Miriam discarded the photograph, clipped the index cards together, and moved on to the next candidate. Yehuda, a pale bespectacled young man with a large square chin that provided balance for his wide-brimmed black hat, was a tough customer: He dreamed of a girl who was both sophisticated and learned, worldly and modest, elegant and unconcerned with material goods. With social dating not practiced among Orthodox Jews, who believe in the separation of the sexes, it was up to his mother to find a match for him. Miriam, for her part, wanted to find a family with whom she as well as her son would be comfortable. And, it went without saying, both families would have to make certain that their children's children would be free of hereditary genetic illnesses.

~

Jewish law prohibits a man to marry a woman from a family of epileptics or lepers, to avoid passing illness on to future generations (leprosy is a bacterial infection, not a genetic condition, but since most of the world's population is naturally immune to the disfiguring disease, susceptibility may indeed be inherited). Jewish scholars have interpreted this prohibition as applying to all hereditary conditions. Through most of history, though, that ban must have been difficult to observe. Recessive conditions, such as Tay-Sachs, have a way of showing up unexpectedly, when two mutation carriers from apparently healthy lineages marry. Dominant conditions, such as hereditary cancers or Huntington's disease, most often manifest when a person has passed reproductive age. Then again, perhaps Jews did their best to observe the ban but genetic drift won out anyway, causing high carrier frequencies of several genetic conditions among Ashkenazi Jews.

In the twentieth century, if one were to interpret the ban literally, one might have devised a testing program whereby all couples wishing to be married by a rabbi would have had to undergo genetic testing the same way some American states still require people to be tested for syphilis. Naturally, before a global solution like that could appear, less-observant Jews began to take advantage of prenatal testing. The incidence of Tay-Sachs began to decline after a biochemical test (not yet a DNA test) became available in the 1970s, but among observant Jews, virtually none of whom would consider a therapeutic abortion, nothing changed: With a carrier frequency of one in twenty-four Ashkenazi Jews, one in five hundred and seventy-six couples was statistically likely to consist of two carriers. With as many as twelve children in a family, three of these, statistically, would be affected. Kingsbrook Jewish Medical Center in Brooklyn maintained a sixteen-bed Tay-Sachs unit, and there was a waiting list. With most Tay-Sachs children not living past toddlerhood, that means that hundreds of couples were watching their apparently healthy babies suddenly become less responsive, turn into vegetables, and die.

Rabbi Josef Ekstein had it happen to him four times. Three of his children died of Tay-Sachs while he was living in Argentina. He was so terrified of stigmatizing his six healthy children that, when his family moved to New York State in 1981, he tried to conceal his wife's pregnancy in case this was another affected child. It was. Ekstein decided to pay a foster family to care for the boy, as is sometimes done among the Hasidim.

Ekstein described the experience to a journalism graduate student twenty years later. He recalled going to visit the foster family. "The child was not kept clean. I saw the child was not taken care of properly. I saw the child had a rash and was not diapered. Then I thought, 'Enough is enough. Everyone knows anyway.' I had to bring him home." All four of Ekstein's affected children died between the ages of two and four.

There is no grief more intractable than that of a parent who has lost a child. When the child's death is preceded by months or years of exhausting caretaking, when there are healthy children competing for needed parental attention and time, when the wrenching experience is repeated four times over, the burden grows greater and greater still. Psychologists have identified various cognitive, spiritual, and "distractive" coping strategies that parents use to survive the loss. Some manage to see the value even in a short life deprived of experience and sensation. Some find support in a community of others who have faced the same sort of grief. Some—these are more often the fathers—create unrelated challenges to shift their thoughts from the child. It was Joseph Ekstein's emphatic inability to cope with his sick and dying children that made him decide to go to war. He decided to eliminate the kind of misery with which he could find no peace. His initial suggestion was to create a straightforward premarital testing program in the Hasidic community to keep carriers from marrying each other. Rabbinical and community leaders rejected the idea: Such a program would stigmatize the carriers, they argued, causing more psychological and social harm to living young people than medical good for children yet unborn. "Those families that had the problems with Tay-Sachs, they were in tremendous fear that I am taking out the problem to the streets," Ekstein told a television interviewer years later. "They were very, very nervous about it that it was going to damage them. Even my own wife was totally against the entire idea."

Rabbi Ekstein, then in his midthirties, was perhaps less equipped to devise a genetic-disease prevention program than most American Jews. The title "rabbi" in his case indicated only that he had completed a course of postsecondary religious education, not that he was a community leader. Indeed, he had always worked as a scribe—in Hebrew. Born in Hungary and educated in Argentina, Ekstein spoke little English. The Hasidim, who believe in preserving the shtetl way of life, generally communicate with one another in Yiddish. That would not do, however, for communicating with medical laboratories or reading papers on genetics. When it came to English, Ekstein could not even use a phone book: He did not know the order of the alphabet. But by all accounts, he was extraordinarily stubborn and persistent. He refused to talk to me; a younger colleague explained that journalists had often mangled the rabbi's words, misrepresenting the program he founded. My own persistence, over months, failed to break through his defenses: The rabbi fed me vague promises and a fair amount of written information, but never met with me or allowed me to visit Dor Yeshorim, the program he ultimately founded.

The convoluted design of the program follows an unfailing paternalistic logic: Knowledge is dispensed with extreme care, and only when absolutely necessary. Young women and men are tested while still in high school—private Jewish schools for the girls, or rabbinical seminaries if they are men. A Dor Yeshorim representative visits the school to talk about genetic diseases and pass out consent forms. On a designated day, most of the students—roughly 90 percent, by

Dor Yeshorim's estimate—come to school with consent forms signed by their parents and the $120 fee—not enough to cover the full cost of the test, but it helps to offset Dor Yeshorim's costs on the one hand and, on the other, to communicate the message that the test is both worthy and important. In return, the students receive small cards with Dor Yeshorim's phone number and the testee's unique identifying number.

The blood samples go out marked only with the identifying numbers, not with names. Each shipment includes blood samples intended for quality control—blood analyzed earlier by a different lab, to check that the results match. Results themselves are entered into the organization's database twice, by two different people, again to ensure that no mistakes are made.

The young people who are tested never see their results. Indeed, individual results are never dispensed. A couple must call with two Dor Yeshorim identification numbers and ask whether they are compatible. The call will be returned—an average of eight minutes later—to either of the home phone numbers the members of the potential couple put down when they took the test. If neither member of the couple is a carrier of any of the Ashkenazi diseases for which Dor Yeshorim tests, or if only one is a carrier, or if both are carriers but for different conditions—if there is no risk that their children will be affected with a hereditary recessive disorder—the voice on the phone will simply tell them they are "a match." If both are carriers of a mutation for the same disorder, they will receive genetic counseling—over the phone. Frances Berkwitz, a genetic counselor who began work at Dor Yeshorim in 1983 at the age of sixty, will tell them that, should they get married, they would have a one-in-four chance of having a child with a severely debilitating or fatal disease. Berkwitz will never know to whom she is talking: She will never see the faces of the couple, and she will not hear their names.

Dor Yeshorim, in other words, is designed to minimize knowledge. Dor Yeshorim staff know the test results but not the people whom they mark as carriers. Most of the carriers will never know their status: The chances of dating another carrier are never more than one in ten. "You don't need counseling, we do the job for you," Ekstein told the journalism graduate student. "You don't get a letter, 'You are a carrier and just do whatever you want.' When you go out and explain to your potential suitor that you are a carrier for this and for that, it's not easy." Indeed, Dor Yeshorim saves both you and your potential partner from unnecessary knowledge. Only the unlucky few—statistically, one in a hundred couples—will learn of their results. But, unlike other recipients of bad genetic news, they are offered the opportunity not to struggle with a decision: In Dor Yeshorim's framing, they will already have been deemed "incompatible."

Dor Yeshorim has no way of tracking how many of the roughly seven hundred couples they had, as of this writing, informed of their incompatibility actually broke up. By all indications, the vast majority do seek a different partner. For one thing, they tend to return to Dor Yeshorim with a request to check their number along with a new one. For another, the number of children born with Tay-Sachs in the Orthodox community in North America has gone from an average of fifty to sixty a year in the 1980s to between four and six a year in the 2000s. Some of the couples may marry and then use pre-implantation diagnosis—in vitro fertilization followed by the testing of embryos to weed out affected ones—but this would be prohibitively expensive for people. In any case, Kingsbrook closed its Tay-Sachs unit in the late 1990s.

~

1) My sister is engaged! She's the sister right above me . . .

2) Ap's, papers. Aaaaahhhhhh! . . .

3) Dor Yeshorim. Let it be known that i am a huge scaredy cat. But I plan on improving. So I will not think about it, and look the other way, and hope for the best.

4) Graduation...

5) Senior Appreciation Dinner.

Jewish teenagers' blogs made it apparent that Dor Yeshorim had become an accepted rite of passage, as banal and momentous as high school graduation. Miriam could not quite recall when Dor Yeshorim became a fact of life. She and her husband had not been tested: The program was younger than her oldest child. In fact, she recalled, they and everyone else in their summer community had given money to Dor Yeshorim in the early days. And now she could not see any of her children marrying without checking "genetic compatibility," in Dor Yeshorim's language.

Normally a budding couple in the Orthodox community would go on eight dates before becoming engaged (for the Hasidim, the average number of dates is more like two, and the dates are conversations in the girl's parents' home, while the Orthodox would likely go out to a lounge). So a courtship might last a month or so—and Dor Yeshorim would urge the young people to check their numbers against each other earlier rather than later, to avoid bitter disappointment. But Miriam's eldest daughter was engaged to her husband within a week of meeting him.

"It was a story for the books." Miriam beamed. "It really was. He had dated for many years. He was dating for seven and a half years. He was very much sought after. He was driving to Baltimore to meet a girl, making like this huge effort—and nothing! He even went out with my daughter-in-law, my oldest son's wife. It wasn't what he was looking for." Miriam's husband had tried to help to find a match for this young man, who had no family in New York, but nothing worked—until he met Miriam's daughter at a small recital in their apartment. "Somehow it was so natural and it was so obvious," said Miriam.

When did they check with Dor Yeshorim? Miriam called to the other room: "Nachum? When did you check with Dor Yeshorim? In your whole dating career? Tell Masha how it happened." In the other room, a baby screeched and a man's voice mumbled something unintelligible. Instead of Nachum, Miriam's daughter, radiant and disheveled in the way of new mothers, emerged from the other room. "We went out on our fourth date," she said, "and he asked me to marry him. But first he had to check with Dor Yeshorim the next day. So he didn't really ask me fully, 'Will you marry me?' but the next day he found out in the morning and he sent me a text, he messaged me, and that night he asked me to marry him.

"The first time he just like gave me a smile, and it was, like, understood. And the second time—" Mother and daughter laughed, and I never got to hear how precisely Nachum proposed, and what exactly he put in the text message that morning. It was all so well understood by everyone—it seemed it was fate. So, what if Dor Yeshorim had said, "No match"?

"Then they would not have married," said Miriam.

"No question about it?" I asked.

"No question about it," she answered.

Just then her daughter, who had been tending to the baby, came back into the dining room, where Miriam and I sat at a very long dining table, Miriam's index cards spread out on the tablecloth before us. Miriam's daughter said she knew a man who was about to get engaged to a young woman but got bad results from Dor Yeshorim and broke it off.

"In Vienna," the young woman continued, referring to the city where her husband, Nachum, had grown up, "there is this couple who never checked Dor Yeshorim. She is from Australia, and I don't think they even knew about it."

"They just didn't think about it," added Nachum, a balding, flushed young man who had now joined the conversation, the screeching four-month-old baby in his arms.

"They just didn't think about it," said his wife. "And they have a kid that has—what does he have?"

"Tay-Sachs!" exclaimed Nachum.

This was the best evidence I had seen of Dor Yeshorim's success. The institution's existence was now so much a given that people who did not use it were, in the new generation of Orthodox Jews, becoming the stuff of legend. And Tay-Sachs was clearly becoming an abstraction: a disease with a forgettable name, affecting faraway people.

Miriam, meanwhile, was recalling a long-ago time, when she herself was looking for the perfect match. It was a long process, and she kept a meticulous record of it, two thick notebooks in which she noted down the details of every date. Her first one took place on February 16, 1975, and it was not good. "Uncle Lazar and Cousin Laishe suggested him," Miriam read aloud to me. "An accountant... It was pouring out. Sunday afternoon. I was so nervous: my very first date. I wore Mommy's blue dress with red trimming. Mommy wanted me to wear my good girdle. I just couldn't. I wore Mommy's good black boots.... Not too bad. Nicely dressed, not horrible-looking. We finally left. Went by train to Madison Square antique show. He just wanted to look at the Yiddishe things. Not my speed. He spoke badly of someone that I knew, which upset me. He never had heard of the *Titanic,* which to me was ridiculous. He bought me sheet music of Freylach. Came home by train, discussed uncomfortable seats in the old trains. Finally: home, sweet home."

"Critique," Miriam wrote at the end of the description. "Not for me. Too religious and narrow-minded. Uncultured, uneducated, not good-looking at all. Took me by train."

Miriam proceeded to meet about one new young man every month. Number thirty-nine came on December 22, 1977. "Shalom is in first-year medical school. If I had known, I never would have agreed to meet him: I'd never marry a medical student." Miriam paused to explain that this was because medical students were never home. "But Uncle Leon didn't even ask me: he just called one night and told me that this Shalom will call. And so he did. He called to speak to me and we had a very nice talk about thirty or forty minutes about euthanasia, abortion, etc." Miriam laughed.

It went downhill from there. "He is probably one of the ugliest boys I've ever met. He had this terrible case of acne. Emaciated, tall, extremely thin face, a mark on his nose from his glasses." A painstaking description of a miserable date followed, in Miriam's fine handwriting.

In another year, Miriam finally went on a date that she rated "Terrific. A+." Like all descriptions of love, the one that closed out her second yellowing and fraying notebook was predictable: "Really gorgeous, unusually handsome boy. Charming. The most charming boy I've ever met. Outgoing, friendly, talkative, and what a personality! I think he is perfect." From all appearances, Miriam's husband, who was baking chestnuts in the kitchen with the younger boys while we talked, *was* perfect.

Miriam shut her notebook. "In those days, we didn't do research," she sighed. "See, look, 'Uncle Leon called'—he called, that was it! It was much different then. No one did proper research. I would never do that to my girl. Today it would never happen like this. Never! Never! Thank God we have come a long way since then."

Here was the explanation of why Dor Yeshorim worked as well as it did. Matchmaking, in the Hasidic and Orthodox communities, was a science. Miriam's index cards, and even her premarital notebooks, were simple databases (her daughters will surely use computers, quite possibly with custom matchmaking software, when it comes time to marry off their children), to which Dor Yeshorim had added just another field. In addition to information on family background, education, extent of religious observance, community activities, other interests, personality quirks, world outlook, appearance, and manner of dress, there was now genetic health—a category that had in fact always been considered but that had now been refined and given its own procedure. This also explained why, though Dor

Yeshorim quickly grew to serve the ultra-Orthodox and the Hasidim in North America, Britain, and Israel, it had the most difficult time reaching into the Modern Orthodox community.

~

"I'll wear a blue shirt, I'll wear a white shirt, I'm not going to go to work in a T-shirt. I haven't worn jeans since I was like ten or twelve years old." Rabbi Howard Katzenstein was trying to explain Jews to me. We were sitting in his cramped office in Lower Manhattan, at the Orthodox Union, where Katzenstein was in charge of the kosher compliance program and also served as business manager. He had a bookcase stocked with cans of Pepsi, jars of canned goods, and various other packaged foodstuffs. This was the kosher hall of shame: items that were marked as kosher—either by accident or oversight or devious design—but were not. "When we have something that is potentially not kosher, I drop everything: It's Priority One," Katzenstein explained when he interrupted our interview to fire off an e-mail acknowledging that another kosher snafu had been resolved. Here was another reason Dor Yeshorim had worked: It worked in a community every one of whose members was committed to observing strict dietary limitations even in this overstuffed time in American history. This they did in the name of community and identity: Faith, after all, is a private affair in Judaism, while observing tradition is a community effort. Dor Yeshorim offered a new tradition, which, the rabbis confirmed, grew logically out of Jewish law, and the community absorbed it easily.

In its first year, Dor Yeshorim convinced all of forty-five young people to be tested—and "they were people who felt bad because of Rabbi Ekstein's personal story." Katzenstein smiled. He himself came to work at Dor Yeshorim in 1993, ran the operation for three and a half years, and has remained on the board since. He was the man Rabbi Ekstein put forth to inform me on the one hand and to keep me away from the organization's offices on the other.

A Jewish real estate family subsidized the organization at first. The family insisted that Dor Yeshorim charge for the test—mostly to ensure that young people place at least some value on their Dor Yeshorim cards. The charge was five dollars—a tiny fraction of the cost of the test. The organization worked out of a one-bedroom apartment in Williamsburg and shipped its samples via car service—to someone who lived near the lab and would hand-deliver them. Incredibly, all the precautions and controls of Rabbi Ekstein's invention were in place even at this informal stage of the organization's existence: Every shipment contained control specimens, every vial of blood had a backup, and every result was entered twice. There were printouts of the database, too, so that Rabbi Ekstein, who did not use a computer, could thumb through the printouts to find the eight-digit number he needed when he answered the office phone in the evenings or on Sundays.

By the second year of Dor Yeshorim's existence, several rabbis had endorsed the program and the number of tests grew to 250. Over the next twenty years, Dor Yeshorim screened 170,000 people. Of that number, 700 couples tested "incompatible," which meant that 1,400 people learned that they carried a defective gene. The roughly 17,000 young people who are tested each year represent 90 percent of the students in participating schools (Jews of Sephardic descent and converts probably account for a majority of the remaining 10 percent, making the number of holdouts negligible).

As of this writing, Dor Yeshorim was testing for genes that cause ten recessive diseases: Tay-Sachs, cystic fibrosis, familial dysautonomia, Canavan disease, glycogen storage disease Type I, Fanconi's anemia Type C, Bloom syndrome, Niemann-Pick, Mucolipidosis Type IV, and Gaucher's disease. Four of these diseases—Tay-Sachs, Canavan, Fanconi's anemia, and Niemann-Pick—kill in childhood. Mucolipidosis Type IV patients may live to middle age but are invariably severely mentally retarded. Bloom syndrome patients are extraordinarily susceptible to many kinds of cancer, which is what often kills them at or before middle age. But cystic fibrosis, which was the second disease Dor Yeshorim added to its panel, is trickier. In the early 1990s, when Dor Yeshorim was sorting the "Ashkenazi" mutations from a long list of mutations that cause this relatively common genetic condition, cystic fibrosis was a disease whose victims died in their teens. A dozen years later, it had become a manageable disorder with a life expectancy reaching into the thirties, forties, and perhaps beyond—albeit at great cost, and requiring constant commitment. Familial dysautonomia, a systemic disorder whose victims used to die young of pneumonia, requires a similar sort of maintenance regime of medication and physical therapy—but affected individuals are now living well into adulthood. Early diagnosis and proper treatment, including a special diet, has allowed children affected with glycogen storage disease Type 1 to become reasonably healthy adults. Finally, Gaucher's disease, an enzyme deficiency, varies greatly, and does so unpredictably: Some people's symptoms are very mild, while others' are debilitating. Dor Yeshorim resolves this issue by offering testing for Gaucher's disease only by request, which means that only people who have reason to suspect they are carriers will likely ask for it. That may be a self-selecting group: The relatives of a very mildly symptomatic person would presumably be less likely to worry than the relatives of someone who was severely disabled.

Speaking to a *New York Times* reporter in 1993, Rabbi Ekstein actually argued that testing for Gaucher's was even more useful to the potential couple than testing for a less ambiguously horrible disease. "With Tay-Sachs, there may be ethical reason to abort," he said. "But there is no ethical reason to abort a Gaucher's baby." Hence an affected couple would probably raise several disabled children. Therapeutic abortion is not generally an option for the Orthodox, but some rabbis—who, in the Jewish tradition, are entitled to resolve such issues—may take some arguments into consideration and make some allowances in some cases. The traditional view is that a fetus is not a human life before the age of forty days, which makes early abortion not a murder but a onetime violation of the order to be fruitful and multiply. Practicing birth control, on the other hand, would be a systematic abdication of the responsibility to multiply, and is therefore not an option for Dor Yeshorim couples.

Dor Yeshorim's stated position is that it will test only for recessive disorders that are severely debilitating or fatal. Depending on one's definition of severe disability and one's understanding of life span, that may or may not be true of cystic fibrosis, familial dysautonomia, or even Bloom Syndrome. After all, Huntington's disease or BRCA1–related cancers also often end lives in great suffering around middle age, but since these disorders are dominant, testing for them would put Dor Yeshorim in the position of marking certain people as entirely unmarriageable, which might correspond to the letter of Jewish law but would never go over in a modern community, even one as antimodern as the Hasids. So why does Dor Yeshorim test for Bloom and cystic fibrosis? Its booklet skirts the issue by stating that patients have "a significantly shortened life span with great pain and suffering," which again begs the questions of how short is short and how great is great. For most people who are tested by Dor Yeshorim, this will never be an issue because they will not be in the position of being a carrier dating another carrier. But roughly one in six hundred and seventy-six couples will learn they may have a child with cystic fibrosis. Dor Yeshorim will do them the favor of perhaps exaggerating the potential disability without discussing the possibility that the child might lead a normal life at an expense and effort that a family with seven or ten or twelve children could not bear. They could, of course, do their own research. They could even forgo Dor Yeshorim altogether and seek genetic testing and counseling at a hospital. But that would mean taking on the full burden of knowledge and decision.

"The Hasidic community are very much followers of what the Hasidic rabbis and leaders suggest." Rabbi Katzenstein was still explaining the Jews and Dor Yeshorim to me. His office phone kept ringing, his cell phone kept buzzing, his e-mail program kept binging its notifications, and he kept talking. "It doesn't mean they are sheep, but they'll certainly consider something. And Dor Yeshorim got the endorsement of many, many of the leading rabbis of the generation, the previous generation, because it's been in business since '84 or something. Then there is what we call the yeshivish community," Katzenstein went on. "I'm from that segment, meaning, probably wearing hats but not as Hasidic. Many of us go to college. I graduated from City College. Likelihood is, they'll become some sort of professional—you know, an accountant—or do things like storekeeper, that kind of stuff." Dor Yeshorim had done well in this community, both because it was well organized—virtually all girls attended private Jewish high schools and virtually all boys, rabbinical seminaries—and because this community did not permit social dating except as arranged by their families, through meticulous research. This was where Miriam came from.

"Within the yeshivish community, many of them will read the *Times,* listen to the radio," continued Katzenstein. By "radio" he meant an AM news station. "Although many of them will not have TVs. The next segment, the Modern Orthodox, many of them will have TVs. And they'll differ slightly in dress." Just then a young man appeared in the open door of Katzenstein's office. He was rolling a cart piled with overstuffed manila envelopes, and he wanted to discuss the cost of sending them first-class. He was wearing pressed jeans, a crewneck sweater, and a knit yarmulke. This, Katzenstein pointed out, was a classic Modern Orthodox appearance. There, in the gap between Katzenstein's white shirt, black pants, and black yarmulke, and the young man's collegiate dress and knit yarmulke, lay a slight but key difference in attitude. There were fewer givens in the life of the Modern Orthodox. Or, as Katzenstein put it, without a hint of derision or envy, "That community is more likely to make decisions for themselves. And that's the challenge: to sell them on the concept of not getting their own results."

A challenge indeed, considering that the most outspoken Jewish opponent of Dor Yeshorim was actually teaching medical ethics at Yeshiva University, the brain trust of Modern Orthodoxy in New York City. Moshe Dovid Tendler was trained both in Talmudic law and in microbiology, and he was a fairly consistent advocate of medical progress, including embryonic stem-cell research and research cloning. So he criticized Dor Yeshorim for its old-world approach. "My grandparents were born in America," he said. "The American ethical and moral values are very important to me. The idea that Dor Yeshorim has genetic information and refuses to share it with the person it belongs to is unfair, irrational, and almost anti-American. If you submit blood, you should be able to have the results." The argument was fragile—after all, anyone who wanted to have his complete results could obtain them elsewhere, and Dor Yeshorim never knew the identities of people whose genetic information it held—but it summed up the suspicions of the more modern among the Orthodox.

In the spring of 2006 Dor Yeshorim undertook another effort to sell itself to the Modern Orthodox community. "Don't Wait Until It's Too Late," screamed flyers posted around Yeshiva University. "It has happened to people here at YU," the flyer added in smaller print. "Don't let it happen to you." Whether "too late" referred to the birth of a sick child or to that stage in the dating process when finding out that the couple were "not compatible" would cause heartache, was unclear, perhaps intentionally so. The flyer offered a special price of one hundred dollars to Yeshiva University students—an almost 20 percent discount. Students discussed the flyer online. "I know many people with various genetic related diseases and none of them nor their parents are sorry about their existence," wrote one student blogger. "I wonder if any couples who have suffered tragedies due to genetic disease would say they would not have gotten married.... The Dor Yeshorim craze has become a bit fad—especially as a new rite of passage for coming of age Bais Yakov girls." The last sentence referred to a girls' high school in Brooklyn.

Dor Yeshorim activists may have derived some satisfaction from the blogger's referring to the organization as both a "fad" and a "rite of passage," but the rest of the comment summed up the problem the program faced among the Modern Orthodox. A student calling himself Q Jew commented, "Why do anonymous Dor Yeshorim instead of going to a hospital and getting a full list of all possibilities?"

Given the chance, Dor Yeshorim would answer: Because you won't, or you will wait until it gets late in the game. And because you will have good reason for procrastinating: because the "full list of the possibilities" is too hard to bear, when it actually contains possibilities.

"One theory is that the more educated believe they have the tools to deal with self-knowledge of carrier status," rabbis Ekstein and Katzenstein wrote in an article they prepared for a Jewish journal. "In reality, even the most sophisticated of individuals, being human, have the same difficulty using intellect to dominate emotions. A good example of this phenomenon is the physician who recently called the Dor Yeshorim office to indignantly exclaim that 'it cannot be that my son is a carrier!' For when it comes to oneself and one's family, science is left at the office."

To bolster their point, the rabbis cited a long-term study showing that people who learned of their carrier status experienced anxiety, embarrassment, and hopelessness. The parents of noncarriers often opposed their children's marriages to carriers, even though, medically speaking, no one had anything to fear.

"People want to know their results as long as they are a noncarrier," explained Rabbi Katzenstein. "Once they are a carrier, they may not be so eager to learn their results. And people believe, before they are tested, the same way that people buy a lottery ticket—they believe they are going to win. But there are those who are definitely carriers and don't know about it. And then suddenly they are faced with that burden of knowledge, the potential stigma and all sorts of other issues, which, if they can be spared of that, why not. You don't have to be quote-unquote know-nothing Hasid in order to have stigma and psychological trauma. I'm not saying the person is devastated, but it can present problems. Why advise people if they don't need to know?"

~

One day in the early 1990s two short men in white shirts and black hats showed up on the doorstep of a DNA-testing laboratory in Framingham, Massachusetts. They wanted a tour. If the director of the lab, a tall blond woman named Barbara Handelin, welcomed them politely, this was a function of her good upbringing and not of any genuine desire to see Rabbi Josef Ekstein and Rabbi Howard Katzenstein of Dor Yeshorim. As Handelin recalled it when she talked to me a dozen years later, she had told Rabbi Ekstein on the phone, "With all due respect, you have to understand that the way you are doing genetics testing, you break some basic tenets of my world. Like full disclosure." She laughed. "I also said, 'It's not clear to me that you really are doing informed consent before testing.'" If this was really the way she phrased her reservations—and that seems likely—that, too, was a function of her good manners, because what she really meant to say was that the way Dor Yeshorim conducted its business was anathema to geneticists, whose basic principles were: never test minors; obtain full informed consent; tell the patient everything you learn but never tell him what to do.

But Barbara Handelin ran one of the few commercial laboratories offering Tay-Sachs carrier testing, and Rabbi Ekstein was by this time ordering thousands of such tests a year and was looking to drop one of his two labs, which had failed too many of his random-control tests. In business terms, they were made for each other, so Handelin gave the rabbis a tour of her operation. Then they gave her a tour of theirs. They shook hands. They became friends: After finding out that Handelin, who was of Finnish extraction, was married to a Jewish man, Rabbi Ekstein took to sending her Israeli matzo every Passover. "They kind of took me on as their pet project, I think." He also sent her his papers for review—for by this time the man who could not use a phone book was publishing papers on genetics.

So how did Handelin, who was a geneticist and an ethicist by training, put to rest her reservations about positively everything Dor Yeshorim did? "I went to schools, heard them talk about testing with the students, at the yeshiva—or is there a different name for the girls' school?—anyway, after spending all that time there talking to Rabbi Ekstein and staff, seeing how they conduct themselves. They demonstrated to me that they were operating by a set of rules and a set of beliefs that at their core are good and noble—so they are different from ones that I grew up with professionally. In the end I said, 'It's very clear that you are preventing great suffering.' And that the way they are doing it works for them."

What makes it work? "It is the absolute comfort with the fact that they put off what many cultures would consider to be very private individual decisions—they are very comfortable with putting that off to a greater community and a greater God, actually. You say, 'Well, I give up my inclination to make my own choice about my spouse to a greater system, or a greater good.'"

So it was that a greater good and a greater God got momentarily conflated in conversation, and this seemed appropriate. In the Orthodox tradition, adherence to Jewish law as originally interpreted is held to be the greater good. Not marrying a person with whom one might bear sick children is fealty to Jewish law and is therefore the greater good. The role of the greater God in the making of children is also clearly described in the Talmud. The man provides the white (sperm), which yields bones, sinews, nails, the brain, and the white of the eye. The woman provides the red (menstrual blood), which yields skin, flesh, blood, hair, and the black of the eye. God provides the spirit, the soul, the beauty of features, the sight of the eyes, the hearing of the ears, the speech of the mouth, the ability to move the hands and to walk with the feet, the understanding and discernment. One might conclude that where a child is deprived of sight, hearing, speech, movement, understanding, and discernment—as children with Tay-Sachs, Canavan, and Niemann-Pick certainly are—God has declined to do His part. Dor Yeshorim's testing provides a peek at which couples will have these God-forsaken children.

The information we gain from genetic testing has a way of filling us with awe and a sense of having touched the forbidden: knowledge of the invisible and the intangible, and, more important, knowledge of the future. This is why the field of medical genetics has produced so many rules—though there are those who would argue that DNA testing is just another diagnostic tool and requires special treatment to the same extent as, say, the use of X-rays, ultrasound, or magnetic resonance imaging, all of which show us what we could not previously see (including the very early, previously invisible stages of human life). The rules that surround genetic testing are counterintuitive: with their focus on counseling and full disclosure, they mandate the dissemination of knowledge, not the protection of it. The guiding principle of Dor Yeshorim's work—Do not share information unless absolutely necessary—corresponds much more apparently to the dread with which genetics can fill its subjects and practitioners. Indeed, the more disorders appeared on Dor Yeshorim's menu—the greater the volume of information a young person might potentially receive—the more sense the policy made.

I remembered reading an article by the philosopher Colin McGinn, published in the magazine *Lingua Franca* about a dozen years before I learned of Dor Yeshorim. McGinn proposed that the mind-body problem, which had occupied generations of philosophers, was unsolvable. In McGinn's view, our inability to understand how our physical selves produced consciousness was integral to the existence of consciousness. What stayed with me was McGinn's description of waking up in the middle of the night to the revelation that the thing he had been trying to understand was unknowable. The discovery relieved him—and, he believed, other philosophers—of the burden of having to learn or invent a solution. He was ecstatic.

I would have been, too. I had spent months falling asleep with a heaviness in my breasts and a fog in my mind. I cycled. In the days immediately following my quarterly visits to the breast specialist, I felt reassured that she had felt nothing suspicious in my breasts. Then the anxiety would build. The frightening facts were all there: My breasts were dense; my breasts were lumpy; my breasts hurt; the kind of cancer I risked developing was so aggressive that a virtually invisible, unpalpable speck could turn into a life-threatening tumor between exams. I was too anxious to perform self-exams, which, new studies warned me, were ineffectual anyway (my mother had found her tumor herself, but she found it too late). A month or six weeks after one of my breast exams, the amount of premenstrual breast pain I felt would grow unbearable. Once I scheduled an emergency breast appointment because I felt a huge lump in my left breast. "Your left breast is larger than your right," my doctor said. "I know this. Why don't you?" Because I could no longer look at them. In essence, my breasts had already become cancerous: My body had turned against me. All I could do now was declare war on it myself.

The economist had told me I should have both a mastectomy and an oophorectomy. The psychologist had told me breasts were essential to my identity as a woman but I could get over losing them. The genetic counselor had told me to sacrifice the ovaries and, she implied, keep the breasts. I had read hundreds of papers, conducted dozens of interviews, attended one conference, and spent countless gray mornings staring at the pine-paneled ceiling at our dacha. I learned every crack and knot in the wood over our bed, and I hated every single one of them.

Karen, the woman who had organized the cancer-family reunion I attended in Green Bay, Wisconsin, told me she had asked her breast surgeon what he would tell his wife to do if she had the mutation. "That will be the easiest question I will have to answer today," Karen remembered the surgeon saying. "I would tell her to have a double-sided mastectomy and a hysterectomy with an oophorectomy, and I would tell her to get it done tomorrow." That, Karen said, "was one-half of my decision." The other half came from talking to her cousin who had already had the surgery. None of that was directly related to studying probabilities and weighing risks: Karen chose to get her information from intermediaries, much as Dor Yehorim's clients do.

When my cousin Natasha tested positive for the mutation in Jerusalem, she found herself scheduled for ovarian surgery before she had a chance to consider the idea, much less get used to it. Once she thought about it, she jumped off the conveyor belt, joining what may be a minority of Israeli women who choose not to have surgery following a positive BRCA mutation result. "I think it is my karmic challenge," Natasha told me when I visited her in Jerusalem a couple of months after her test. We stood on the side of a mountain road, the wind filling our hair with sand: My two children and Natasha's two grandchildren were in the car, and we had to talk about this quickly, before getting in. "If I cut off some part of me that may develop cancer, the cancer will come through in a different organ. I have to confront it with my own spiritual force." I was in no position to argue with her: She had outlived my mother by four years and her own mother by a year, which, in our mutant world, seemed significant. Natasha hugged me and said softly, "But you do whatever you need to do."

I needed to do something. I did not believe in karmic challenges that could be defeated with the strength of one's spirit. I did not even really believe in a God who might have chosen to deprive me of the 187th allele in my BRCA1 gene. I believed in action, and, most of all, I believed in knowledge. But I would have given all my breasts and ovaries to wake up one night, like Colin McGinn, with the understanding that this knowledge was not meant to be held by me. In the absence of such a revelation, all I could do was reduce the amount of knowledge that drove me crazy. In my mangled internal math, the 40 percent risk of ovarian cancer was knowledge I could live with, and the 87 percent risk of breast cancer was not.

Chapter 9

THE OPERATION

I spent the night of August 22, 2005, in front of the computer—editing someone else's story for a magazine where I worked—and chatting by instant messenger to make sure the time until I left for the hospital was filled with words. That I would have my surgery on the anniversary of my mother's death was a fateful scheduling accident: The breast surgeon who would do the mastectomy and the two plastic surgeons who would do the reconstruction bounced the days of August back and forth among them until finally they informed me that August 22 was my only option. I very nearly canceled the surgery several times before finally deciding that assigning the worst days of my life to a single date was perhaps not such a bad idea; a friend suggested that ultimately I might just choose to cross the date off the calendar altogether.

At seven in the morning, exactly thirteen years after my mother last woke up and not ten miles away, I would go to sleep—to wake up with a different body, one that would be divorced from her legacy. While I slept, a breast surgeon, a lovely young woman with the ingratiating manner of a Midwestern saleswoman, would cut off my nipples and make slits under my arms to scoop out all the breast tissue. At the same time, a plastic surgeon would cut open my belly and scoop all the fat tissue he could find there, to scoot it up into the empty space under the skin that used to cover my breasts (the "breast envelope," as they call it). A third surgeon would perforate the muscle tissue to get at the blood vessels that would need to be pulled up to my chest to feed the transplanted fat. I would wake up with small nipple-less breasts that should look and feel real to the touch but that would probably never have any sensation. A few months later, once everything had healed and settled, I would be able to have decorative nipples constructed and tattooed to look more or less like the real thing.

I had turned out to be a coward on the nipple issue. American breast surgeons generally insist that the nipples must be removed in a mastectomy—because it is impossible to scrape them clean of breast tissue the way it can be done with the thicker skin of the "envelope." Some European surgeons, as well as a few Americans, have, however, been performing what are called subcutaneous mastectomies, which preserve both the nipples and breast sensation. But I could find only a couple of studies analyzing the safety of this procedure, and what they said was that there was not enough information to know whether a subcutaneous mastectomy was as effective in preventing cancer as what surgeons call "a simple mastectomy." In fact, the studies suggested it might not be: The few women who went on to develop cancer following a preventive mastectomy had all had the subcutaneous kind. And I knew it would be an uphill battle, trying to convince the breast surgeon to spare the nipples. I also knew I could probably win—unfair as it might be, I could trump most objections with the argument that I was writing a book about my experience and my research—but I could never quite convince myself. At best, I would feel I had been reckless. At worst, I would come to fear and resent my nipples the way I had come to fear and resent my breasts in my year of mammograms and biopsies. So I decided to sacrifice them.

The operation lasted a staggering thirteen and a half hours. I woke up in the recovery room—a space the size of half a football field, designed to sustain a dozen simultaneous crises. Everything in it was makeshift: beds that were wheeled in from the operating rooms, and the bodies in them, held together with string and tape; movable monitors, X-ray machines on wheels, curtains that created that hospital sort of falsified privacy. I spent the night diving in and out of the anesthetic darkness. Twice a nurse woke me up rudely by screaming, "Do not forget to breathe!" I had apparently forgotten. I remember resenting her intervention.

By morning, the thickest of the anesthesia had worn off and I was able to take in my surroundings. A man in his fifties was wheeled in to a space diagonally across from me, woke up, and began to stare me down flirtatiously. Then a nurse came, and he started hitting on her with lines so hokey I had never thought anyone might actually say them: "That's a pretty name. Are you married? Good." Then he discovered he was missing his right leg.

I felt like I was in a bad movie. A fat woman directly across from me woke up in extreme pain and howled and laid into the nurses, who made disdainful comments. I had known this about nurses from earlier experiences of giving birth and visiting friends in American hospitals: Nurses tend to arrange patients according to a hierarchy of pain tolerance. It is an entirely unreasonable value system: For the most part, people can do nothing about how and when they experience pain. As it happens, though, I have an extremely high pain threshold and a high sensitivity to painkillers: One milligram of morphine knocks me into space. That makes me a virtual star with medical personnel. So now I was playing the strong, silent, stoic type. I put on my headphones, turned on the CD (is there another song quite as underhandedly dark as Leonard Cohen's "Everybody Knows"?), pressed the button that sent another milligram of morphine through my veins, and sank into despair.

I would spend the next week in this state. I had miscalculated—or misimagined—something. I had asked the doctor whether the recovery would be like that from a cesarean section, and he had confirmed that it would be. I had had a difficult time giving birth—fifty-six hours of labor followed by emergency surgery—so it seemed like a fair comparison. I remembered being in pain the first few days, but more than that, I remembered the recovery being not only easy but automatic: It required no conscious effort. I expected something similar this time.

I had been very, very wrong. This time I woke up in a strange body—not my own badly battered one, as after the cesarean. The front of my torso had been cut up and rearranged and was now fragile beyond imagination. A sign at the top of my chest, in black Magic Marker (the permanent kind, as I discovered later over a series of showers), said, "No pressure here." This was where the microsurgeon had fused two blood vessels together. I joked I would have the sign permanently tattooed, but it terrified me. Then there were my so-called breasts, two small, badly bruised mounds. Every hour a nurse would come by to see if they were still alive: She would touch them to see if they were warm and listen for blood flow using a tiny Doppler—the tool usually used to listen to the heartbeat of a baby in the womb. Then there was the rest of me, bruised, bloody, and absolutely numb. Plastic tubes with pear-shaped vessels at the bottom—drains to absorb excess fluid—came out of my underarms and my abdomen: four in all. The breast surgeon talked me into looking at this horror once: Now I realized her saleswoman's voice was more like that of a guardian angel, and I clutched her hand while I looked down. I acknowledged that the reconstruction was masterful, if you discounted the bruising. And then I could not bring myself to look at myself for over a week.

Every day I woke up with a sense of pervasive regret. What had I done? Physically, I felt like I would never bounce back. Psychologically, I felt disgusted.

This was not the least bit like the period following childbirth. My body was not regenerating. It was not producing milk, it was not rejoicing in the presence of a baby. All it was doing, it seemed, was recoiling in horror. I did all the things a good patient was supposed to do: I got up on Day Two, I took a lap around the hallway on Day Three, and I was discharged on Day Four. But every time I had to shower, it took me hours to get ready and then finally I would ask Svenya to help, because I was terrified of myself, so I needed to shower with my eyes shut. I began to think that there was reason to surgeons' refusals to operate on young women. I even thought my own counterargument—that if these same twenty-something women had come looking for a sex change, they would have gotten the surgery, chased by hormones—was shortsighted. The women who wanted their breasts cut off because they believed they were men would experience the change as natural, as releasing their true bodies into the world. What I was experiencing was pure violence.

Three of the drains came out after a week. I began doing exercises to regain the ability to raise my arms: For now I could barely get them to shoulder height. I also resumed working. The pathology results came back. They were a few days later than expected, and I had grown nervous: I had talked to enough women who had had what they thought were prophylactic mastectomies, only to have tumors discovered following the surgery—and to have to go on to chemotherapy. But I did hope the pathology would reveal just a single cancer cell—or rather, noninvasive, very early-stage cancer—so I would not have to undergo chemotherapy but would still know I had done the right thing. No such luck: The pathology results were clean. What had held them up, my breast surgeon explained, was a bunch of benign granulomas—small growths that are usually associated with a disease called sarcoidosis, of which I had no other symptoms. I could hear the surgeon shrug over the phone: Whatever that disease was, it was not her turf. Privately, I thought I knew exactly what caused it: fear. I had spent so much time imagining what might be growing in my breasts that, ultimately, something grew.

A few days later I bought some clothes: Because of the scar that went all the way across my lower abdomen, I would have to wear hip-hugging trousers for the foreseeable future. The last drain came out after two weeks. I got behind the wheel. The next day, I went to visit an old friend, and we spent the evening talking about love and not talking about surgery or cancer. I had to ask her husband to back the car out of the driveway: I still could not turn my torso. But on the way home, I felt in awe of my own body, which had somehow leaped over the violence I had done to it and was back to serving me so soon. When I got home, I finally looked at myself.

I was still yellow and black in places, and the inches upon inches of stitching were still red and raised. But, for a bruised and bloody torso, it did not look half bad. I liked the small breasts and the taut stomach. They were unfamiliar, but I could learn to live with them. Most of the front of me still felt leaden—numb and, it seemed, hard. I was soft to the touch, but the touch echoed so unpleasantly—like pins and needles amplified—that it was clear it would take a long time to learn to be touched. For now, I felt like I was carrying something like a shield on my front, and this seemed fitting enough.

The next day, I put on my new clothes. I looked in the mirror: ridiculously low-riding jeans and a T-shirt stretched over a preternaturally flat stomach and small braless breasts. It looked like a teenager's body with a thirty-eight-year-old face pasted on top. In an odd way, this had been precisely the point. I laughed.

PART 3

The Future

Chapter 10

THE FUTURE
THE OLD-FASHIONED WAY

"*She is just* an ordinary-looking baby, isn't she?" the pediatrician asked, looking at the blanketed lump placed in a plastic cradle beneath a lamp. "Except that they shaved her head to put IVs in her at one time or another," he continued. "She has this colostomy. That's just a bowel hanging out there, with some stools coming out, and they'll just close it up eventually and rehook it back up to the rectum." Dr. Holmes Morton poked his short fingers at the baby. She looked good to him.

Marlene was one month old. She and her twin sister, Arlene, had both tested positive for maple syrup disease, also known as maple syrup urine disease (MSUD), so named because the urine of babies born with this recessive disorder smells like maple syrup. In fact, if the disease is allowed to progress unchecked, everything smells of maple syrup—not only the child's urine but also the earwax and, indeed, the child's whole room. And starting at about four days of age, the child will suffer irreversible damage to the brain and the central nervous system. The disorder is an enzyme deficiency that leads to a buildup of certain kinds of amino acids and a deficit of other kinds in the brain and elsewhere. A technical term is "metabolic derangement." This deranged state of the body can lead to disability or death resulting from cerebral edema or other equally horrifying causes. Most places in the world, where only something like 1 in 185,000 babies is born with the disorder, the child is doomed. In Lancaster County, Pennsylvania, where roughly 1 in 200 babies born to Old Order Mennonites is affected with the disease, babies are tested and put on a diet—a specially designed formula that will be their only food for the rest of their lives—and proceed to develop and grow normally, or as normally as any person with a serious chronic condition requiring constant attention and intermittent medical care. Morton is the man who made sure newborn screening for maple syrup disease was instituted in Pennsylvania, helped design the formula for feeding the affected babies (and children and adults), and proceeded to treat them all at his Clinic for Special Children in Strasburg, Pennsylvania. No one had ever treated adults with maple syrup disease, because affected individuals did not generally live to become adults, so Morton and the one other doctor at his clinic, despite being pediatricians, treated the adults too.

Morton looked like a country doctor should look. He was not tall, and his slightly stooped posture made him seem even shorter. He wore a gray tweed jacket, a blue oxford button-down shirt with noticeably frayed cuffs, and a bow tie. He talked a lot, displaying a boundless willingness to discuss all relevant topics with all comers, which caused him to take detours—to a hospital waiting room to collect a family's medical history, down to the hospital basement to study a series of MRI shots of the brain of a girl with a mysterious seizure condition—and the detours in turn caused him to run habitually late. On the first day I spent with him, he had a single snack, in the late afternoon: two saltine crackers snatched from a packet behind the registration desk at his clinic.

The day had begun with a visit to Children's Hospital in Hershey, a taxing forty miles from the clinic, to check on month-old Marlene with the colostomy bag and an IV drip of Morton's special maple syrup disease formula. The maple syrup was not the problem: Both Marlene and Arlene had been diagnosed within forty-eight hours of birth, put on the special formula, and sent home. But then Marlene began to show signs of a bowel obstruction. She returned to the hospital, where she was diagnosed with Hirschsprung's disease, another apparently recessive disorder. Babies with this condition lack normal nerve cells in a part of the intestine, which means stool is not moved forward, causing an obstruction to form. Surgeons had removed the part of the girl's intestine that was missing the normal nerve cells—"about most of her left colon," as Morton put it—and she was now recovering. Morton was pleased. He talked up the merits of her diet, too: "It's actually an unusually good formula, because we made it," he said and told me the story of working with a company called Applied Nutrition to design a food suitable for maple syrup babies and children, on the condition that the company would provide Lancaster County children with the formula free of charge for a period of time—and get data on their nutrient levels in return. Then he had one of those meaningful bedside conversations with the baby's attending physician.

"Looks like she just advanced on her feeds?" the country doctor asked.

"Yep," the man in hospital scrubs answered.

"Doing all right?"

"Yep."

"So how much faster do you want to advance her?"

"Well, that was our question for you."

"Well, the goal is three ounces every three hours."

"All right."

With that, Morton gathered up his bulging pigskin briefcase and readied to go, but a group of doctors in scrubs stopped him to talk about an article that had just come out in *Smithsonian* magazine. Titled "Medical Sleuth," it told the story of Morton's investigation into the sudden death of an Amish infant. Her parents were charged with abuse, but he ultimately proved that the baby had died as a result of a genetic condition. This was the second article about Morton in a major national magazine in a couple of months: Long revered locally, he was now becoming a national entity, if not yet a celebrity.

He led me to the elevator, where he lectured me on the management of maple syrup disease, down to the parking lot, where he expounded on the business of medicine—or rather, the way medicine had been turned into a business—and into the car, where he talked to me about the need for genetic testing for everything, the whole thirty-mile drive to Lancaster, Pennsylvania. We were on our way to Lancaster General Hospital to see the little girl's older brother, thirteen-year-old Henry, who had been hospitalized thirty-six hours earlier with a maple syrup disease crisis. These children have episodic crises, often brought on by an otherwise innocuous viral infection that causes their metabolism to go out of whack. Morton's treatment protocol prescribes a special "sick-day" diet for maple syrup children, but that does not always help. Their metabolism loses its mind, and so do they: Their brains swell.

Henry's father had been unable to wake the boy up, and when Henry did wake up, he had been absent. "When he came in the other night, he didn't know who he was or where he was," Morton explained. "Wasn't particularly sick, he doesn't have like pneumonia or gastroenteritis or anything, he just had an upset stomach and started vomiting, and how much of that was just due to the metabolic disease and how much was the underlying bug that provoked it, we don't know, except it was a fairly ordinary illness that in another child might make him feel bad for a day or two but in these kids just sort of cycles into this state where they just can't get themselves out of trouble. They can't drink enough formula to get the lucene [an amino acid] level down. And basically, as they become fasted and ill, they begin to break down protein, which is part of the normal metabolic response to the illness, and their blood amino acid levels go up, or their lucene level goes up, and the higher the lucene level is, the more encephalopathic they become and the more they vomit, and it's just sort of a vicious cycle."

There was a sickening, treacly smell in the boy's room: It smelled of maple syrup, and this meant he was still in trouble. His lucene level had been 17 when he arrived. Per Morton's standard instructions in these cases, the hospital had put Henry on a special IV solution that was pumping four thousand calories a day into his body, forcing his metabolism into overdrive, leading him to synthesize protein at a very high rate, restoring a normal balance. Within twenty-four hours, Henry's lucene level had gone down to 13, and Morton was predicting it would be down to less than half that in another twenty-four hours, which would mean Henry could be weaned off the IV solution and return to his usual formula.

Henry was a pale boy, very small for his age—one would expect that, with a diet that restricted him to sustenance levels of proteins—but otherwise he looked just like any child in a hospital room: simultaneously bored and pleased to be able to watch television all day long. "Gonna ruin you, watching that thing, you know," said Morton, shoving his fist into the boy's slightly distended stomach. "Does it hurt when I punch there? Is it sore? These guys get pancreatitis," said Morton, turning to me now. "They always have this epigastric tenderness and vomiting, and it's—we think the pancreas swells just like the brain does. It hurts down there and up there, and they just stop thinking straight. So," he was facing down at Henry again. "You look a lot better."

Before leaving Henry's room, Morton swirled a Q-tip in Henry's ear. "What people don't realize," he explained, "is that the smell of maple syrup isn't from the tree but from the bacteria. So when people come in, I tell them they can send an expensive test to the Mayo Clinic and wait two weeks or use a Q-tip and know." Morton stuck the Q-tip under the nose of Corinne, a new nurse on the ward. Henry's earwax smelled like maple syrup. "Henry can teach you all about MSUD," he said.

Henry could also teach her all about genetics. Henry's parents were double first cousins—their mothers were sisters and their fathers were brothers—which meant that roughly 25 percent of the children's genome was homozygous: Both copies of the gene were identical. Like most Mennonites, the couple had a lot of children, though Morton could not be sure how many exactly. "About eight, maybe nine," he said slowly, as though struggling to remember. "There are a couple of kids I don't know." He had reason to know only the children who had any of the two or three genetic disorders that ran in the family.

Four of their children were affected with maple syrup disease. They also had two children with SCID, severe combined immunodeficiency, or "bubble boy," disease, a recessive disorder that handicaps the immune system, making children catastrophically vulnerable to all sorts of bacteria and viruses. Babies with untreated SCID rarely survive their infanthood, but treatment—including bone-marrow transplants—can allow them to develop normally. One of Henry's siblings, affected with both SCID and MSUD, received a bone-marrow transplant, which not only cured the SCID but significantly lessened the symptoms of MSUD. "So there was another one of those little experiments you don't get to do every day," said Morton. The beneficial effect of a bone-marrow transplant on the symptoms of maple syrup disease is the sort of thing that would make for a nice article in a medical journal—"it tells you that bone-marrow transplant is a kind of gene therapy"—but Morton did not have the time to write articles about every one of his "little experiments."

Were he so inclined, Morton could mine the family for a good number of scientific reports. Three of the kids—including the newborn Marlene—had Hirschsprung's disease. The intestinal disorder is believed to be what is called a complex trait, a condition caused by a combination of mutations in several different genes, with one mutation the main culprit, while specific variants of other genes act as modifiers, determining the severity of the condition. Three genes have been identified as the main culprits in different populations: RET, EDNRB, and EDN3. It was a mutation in EDNRB that was believed to cause Hirschsprung's in the Mennonite population. Back in the mid-1990s, when few disease-causing mutations were known by name, a geneticist named Erik Puffenberger wrote his doctoral dissertation on this gene and its role in causing Hirschsprung's among the Mennonites. But here was the thing. Baby Marlene's two older siblings with the disease carried the mutation described by Puffenberger, but Marlene did not. And, said Morton, she seemed to be homozygous for an area of the RET gene, suggesting that might be the cause of her Hirschsprung's, which, in turn, suggested two things: that the unlucky parents were actually both carriers of two separate mutations that caused Hirschsprung's; and that, more generally, both of these mutations were found in the Mennonite population. "This is cause for him to redo his Ph.D.," chuckled Morton.

~

If Morton looked like what a country doctor should look like, then Erik Puffenberger looked like what a scientist should look like. Not the absentminded professor type, but the young, anal, collected type, the type to whom you would entrust the mapping of your child's genome. One look at Puffenberger told you this was a man who was never late, never forgetful, who would never spill one drop of fluorescent marking solution or misplace a single DNA chip. Puffenberger wore belted black Levi's with a red shirt neatly tucked in, and he smelled of Old Spice. Puffenberger made very certain I understood everything he told me. To this end, he tended to rephrase his explanations, which I recorded on my Dictaphone, several ways. Then he made some notes on a sheet of paper, folded it, and handed it to me. Then he showed me a PowerPoint presentation, printed it out, and e-mailed it to me. To be sure, I was a willing recipient of all these explanations: The points made by geneticists can be hard to understand, and this was an important point. The point was, inbreeding does not cause genetic disease but it does increase the probability of its occurring—slightly.

Puffenberger pointed out that Morton had grown up in the coal-mining region of West Virginia, the national butt of inbreeding jokes. (In 2004 the governor of Virginia publicly protested the clothing retailer Abercrombie & Fitch's marketing of a T-shirt that read, IT'S ALL RELATIVE IN WEST VIRGINIA.) Puffenberger himself grew up in Lancaster County. The Old Order Mennonite population he had used for his doctoral research was a geneticist's dream, because it had stayed essentially isolated for roughly three hundred years, since two Anabaptist groups—the Old Order Mennonites and the Amish—had accepted William Penn's invitation to immigrate to the United States. They lived on their farms and eschewed most of what civilization had to offer, such as television, cars (not all of them; some Mennonite groups now allow their members to drive, provided they paint their cars, including the normally chrome parts, black like the buggy, which has remained the privileged vehicle), and telephones—which meant that many of them married not just within their religious community but within the immediate vicinity, which is one of the ways they ended up with marriages between double first cousins.

Contrary to all those West Virginia jokes, none of this meant that people had to be sick. If you do not carry the gene for any recessive disorder, marrying your double first cousin will probably not endanger your future children. But in a population where—as in most or all populations—some people carry some mutations for some diseases, those people's cousins and other relatives are more likely than others to carry the same mutations.

Take maple syrup disease, said Puffenberger. The carrier frequency among the Old Order Mennonites in Pennsylvania was roughly 10 percent—extremely high for any genetic disorder anywhere. The disease is found elsewhere—say, in Northern Europe, where the Mennonites originated—but the carrier frequency in the old country is roughly one in 300,000. The very small "source population"—the group that originally came over to the United States—must have included one or two people who carried the mutation, which reproduced efficiently over the following three centuries. "In that case the allele frequency in the population is about 5 percent," he said—meaning that in all the genomes of all the Pennsylvania Mennonites, about 5 percent of the maple syrup genes would be abnormal: one abnormal gene per carrier, one carrier per ten people. "What inbreeding does, because inbreeding is the marriage of people who are closely related, you tend to share alleles, you tend to share more DNA than somebody who is not as closely related to you. What does that do? That increases the number of homozygotes at both ends of the distribution." That is, genetic bad luck becomes unevenly distributed: Where in a population with little consanguineous marriage the 5 percent of "bad" alleles would be spread more or less randomly, in a population where relatives tend to marry, the bad genes will bunch up at one end of the spectrum while the vast majority remain homozygous for the healthy version of the gene. "The allele frequency is still 5 percent," Puffenberger stressed. "But it means in the long run, in an isolated population with inbreeding, you can create more affected individuals, or, said in a different way, given a fixed incidence, you need a lower carrier frequency in that population to give you the same number of affected individuals. So as inbreeding increases in the population, you are going to increase the number of affected individuals. But the underlying problem is the high mutation frequency. One in ten is just too high. But it's a random chance event that this mutation got in high frequency."

Puffenberger was on to his other topic of passionate debunking. He was a convinced opponent of the selective advantage theories of genetic disease. "You read a lot in medical literature about postulating that a mutation, for instance cystic fibrosis, got in a high frequency because of a carrier or heterozygote advantage," he said. "But I find those explanations to be rather implausible at times. If there was a beneficial effect of being a carrier for maple syrup urine disease, why don't we see carrier rates higher in other parts of the world? Is there something specific to Lancaster County that gives them a benefit to being a carrier for MSUD? Or is it just by chance, bad luck essentially?

"I think you can make those arguments about any mutation, in any population. Particularly, if you think about something like cystic fibrosis, there is one mutation that's in fairly high frequency— probably 75 percent of mutations in cystic fibrosis are one mutation— everyone who carries that mutation is related to one another, because that mutation happened one time on one chromosome a long, long time ago, about two thousand years ago. So how does that mutation rise in frequency? You might make the argument, if there is a selective advantage to being a carrier for cystic fibrosis, why aren't other mutations equally high frequency? In fact, we know of hundreds of different mutations on that gene. All the rest of them have risen to fairly low levels. Why is that? Maybe propped up in a geographic isolate in Europe, a group of people who were isolated and had a high frequency of that, were then subsumed into a larger population and the mutation spread in that way. You can imagine that if the Mennonites suddenly decided to dissolve themselves and joined the general population, our carrier frequency of MSUD in Lancaster County would suddenly go up. So you can imagine that two thousand years ago they didn't have a very big population running around Europe. There would be essentially lots of little populations, genetically distinct little populations, any one of which could have had a rise in some mutation, which then was spread as populations sort of merged together. And you can imagine, and you can make a plausible argument that mutations got to their high frequencies just by chance.

"Just look at the surnames in a population. The distribution of surnames is going to be based on what? A random event: how many boys a family has. The most common Amish name in Lancaster County is Stoltzfus. One in four among the Amish in Lancaster County has the last name Stoltzfus. What's the selective advantage there? Unless you argue that Stoltzfus men are inordinately attractive compared to other men, or had an inordinate amount of wealth and all the Amish women just want to marry Stoltzfus men, you'd have a hard time explaining why a single surname had risen to such a high frequency. It's chance. They had a lot of boys, probably in an early generation, probably had a bigger family. Maybe more of the boys stayed in the population."

The surnames made for a great analogy. I assured Puffenberger that I agreed with him, but that was not enough. "It's a view that many people would agree with," he granted me that much. "But then you see all these papers. There was a rather interesting article in *Scientific American* just a couple of months ago about founder mutations, and much of the article spent time talking about heterozygote advantage for all these mutations. We just don't see that here." Puffenberger had tested the theory several different ways. One was to look at areas of the genomes of Old Order Mennonites with an eye to areas of decreased diversity—parts of the genome where many different people would be similar to one another. Suppose being a carrier for maple syrup disease did not confer a selective advantage: That much seemed obvious, since the disease remained so rare in Europe. But perhaps there was a gene somewhere in the general maple syrup region that played some sort of important role, and among the Pennsylvania Mennonites, the maple syrup mutation just happened to tag along. Puffenberger was giving the selective advantage theory as much benefit of the doubt as he could muster, but it still did not work: Neither this part of the genome nor any other looked "selected for" in the Old Order Mennonites.

He went further. He looked at a region of mitochondrial DNA in about 250 Mennonite and Amish individuals to see how much variety he could find there—specifically, to see if certain haplotypes were significantly more common than others. For the mitochondrial DNA, which is passed on through the maternal line, it worked roughly the same as with surnames, which are passed on patrilineally: "There are a couple of mitochondrial haplotypes that are very common in the population that were only introduced by a single woman two hundred years or two hundred and fifty years ago. And how did that get in such high frequencies? Was that mitochondrial DNA so superior that it is spread in the population by selection? You look at the genealogies, and your answer is right in the genealogy: They had a lot of girls. These events can be random. It's all right to be random!"

Puffenberger was now finished with articulating this part of his scientific platform and willing to entertain questions. He was master of the Mennonite and Amish genome. He had identified the genetic causes of over one hundred diseases and will probably have identified a dozen more by the time this book is published. In the previous generation of geneticists, scientific reputations and entire careers were made on the basis of a single such discovery. Puffenberger was one of the first representatives of a new generation, one that treated genetic analysis as just another diagnostic tool. This made him the perfect partner for Morton, who dealt with rare genetic disorders as just another illness to be treated and managed.

~

The drive to Susie's house in Lancaster County took me past picturesque Amish homes, the ones that stand pure white against the green hills, the ones with the large barns and tall silos that, when the narrow road was empty except for a horse-driven buggy or two and my rental car, made me feel like I was time-traveling. Susie's house belonged, unmistakably, to the second half of the twentieth century.

There was linoleum on the floors, and a ramp for the disabled up to the porch. As in other Amish homes, there was no electricity—or, rather, no hookup to the municipal power lines: There were several electrical devices in the home, including a wheelchair, but these operated on batteries that could be charged using a generator planted out in the small barn. Susie talked to me while she sewed—she was working on a simple quilt, laid out on a board by the window, to make sure she used as much light as there was in a day. She worked a large needle at an even pace, looking up at me through her large aviator glasses most of the time. She was a rare sort of storyteller: even-voiced, as though detached, but so true to her narrative that, whenever she came to the birth of one of her children, I could not guess whether this would turn out to be a healthy child or an affected one.

Susie and Amos's first son was born in 1970, two years after they were married. That made Elmer thirty-five by the time I met his parents. He was grown-up and married: He was fine. The next son, Levi, was born in 1973. "He was a fussy baby, and we could see that he was not developing so fast," Susie said. "He gradually lost muscle control. Eventually it seemed that some of his muscles were doing the opposite of what he wants them to do." When this baby was nine months old, the family began making the rounds of doctors trying to figure out what was wrong with their son. "At the end of all those tests they said there wasn't much they could tell us. And I asked about cerebral palsy, and they said, 'That's probably what he has.' And I admit I wished they would just come out and say it." Whatever it was, though, Levi would never be like other children. The Amish have ways of thinking about children like that. They call them "God's special children" and similar names. They believe there is a separate heaven reserved for these children, who usually die quite young. They talk about this, and they do not usually talk about the crack that appears in a young couple's world when they realize that their child will not walk, talk, or feed himself.

"We had our time of acceptance," said Susie. "We realized he was a sweet little boy, even though he could not talk like other children." The hardest thing about taking care of this baby was feeding him: His muscle problems meant that he could neither swallow nor easily keep food down. Keeping the little boy nourished was a round-the-clock occupation.

"In 1975 we were blessed with another little boy," said Susie. She talked with the particular Amish accent: throaty and deliberate. The Amish speak to one another in a language they call Dutch, which is the local way of pronouncing "Deutsch": The language is in fact a variant of German. Since most Amish are educated in Amish schools, their English, while fluent, bears the mark of a second language: Their speech is exceedingly literary. "Everything appeared fine at first," she continued. "Then, when he was six and a half months old, he got a cold. Just a regular cold. When we thought he was recovering from the cold, we put him down for a nap. And he slept and he slept, and we finally woke him up, but it was quite a different little boy than the one we had laid down. He was completely helpless. He had been just the most playful little boy, and now he couldn't use his hands or hold up his head."

It was winter. The roads were icy, which made it very difficult to get to the hospital using a horse and buggy, with its thin wooden wheels (calling a doctor was not an option, since there was no phone). They did make it to the hospital, though, and got another confused bunch of diagnoses, including pneumonia, which was not confirmed by an X-ray, and encephalitis. When Susie went home with baby Sylvain ten days later, all the doctor could tell her was that her son was very sick and might never recover.

"So we had two little boys to take care of now," said Susie, and, uncharacteristically, looped back in her story to her memory of her third son's awakening from that nap. "When he first woke up, he clearly wanted to hold up his head and start playing, but he couldn't.

And he just cried, heartbroken, and he would cry every time. Apparently, somewhere in his brain he remembered playing." Babies can feel heartbreak, just like their parents. After a while Sylvain seemed to give up his memory of playing, and later he learned to lift one of his hands and one of his feet, and this made him smile. "Oh, he was just the sweetest little boy. People were afraid to hold him, but everybody loved him."

Susie and Amos's fourth son came along two years after Sylvain's birth. "The doctor assured us he was okay. The doctor thought I was just overanxious, but by this time you start to wonder." When Steve was eight weeks old, he was diagnosed first with an ear infection and then with meningitis. By this time Susie knew enough to be leery of diagnoses: "He just didn't appear as sick as you'd expect with meningitis." After they left the hospital, Steve seemed recovered but not quite: The parents noticed that his fontanel was inflamed. In the late 1970s CT scans were just coming into use, and baby Steve got one. "The doctor said, 'The news is good and bad. He doesn't have hydrocephaly. But we found a lot of brain atrophy. I'm afraid he is going to be just like your others.'

"I don't know if you have any idea what we were going through by this time," said Susie. I did not answer and waited for her to continue. "I can tell you straight out," she said. "We had a hard time. With ourselves. But we had our hands full." They had four small children, three of whom were severely disabled.

"That winter we went to visit grandparents," Susie said. "I had the baby in my lap, and the two handicaps were on a mat in the back. All of a sudden my husband stops the carriage. He drove the horse into a snowbank to stand better. And I said, 'What's wrong?' He said, 'Something is not right back there. Sylvain is not lying there like he was.' He grabbed him—and there seemed to be nothing there, no breathing, nothing, and I quick gave the baby to Elmer—he was seven years old at the time—and tried to do mouth-to-mouth. I didn't really know how to do it, but I tried." Eventually they knocked on a door, someone called a doctor, the doctor came and said that there was nothing to be done.

Susie told me in detail about the days that followed. There was a record snowstorm. It was nearly impossible to get to the cemetery. The body lay in their home for four days. It took the efforts of the whole local Amish community, and a tractor and plow, to enable them to bury Sylvain. At the cemetery, they had to leave the tractor, and shovel by hand because they could not see the gravestones for the snow. "It looked like a white room, because the snow was over our heads," said Susie. The blank whiteness seemed the perfect image of her grief. Without pause, she went from describing the cemetery to describing the way she and Amos felt. "And when he passed, it seemed as though a part of us was left there at the cemetery. But people were nice and they came to visit us, and eventually we came to accept it. Levi had a hard time: It seemed he missed his brother so much, and he had a hard time because he couldn't even speak."

In the summer of 1979 Susie gave birth to a daughter, Livanna. Four days later Levi died suddenly.

"When she was about three and a half months old, I told the midwife that I'm not quite sure about her, I don't think she is quite as active as other children her age. She looked at her and the doctor looked at her, and they both told me I was just overanxious." Susie had been overanxious like this before, and she had been right. "A month later I came for a checkup, and the doctor said, 'How do you feel about her now?' I said, 'Now I don't need you to tell me, I know there is something wrong with this baby.'" Susie showed the doctor just what was wrong: The little girl could not kick with her legs; when she reached out, her movement was slow and her arm stiffened; if she wanted to grab a toy, she visibly had to struggle to get her hand into position. By this time Susie was ready to try something new. She told the doctor she was switching the baby to soy-based formula. This proved a good idea: Livanna started gaining weight faster.

"So. We accepted her. She was a cute little girl. She wasn't as helpless as Levi and Sylvain were. By this time Steve had started walking—it wasn't the normal kind of walk, he fell a lot." But he was surprisingly self-sufficient for a baby who had made his parents feel so hopeless. He learned to feed himself at the age of twenty months. At two, he started speaking. Physical therapy seemed to help him learn to walk. Livanna compensated for her lack of muscle control as well: At six months, she started pulling herself around by her hands, which would serve her well in the future, too, since she never gained control of her legs. Though the children had some difficulty making themselves understood, their comprehension was clearly no worse than that of other children their age—confirming Susie's sense that the two children she had lost, Levi and Sylvain, had been quite aware of all that went on around them but simply lacked the ability to express themselves in any way.

When Livanna was three years old, Susie gave birth to Ruben. "I thought there was better motion in his movements: He started holding his head well at six weeks, he squirmed well. I thought he was okay. When he was four months old, I got a walker at a yard sale. My husband said I was tempting fate. He got whooping cough, but he maintained muscle control. Everything looked fine. And the wonder of it stays with me to this day, because it stayed this way!" This was Susie's one story with a surprise ending: Ruben was a healthy child who grew up to be a healthy adult. "I think everybody rejoiced with us that we finally had a child who could be okay."

Alvin was born three years after Ruben. "He gained slowly, but doctors felt he would be okay, and it looked promising." With Ruben healthy and Steve and Livanna doing well in grade school—Livanna never learned to walk but picked up reading easily when she was five—Susie may have lost some of her vigilance. "When he was one year old, he got diarrhea and got over it. Six months later he got it again, and that time it didn't turn out so well. He lost his muscle control. When we got to the doctor's office, we could barely wake him up, and when we did, he couldn't hold up his head. He'd been starting to talk and he was feeding himself, and now he could not hold up his head." A familiar nonsensical sequence followed: a misdiagnosis of meningitis, a referral to Johns Hopkins, where the doctors, who were by now familiar with the family, had nothing new to say. "We all cried, as though he'd died."

Little Alvin cried and thrashed around constantly. Phenobarbital only made him worse. "It almost got the better of us for a while. We tried to bottle-feed him, but he was putting too much energy into sucking, getting just one ounce an hour. When we spoon-fed him, he tried to grab the spoon, because he'd been feeding himself, and he could do nothing with it and it broke his heart. We put him on the floor: He tried to crawl and fell on his face." Susie remembered she had some medication left over from Levi, who had been dead for eight years. It was Valium, and it worked by settling Alvin down enough to enable him at least to eat.

In 1988, when Alvin was three, Livanna was nine, and Steve was eleven, and Levi and Sylvain had been dead nine and eleven years respectively, a beat-up car pulled up in Susie and Amos's driveway. "Amos first thought he was a feed salesman," said Susie of her first encounter with Holmes Morton. "And he came in and told us that he thought he knew what was wrong with our children. Now this just doesn't happen, a man doesn't come to the door and tell you what's wrong with your children."

~

Holmes Morton had been working at Children's Hospital in Philadelphia. He had graduated from Harvard Medical School—this after forgoing college to work in the Merchant Marine, then serving in the navy, then deciding to study developmental neurology and entering college at twenty-five. So by the time he came to Children's Hospital, Morton was in his late thirties, with a wife and three children. He came across a blood sample drawn from a boy in Lancaster

County who suffered from a mysterious debilitating condition. "His story was—and what had fascinated me about the case—was that he had learned to sit up and walk and then he became ill and within a few hours he was disabled." The boy was now four or five years old, and his diagnosis was "cerebral palsy"—in his case, an umbrella term that simply meant he had suffered neurological damage.

Morton used some technology that was new at the time and diagnosed the boy with glutaric aciduria type 1 (GA1), a disease that was considered so rare in the late 1980s that even most specialists had never seen a patient with it. It is a recessive disorder that causes the body to be unable to process certain amino acids, which build up. Much like with maple syrup disease, the body is set up to go into overdrive and fail when anything upsets its metabolic equilibrium. When a child with GA1 develops an ordinary bacterial or viral infection, he is likely to suffer a stroke that will do irreparable damage to the basal ganglia, the parts of the brain that are responsible for controlling movement and perhaps for other functions (the caudate nucleus, which is affected by Huntington's disease, is a part of the basal ganglia).

Almost no one had seen a person with GA1, but Morton decided to go take a look. He took a day off, drove the couple of hours to Lancaster County, met the boy and his parents, and asked them if they knew of other children with similar symptoms. They said they knew a lot of them. "So I just came out and started finding these kids," Morton said. "And by the end of that summer I had diagnosed twenty cases, within all the major family groups of the Amish." GA1 would turn out to be as common among the Amish as maple syrup disease is among the Old Order Mennonites: Roughly one in ten people is a carrier.

The Amish had long held a fascination for geneticists: They were a closed population, they obviously had a number of genetic disorders common among them, and they were generally willing to cooperate with researchers. Most people I met in Lancaster County explained this surprising acceptance of modern medicine by saying that the Amish culture is, first and foremost, pragmatic: While something like television is rejected because it may distract people from their work, their faith, and their community, modern medicine, which can potentially help keep Amish people healthy, does not fall into the same category. So for years genetic researchers had taken blood, urine, and tissue samples from the Amish, especially those who were apparently affected with hereditary conditions. Holmes Morton found out that Victor McKusick, one of the world's best-known geneticists and author of *Mendelian Inheritance in Man,* the bible of medical genetics, had visited the same households Morton was visiting now and had not figured out what was wrong with the children. Glutaric aciduria type 1 among the Amish became Holmes Morton's own project.

Children's Hospital did not encourage his research, so he left, finding a temporary home at the Kennedy Krieger Institute in Baltimore under Hugo Moser, a patriarch of neurological genetic disorder research who was portrayed by Peter Ustinov in the 1992 film *Lorenzo's Oil,* about a little boy who suffers from adrenoleukodystrophy, another inherited metabolic disorder that destroys the central nervous system. "Hugo gave me a little bit of money to feed my family for the year," Morton told me. "And at the end of the year Hugo put in a competing grant with mine with the National Institutes of Health. He is a pretty famous guy, so it was pretty obvious to me that I wouldn't get the money."

By this time Morton had spent a year tracking down Amish families affected by GA1. "He told us what he knew about glutaric aciduria type 1," Susie told me of the day her husband mistook Morton for a feed salesman. "He came back that evening to collect urine samples. He came back after the weekend and said, 'Susie, I found what I expected.' Over the weekend, I had cooled off. I thought, now this is just another doctor who thinks he knows. But now, I tell you, I had to sit down."

Morton put Susie's sick kids on low-protein diets and riboflavin (vitamin B_2). They improved little, if at all: The damage to their brains was indeed permanent. But he started testing babies and newborns in Amish families, sometimes managing to warn the parents in time to restrict the child's diet sufficiently to prevent a stroke. By the time he failed to get the federal money, he was already treating children in Lancaster County and could not stop. So in the spring of 1989 Holmes Morton and his wife, Caroline, set up a nonprofit organization called the Clinic for Special Children, which was a little grand for an operation that was limited to Morton's car. They figured the Amish and the Mennonites, who had a system for raising money to pay large medical bills—they generally have no insurance, not even Medicaid, even though they pay taxes—would be able to gather some money to help care for their "special children." In September of that year the *Wall Street Journal* printed a front-page story on the effort, drawing more than half a million dollars in donations that essentially launched the clinic. The Mortons moved their office from Holmes's car into a space provided by Lancaster General. In another year, men from the local Amish and Mennonite communities held a barn raising—the sort shown in the movie *Witness,* where hundreds of men in identical black single-strap trousers and primary-color shirts work like ants to erect a building in a day. The clinic was built on land that belonged to an Amish farmer whose granddaughter Morton had treated for GA1. Thereafter the clinic survived on small fees it collected from patients, large and small donations, and regular fund raisers at which local residents auctioned to the larger public their quilts, buggies, and other handmade artifacts to bring in as much as a third of the clinic's budget. The most expensive equipment in the clinic, which had allowed Erik Puffenberger to take his colleagues and patients into a new era of genetics, had been donated by the manufacturer.

The building I visited—the original barnlike construction plus an addition raised in 2000—looked like a Hollywood rendering of the Amish aesthetic: soaring ceilings, exposed wooden beams, shining hardwood floors, spacious offices for the two pediatricians, sparkling-clean exam rooms, and a general sense of friendly calm so rarely found anywhere, especially at a medical institution—particularly one where parents brought their very sick, their inexplicably crippled, their hopelessly unable to communicate, impossible-to-understand children. The secret to the calm, aside from an apparently boundless capacity for acceptance characteristic of both the Amish and Old Order Mennonite cultures, was a belief Holmes Morton seemed to project from the moment he set foot in Lancaster County: He knew what was wrong with the sick children here, and if he did not know, he could figure it out, and then he could cook up something that would help. Cook he did: Morton spent the first twenty minutes of our acquaintance telling me about an experimental pig-brain-bouillon treatment he and his junior partner at the clinic, Kevin Strauss, had devised for babies born with a disorder that apparently robbed the body of necessary compounds called gangliosides, causing children to become severely microcephalic. The process sounded downright bizarre—Strauss had gone to the butcher's to purchase fifteen pig brains, which he boiled down to forty cubic centimeters of "brain soup." And the logic was suspiciously simple—pig brains contain a lot of gangliosides—but the approach was probably quite similar to those that had led Morton and his colleagues to devise ways to manage diseases like glutaric aciduria and maple syrup disease.

The Clinic for Special Children came a few years too late to help Susie's children. If this tardiness caused her particular regret, she had come to peace with it a long time before I met her. From the way she told her story, in fact, it was clear that Morton's appearance in Lancaster County had marked a new era in her life. He named her children's disease. He helped keep other children from becoming similarly disabled—although some 20 to 30 percent still suffered strokes, despite all the precautions. The words would certainly be too grand for Susie's restrained speech, but perhaps Morton had managed to make her feel that her children's suffering and her family's grief had not been in vain. Susie told me of Morton's work in great detail: She followed it in part because her sister, Rebecca, had worked at the clinic nearly from the beginning.

All of Susie's own children were grown now. Steve was able to find work helping a family in whose basement apartment he was now living. Livanna was renting an apartment with two other disabled young women, helped by a home-care aide who came once a day. She visited her parents once every two weeks, arriving on a bus for the disabled. She lived on Social Security—a level of integration into the larger world that pushed the margins of what the Amish found acceptable, but church elders had apparently chosen to tolerate this, perhaps out of consideration for all the hardships the family had had to endure. Alvin, too, had found work out in the larger community, through a workshop for disabled adults, where he had the task of counting screws and packing them into boxes. He had a paycheck—between two and ten dollars for a day of work—which made him very proud.

Susie was showing me around the house, explaining the battery-powered lift used to get Alvin on and off the toilet and in and out of the bath, when a van for the disabled announced its arrival outside with a series of squeals and beeps. Alvin rolled in on his rechargeable wheelchair. He was very thin, dressed all in black, and he sat tilting noticeably to one side, his knees drawn up and his hands hovering close to his long face. He had never learned to walk or talk, so he communicated using jerky gestures and a lot of lopsided smiles. Susie introduced us. Alvin said he had had a good day at work.

~

"So the big joke now is that Hugo got his oil and got to be in the movie and Morton went to Lancaster County." Seventeen years later, Morton still resented the way his mentor had undercut him. Hugo Moser had received his grant to do research into Lorenzo's Oil, so named for a little boy (now a grown man) whose parents educated themselves in chemistry and medicine and devised a compound that seemed to help soften the symptoms of his adrenoleukodystrophy. Moser's initial research had failed to prove that Lorenzo's Oil worked—a fact Morton noted with obvious satisfaction, even though more recent studies of the compound had been much more hopeful. But Hugo Moser got to be in the movie.

It is one of those films with a plot so heart-wrenching as to make perception of its aesthetic qualities impossible. Nick Nolte, armed with a grotesque Italian accent, and Susan Sarandon play the parents of an adorable little boy who suddenly begins to have temper tantrums on the one hand and coordination problems on the other. He begins to lose his abilities to walk and talk before the family finds out what is wrong: adrenoleukodystrophy, a metabolic disorder caused by a mutation in the X chromosome. It is one of several mutations—like the one that causes a kind of hemophilia—that only women pass on to their children, of whom only the males are affected. This is another genetic condition that seems to have a conspicuous built-in cruelty. Since women have two copies of the X chromosome, they always inherit one healthy copy of the gene from the father, and this allows them to remain healthy. It is the boys who stand to inherit the one bad copy from the mother, with nothing to counterbalance it, and get sick.

Boys with adrenoleukodystrophy generally develop symptoms before the age of ten, then suffer from progressive neurological deterioration and die within a couple of years. Lorenzo Odone, whose story inspired the movie, lost his speech and all of his mobility but did not die. Years after he became unresponsive and should have been dead, he began to show signs of consciousness and learned to communicate using his eyelids and one of his fingers to sign "yes" and "no." His parents believed this happened because of Lorenzo's Oil, and this may or may not be true: Studies seem to show that the treatment is possibly effective only as a means of prevention and becomes useless once symptoms set in. Quite clearly, though, Lorenzo survived because his parents refused to let him die, which means they cared for him so closely and so attentively that they certainly did things that could never be measured and published in a medical journal to be reproduced in a patient population and checked in a proper trial. Their best medicine was not pure love: It was a thorough understanding of the disease earned through years of sleepless nights spent at libraries (Augusto Odone, the father, ultimately published several medical papers in major journals), combined with unblinking observations of the way the disease worked in their particular boy. In other words, it was the essence of medicine as it will be practiced in the future.

In the course of researching this book, I interviewed dozens of medical doctors, geneticists, and biochemists who defined their research in terms of "personalized medicine." Some were looking for genetic markers that would allow doctors to identify patients who are likely to vomit when coming out of anesthesia—a common problem that is a major source of complications following surgery. Some were already testing patients with colon and other kinds of cancer for genetic markers that would predict their response to one or another sort of chemotherapy. The idea is that eventually even community-based physicians—those outside academic research centers—would routinely be using genetic tests to tell them which medicine to administer, and how, to a particular patient.

By the time I conducted these interviews, Holmes Morton had for seventeen years been practicing medicine one child, one life, and one illness at a time. He had conducted clinical trials that involved one, two, and three children. Some of these were trials of his own invention, like the brain-soup experiment. Others involved the use of experimental treatments on children who for one reason or another could not enroll in one of the larger academic trials. When I spoke to him, for example, he had begun administering valproic acid to a child with type 1 spinal muscular atrophy, a condition that destroys cells in the spinal cord, robbing a child of the ability to use his muscles, and usually leads to death in early childhood. The parents of this particular baby, a horse-and-buggy Mennonite family from Ohio, had located a large clinical trial of valproic acid in California, but the trial turned out to be randomized and placebo-controlled, which meant that the child stood a 30 percent chance of being assigned to a placebo group. "To make a long story short," Morton said, "the family asked me if I would prescribe valproic acid in the hopes that it would help this child. Well, I prescribe valproic acid all the time for less interesting reasons, so I said, 'Sure, why wouldn't I?' And we monitor liver tests, and we monitor blood counts to make sure the valproic acid is not causing any toxicity. And what's the downside of a trial like that? Not much, you know."

It was just the opposite of what was still medical gospel in the United States at the turn of the twenty-first century: that proper medical knowledge is gained only through evidence gathered from large-scale double-blind placebo-controlled randomized trials. Few people had yet noticed that the medicine of the future could probably at no stage be either blinded or randomized. But in southeastern Pennsylvania, a small group of slightly odd people was already practicing personalized genetics-based medicine.

~

Genetics-based medicine begins with a genetic diagnosis. In the old days—the 1990s—determining the genetic cause of a disease generally required a large group of patients and DNA samples from members of their families. Following the logic that genes generally travel in clumps—meaning that if a person inherited a mutation from one parent, she or he probably also inherited a sizable group of genes in the vicinity of the mutant—researchers looked for areas of the genome where all the samples seemed to be similar. This usually allowed them to focus first on the right chromosome, then on the right neighborhood in the chromosome, and gradually zero in on the gene. The search for a mutation took months or, frequently, years. Often, before a test for the mutation could be developed, a test for a genetic marker appeared. This happened at the "neighborhood" stage of the search and essentially amounted to looking for signs that the person had inherited the suspect region from the affected parent; the test invariably required a DNA sample from the affected person, too, to ensure that both the person being tested and the affected parent carried the marker. Marker tests could also yield inconclusive results.

Technology that appeared in the new millennium allowed scientists to look at more of the genome much faster. The Clinic for Special Children received a donation of equipment from Affymetrix, a company that made machines that could rough out the map of a person's genome in the space of about three days. The technology was based on an array of genetic markers: Erik Puffenberger, when we met, was using chips that contained ten thousand markers, and he told me breathlessly that Affymetrix was about to come out with a five-hundred-thousand-marker chip. By the time I was writing this chapter less than six months later, Affymetrix had announced the coming of a million-marker chip, and, less than a year after that, when I was editing it, the 1.8-million marker chip had already arrived.

Where it used to take months to scrutinize three or four hundred places in the genome for conspicuous matches, the new machines took three days to look in ten thousand spots. The number of markers meant that, instead of using a large patient population, Puffenberger could now take just five, three, or even two people with the same disorder. Alternatively, he could test a single affected person and compare his genome to that of his unaffected siblings, looking for areas of difference that could be significant. Puffenberger had done this in the case of a family that had had three children born with an extremely severe seizure disorder—the babies seemed to start seizing in utero and after birth continue to have thousands of seizures a day. Two of the children had already died, so Puffenberger compared samples from the single affected child to those of the child's six healthy siblings, and located the mutation.

The clinic had had the equipment for all of two or three months by the time I visited. Puffenberger had already used it to map four or five disorders and had grown accustomed to comparing the new era of genetic mapping to the dark old days when one had to labor for years—as though many decades had passed. I arrived on the second day of Puffenberger's three-day mapping cycle. He was arranging tiny vials of samples on a palm-sized tray.

"These two have an unusual disorder," he said, looking at small drops of liquid in a vial. "Mental retardation, and I believe they have seizures. We think this might be a new disorder. They are Amish from out in Indiana, and we think this might be a new disorder.

"Then these are four patients, Mennonite patients from Canada, who have a muscle myopathy.

"These are two patients here at the clinic who have an unknown mitochondrial disorder. They are brothers.

"And here we have a child that died, and we are trying to figure out if they had a chromosomal abnormality. And the next child then had a seizure disorder that we know is not the one we recently mapped, so we are adding him to our big panel of genotype seizure patients."

Puffenberger used a tiny pipette to put a fluorescent "tag" on each sample and left them in the fantastical Affymetrix machine to "hybridize." He then led me into another room, a very large one, where the level of background noise from all the lab equipment approached that of a roaring engine. There on a small desk Puffenberger kept his PowerBook, where a collection of Excel files held the fruits of his labors.

Puffenberger opened one of his files with apparent pride. It contained the data on three siblings with a severe dystonia disorder, and two other children, from two different families, who had turned out to have the same disease. We were looking at tens of thousands of Excel cells with the letters AB, BA, AA, or BB in them. The first order of business, once the ten thousand markers had been mapped, was to look for AAs and BBs. "These patients all have the same disorder," explained Puffenberger, "so we are interested in finding a region where this patient, this patient, and this patient are all homozygous for the same allele." Every so often an area of repeated pairs of letters showed up, but some of these could be easily dismissed—either because they were only very small pieces of DNA or because control samples (people with a different disorder or healthy people from the same population) were also homozygous in that area.

Once Puffenberger identified a large area of homozygosity, he worked on the assumption that the defective gene was found there. The next step was going to the online database of known genes maintained by the National Center for Biotechnology Information, a resource of the National Institutes of Health, and looking at genes that are known to reside in the suspect area. The NCBI data, culled from the Human Genome Project, included a list of genes whose functions had been studied or theorized, as well as a list of genes that were not known to do anything in particular—either because they did not or because no one yet knew what they did. For this disorder, Puffenberger decided to look first at two genes, one of which was known to cause seizures while the other was known to cause a movement disorder when it was knocked out in a mouse. He would sequence both genes. If he did not find a mutation, he would go back to the database to identify other candidate genes and sequence them.

Puffenberger made the entire procedure as exciting as one could make a very long process of chasing numbers and letters. It did seem that the quest for the cause of disease had lost much of its detective-novel luster and turned largely mechanical. The impression was slightly deceptive. The straightforward number-crunching route worked only some of the time. Other times, Puffenberger had been stumped. Indeed, one condition had stumped the entire clinic for years. It was a kind of retinitis pigmentosa, a disorder that causes progressive loss of vision. In the Amish, it was accompanied by a loss of reflexes and position sense: Affected people have an odd gait and trouble gauging their movements. Caroline Morton told me of one young man who would crush a paper cup if he were given one to drink from, because he could not quite tell what his fingers were doing.

Dr. Morton and I passed another affected person, a middle-aged woman, on the winding road to the clinic. "She shouldn't be out here," he said. "She has retinitis pigmentosa, like we were talking about. She has bad night vision and also almost no reflexes. She also has deafness. At some point—I'm not sure exactly how it happened—but Caroline hired her to clean our house. And my daughter Sarah walked in, and she was there with the vacuum cleaner, dragging it around. And she couldn't hear well enough to know that it wasn't turned on, and she couldn't see well enough to know that it wasn't cleaning." Something told me this was not the only time the Mortons tried to supplement their good work at the clinic in ways that were not entirely practical.

It was back in 1999 that a group of scientists, including Dr. Morton, identified a likely location of the gene that caused posterior column ataxia with retinitis pigmentosa, as the condition was formally described. They theorized that a single mutant gene would cause damage both to the spinal cord and to the retina. And now Puffenberger had spent years looking for the gene. "We sequenced every gene that is expressed in the retina, and they are all normal," he complained. "Then we went and sequenced all the other genes that are expressed in the brain, and they are all normal, and now we are sequencing the ones that really make no functional sense whatsoever, but it's got to be one of these! So it's not always easy and obvious."

It may not always have been easy and obvious, and it was profoundly frustrating at times, but on the whole, genetic analysis as practiced at the Clinic for Special Children in Strasburg, Pennsylvania, represented a thorough demystification of the science.

～

The P. family came in with two boys, ages five and six. Morton knew the Amish family well: Its four affected children had allowed him and Puffenberger to diagnose a disease that had not been named, and to find the mutation that caused it. The parents were both thin, very light-haired, very young, and very tired. They had had extremely poor genetic luck: Four out of four of their children were affected. Children with this disease are born apparently normal and develop well, with perhaps slight motor delays, until the age of eighteen months. Then they begin having seizures, their language ability disappears, and social skills deteriorate. "If they were somewhere else, they would probably be diagnosed as autistic," Holmes Morton explained. He had as little regard for this possible diagnosis as for any umbrella term that he believed obscured the fact of a genetic abnormality.

The two P. boys were beautiful, with large eyes and smooth light brown hair, cut roughly—clearly at home—and a little long. They wore maroon shirts and the usual thick black trousers with suspenders. When I came in, they were sitting quietly in their parents' laps. Then Isra, the younger boy, stood up, stumbled over to me uncertainly, and poked his finger at the fly of my jeans, which was directly at his eye level. He then leaned with his elbows on a chair seat, stuck his head against the chair back, and pushed. His brother Nathan was doing the same in his father's lap, butting his head against his father's stomach insistently.

Isra had noticeably better coordination than his brother. He had a vagal nerve stimulator—an electrical device that is implanted in epilepsy patients to help control seizures. His older brother, unfortunately, was not a candidate for this surgery, because he was having seizures on both sides of his brain. Isra's seizures had stopped after his surgery. "We thought we had it licked," said Morton. But after nine months the seizures returned, although they were apparently less frequent than before.

Susie, the boys' three-year-old sister, was affected a bit less severely. "She is talking more than the boys ever did," said their mother. The boys were now both on the floor, Isra rolling around with his diaper sticking out the top of his trousers and Nathan sitting quietly. Their youngest brother, Ethan, may have been the luckiest: He was diagnosed early, because by that time Puffenberger had identified the mutation that caused the disorder. Dr. Morton put him on valproic acid, which is commonly used to treat seizure disorders and some other neurological conditions, and so far, at the age of twenty months, he had not had seizures. He was learning to talk.

The boys had the clinic's disease of the month. The clinic team's article on identifying the mutation was coming out in the next issue of the *New England Journal of Medicine*. They had not only defined the disorder but had found a mutation that seemed to shed new light on a substance called contactin-associated protein-like 2, or CASPR2, which, they now suggested, influences brain development. The genetic mutation in these children meant they had less CASPR2 than is normal.

"Maybe as they grow older we might still see changes?" asked the father. Ethan's apparent health had made him hopeful that valproic acid might help his older sons, too. Morton nodded.

"Interesting thing about valproic acid," he began, and launched into a small lecture involving the spectacular qualities of a remarkably versatile drug, gene expression, and the small but valid hope that he still had for being able to help Isra, Nathan, and Susie. "These children are actually better off than Annabelle was," he pointed out, referring to a girl from a different family. "She just seized until her brain burned out."

The father nodded. This was clearly not the first time he had discussed gene expression with Morton. But today the family had come in with a more pedestrian complaint: Both of the boys had earaches. Morton had given them ibuprofen, and as the visit went into its second hour, it became evident that the pain reliever was working. "What was the name of the medicine you gave them?" the father asked. Ibuprofen, said Morton, and added that Tylenol would also work. "Are they safe to take?" the father double-checked. "Are they available in drugstores?" Thanks to Morton and his colleagues, knowledge of gene expression was more common than ibuprofen and Tylenol in the very separate world of the Old Order Amish.

Morton walked the P.'s to the waiting area, where he told the receptionist, "We had short visits." The family had spent an hour and a half in the exam room. They had had time, too, to talk about other families with similarly affected children, and to agree to get one of them to come in to be tested for the newly discovered mutation. Morton had promised to do the test free of charge.

~

On the evening of the second day I spent with Morton, I finally saw him eat. Lancaster, Pennsylvania, has a perfectly ordinary highway cutting through it, with discount shopping malls, Staples and Holiday Inns along it, as well as a roadside Waffle House. I sat in a booth with Morton, Puffenberger, Strauss, and Charles Hehmeyer, a medical malpractice lawyer who had driven the two hours from Philadelphia to help the clinic's staff teach a class at Franklin and Marshall College, where we would all go after dinner. On the face of it, this was a clear case of overkill: Four grown, professional, even famous men teaching a two-hour undergraduate seminar was probably three men too many, even if the students were the sort most valued by Morton—unspoiled by medical school and the business of medicine as business. But I had already realized that resources around here were allocated very differently from anywhere else I had seen, especially anywhere else in the United States.

The topic of that night's Franklin and Marshall class was new-born genetic screening. The team's basic policy suggestion: test all babies for all diseases, including rare ones. Morton had successfully campaigned to add some rare disorders to Pennsylvania's newborn screening panel. Charles Hehmeyer had been suing hospitals across the country for malpractice in cases where children died of or were disabled by genetic disorders that could have been diagnosed and addressed at birth. A few unlucky undergraduates were assigned to defend the opposing view, and they had done the best they could to advance the argument that the diseases were so rare that testing children for them was, on a large scale, a waste of money. This reasoning, standard for health-care bureaucracies, was indefensible under the circumstances. First, adding disorders to the existing testing panels barely increased the cost of the procedure. Second, by Hehmeyer's count, in the United States a thousand children a year were killed or injured by the preventable effects of undiagnosed genetic metabolic disorders. And third, these four remarkable men had come to tell the students that, basically, everyone should be tested—for everything, if necessary.

The intoxicating sense that anything was possible, which pervaded the work of Morton and his associates, was natural and integral to his thinking. Within perhaps half an hour of meeting, he and I had engaged in a discussion about the pros and cons of testing children for adult-onset diseases. Conventional wisdom, professional consensus, and general policy were all clear on this much: Genetic testing of minors should be performed only if it is of immediate medical benefit. My own mutation was often used as an example of a mutation for which it would be counterproductive and possibly dangerous to test a child. The arguments were that the child would come of age some years before she was likely to develop cancer and could make her own decision about testing then—by which time effective treatments might have been developed—and that informing her parents might result in fragile-child syndrome and otherwise complicate relations. But just a week earlier I had heard Stephen Narod, billed as the world's most-cited expert on breast cancer, tell a conference audience that the course of breast cancer in mutation carriers seemed to be determined in adolescence. For example, a girl with a BRCA1 mutation who had her first period after the age of fourteen was less than half as likely to develop premenopausal breast cancer as a girl with the mutation who had her first period at eleven or earlier. This, said Narod, seemed to suggest that whatever ultimately set off the cancer started happening at puberty. My mind rushed: It was entirely possible to affect the age of onset of menstruation through exercise and diet. Should I have my daughter, who was four and a half at the time, tested for the mutation and then decide, say, whether to encourage her dream of becoming a professional figure skater? Professional-level athletic training would give her a good shot at delaying the onset of periods, thereby cutting her risk of cancer in half. Granted, choosing a demanding and dangerous sport was a complicated decision and I was not certain I would ever really encourage it, but now it seemed there was an important unknown in the equation. Following his talk, I asked Narod whether his study results suggested that prepubescent girls should be tested for BRCA1 mutations. To my surprise, he adamantly defended the professional consensus, saying children should not be tested.

Morton, on the other hand, was in favor of genetic testing, period. His arguments, too, were unassailable. First of all, knowledge was power, and diagnostic knowledge meant the power to heal, or at least to treat. In the case of incurable and untreatable diseases, the right genetic diagnosis would at least save parents tens or hundreds of thousands of dollars that they might otherwise spend trying to figure out what was wrong with their children—something that had happened to a number of Amish and Mennonite families—and might bring a kind of peace, as it clearly had for Susie, whose children Morton had diagnosed when it was too late to help them. Attempting to help even a hopelessly sick child might ultimately yield treatment for others with the same diagnosis—provided there was a diagnosis. The same was true for adult-onset and late-onset diseases, Morton argued. Take Huntington's. Studying people believed to be presymptomatic would yield a better understanding of early symptoms, which would ultimately lead to treatment or prevention. The same argument held for hereditary cancers. Fundamentally, he believed, the more you knew, as a patient or as a doctor, the better. Sure, one had to be honest, caring, and ethical, but special rules created to insulate people from genetic tests in particular seemed only to irritate Morton and his team.

I came to think that the difference between those at the Clinic for Special Children and other doctors, geneticists, and genetic counselors I had met was that here in Lancaster, Pennsylvania, people really believed that treatments, cures, and other breakthroughs were just around the corner. The rest of us talk about living in an age of unprecedented medical progress; these people were actually living in it. Most medical researchers have had the experience of spending several years working on a failed treatment—or even just a failed idea for a treatment that was never so much as tested. Morton, despite his grudge over having been beaten out for a grant two decades ago, had never experienced the sort of devastating failure that irreparably alters a person's sense of time, progress, and his own utility. In fact, he had to be one of the luckiest doctors on the planet. He had built his own clinic. He had revolutionized health care in two populations, the Amish and the Old Order Mennonites. He had developed effective treatments for diseases that used to cripple and kill universally. By all accounts, his less-than-total success with glutaric aciduria type 1 troubled Morton terribly, but on the whole he had one of the most creative and rewarding jobs in all of American medicine. What for other doctors, those anchored in traditional research centers, might have been career-making events were for Morton virtually daily occurrences: identifying new diseases, inventing new approaches to treatment, running experiments at will.

"We have our gene therapy," he told me, laughing. The promise of gene therapy, which had in the 1990s seemed like the imminent and obvious result of all the advances in genetic research, had been receding for nearly a decade. In 1999, at the University of Pennsylvania, an eighteen-year-old man named Jesse Gelsinger died four days after receiving an experimental gene therapy treatment aimed at curing his ornithine transcarbamylase deficiency, a metabolic disorder that had required him to stick to a low-protein diet and a drug regimen. His immune system went into fatal overdrive in response to the adenovirus used as the vector to deliver the therapeutic gene. He died of inflammatory shock: His blood developed clots, and many of his organs failed. The Food and Drug Administration reacted by shutting down a number of gene therapy trials around the United States; several research centers shut down their trials voluntarily. The University of Pennsylvania disciplined the director of its gene therapy program, James Wilson, by restricting him to animal experiments.

The first human gene therapy trial considered successful was performed at the Hôpital Necker-Enfants Malades in Paris in 2000–2002. It involved eleven boys affected with the "bubble boy" disease, the severe genetic immune deficiency. (The patients were all boys because their particular mutation was X-linked.) Nine of the boys were considered cured. Within two years of the trial, however, three of the patients developed leukemia. One died, while two others were helped by chemotherapy. As soon as news of the first case of leukemia got out, the Paris hospital halted its trial, and similar trials were stopped in the United States. The leukemia was subsequently shown to have resulted from the gene transfer—in yet another in a long line of setbacks showing that routine gene therapy is probably decades away.

In 1995 the *Wall Street Journal* predicted that a gene therapy product would be on the market within a year. That may have been hype, but certainly by June 2000, when President Bill Clinton and British prime minister Tony Blair stood side-by-side to announce the completion—technically the near-completion, at that point—of the sequencing of the human genome, gene therapy seemed at hand. This was the basic promise of the era of medical genetics. It would allow doctors to dig down to the root causes of disorders and body responses. It would then allow doctors to calculate the best course of action for treating a particular individual. Best of all, it would allow doctors to tinker with the mechanism itself, treating not the symptoms of a disease or even the disease itself, but the cause of the disease.

This has already happened, in a way. Eight of the boys from that Paris trial, including the two who survived leukemia, were alive and basically healthy at the time of this writing. Normally, children with this disorder could not survive more than a year without a bone-marrow transplant. In fact, the very first gene therapy attempt, performed on a four-year-old girl from Cleveland in 1990, was successful: Ashanthi DeSilva, who suffered from an adenosine deaminase deficiency, another cause of severe immune deficiency, was cured. Gene therapy trials in patients with lung cancer, pancreatic cancer, and malignant melanoma have all shown promise. Several lives have been saved. The difference between the promise of gene therapy and its practice to date is scale. The initial vision, even if no one quite articulated it, was that of an assembly line: All of us become reduced to a long line of letters, all of which can be examined, the causes of disease pinpointed and fixed. It would be like doing a check on a problematic computer hard drive, combing through every byte of information to identify and, if possible, to correct each error.

That may happen eventually, when we learn much, much more about the mechanisms by which genes cause disease. But it may never happen. It may turn out that the uniqueness of each human body exceeds all the computing power humans are capable of developing: There will always be too many variables and unknowns.

The clinical trial that ended with the death of Jesse Gelsinger involved seventeen other people—none of whom suffered a reaction like Gelsinger's. There was talk that if the young man's own treating physician had been involved in his care following the gene transfer, the complications could have been handled better. Knowing his patient, he might have used his experience and his informed intuition to help him. It seems obvious, but in a basic sense it goes against the very premise of gene therapy: that a single genetic condition will have a single genetic cure, which can be administered almost automatically.

What Morton meant when he said "We have our gene therapy" was actually referring to the use of liver transplants. He had figured out that maple syrup disease can be cured by liver transplant. Even though the patients' bodies were genetically programmed to lack a necessary enzyme, the insertion of a healthy liver would fix the problem. Several of Morton's patients—some who had problems that exacerbated their symptoms, and one who had had enough of drinking formula—had opted for the operation. In a particularly exciting experiment, a team in San Diego transplanted a healthy cadaverous liver into a maple syrup patient and the maple syrup patient's liver into a man who was dying of liver cancer. The maple syrup liver was healthy, and the recipient's body managed to provide the enzymes to make it function normally. "That's actually the neatest thing about transplants—what we call 'domino transplants,'" said Morton, and then repeated his almost-joke about already practicing gene therapy.

He was right, as usual. Gene therapy, like genetic medicine in general, will be practiced one patient at a time. And this was exactly what was already happening at the Clinic for Special Children. Erik Puffenberger was already isolating mutations one patient at a time. Holmes Morton and Kevin Strauss were inventing treatment protocols one patient at a time. This was the handmade version of the imagined assembly line of the future. That future may or may not happen, but until then, we all will want a kindly country doctor tinkering with our genes, which he can navigate like they were his own.

Chapter 11

BIOBABBLE

Anastasia Kharlamova, a tall, pretty, and confident woman in her early forties, shed her urban-professional image by exchanging her sheepskin coat for a dirty lilac down number and slipping on a pair of hiking boots that had seen better days. She led me down a gravel path, through a gate, to a field filled with cages holding hundreds of silver foxes. We entered a covered aisle lined with two rows of female foxes. They barked and squealed and made funny quacking noises and banged their aluminum feeding boxes with their paws, trying to draw attention to themselves. Anastasia murmured Russian endearments as she walked along, occasionally stopping to take a fox out of a cage and cuddle her. One of the foxes resisted going back in, wriggling into Anastasia's arms and angling to lick her face.

A few aisles over, we were greeted by ominous silence—until Anastasia touched her fingers to the cage bars. Then the fox in the cage would bare her teeth and lunge at the front of the cage, closing her mouth with a metal snapping sound.

"Careful of your nose," said Anastasia, addressing a fox. "They have no teeth left by the end of their first year," she explained, turning to me.

The silver-fox farm outside of Novosibirsk is probably the longest-running continuous genetic experiment in the world. In the 1950s the Soviet geneticist Dmitry Konstantinovich Belyaev set out to figure out what happened to wolves on their way to becoming dogs. That is, how did wolves, which can be generally, but accurately, described as weighing between fifty and a hundred and thirty pounds, having a narrow chest and a powerful back, possessing a bulky coat that is gray to gray-brown in color, as well as stout and blocky muzzles and pointy perky ears, manage to turn into creatures that can weigh as little as two pounds or as much as a hundred and seventy; have a long stringy white coat or no coat at all; have a barrel chest or a ridge down a slender back; have a square jaw or ears that flop all the way to the ground? Granted, people bred dogs to look a certain way—but any breed had to start with a spontaneous mutation or mutations that changed the wolf's coloring, coat, the degree of ear perkiness, or any of the other components of wolf morphology. Everything science knew about mutation rates said that this could not have happened in the mere twelve to fifteen thousand years that had passed since wolf started turning into dog.

Belyaev was a lucky man: He was working with animals. It had been roughly a dozen years since the science of genetics had essentially been banned in the Soviet Union. Former geneticists were working as housekeepers and street cleaners. Belyaev managed to continue studying animal genetics under the guise of doing physiology research. In the late 1950s, as Soviet policies relaxed following the death of Joseph Stalin, Belyaev was tapped to head a new genetics institute, to be founded in Siberia. He gathered scattered geneticists from around the country—one of the women he picked to head a laboratory in the institute had worked as a preschool teacher until she was exposed as a geneticist and fired, eventually finding work playing piano in a restaurant—and set up the institute and the experiment of his dreams. He picked foxes because they were closely related to wolves and could be obtained in large numbers, from fur farms.

Belyaev's theory at the start of the experiment was that there was something about the process of domestication that threw genetics out of whack. "Dmitry Konstantinovich believed that behavior is regulated through a complex combination of regulatory molecules, including hormones, neurons, neuron receptors, hormonal receptors," explained Lyudmila Trut, a geneticist whom Belyaev had drafted as a college senior half a century earlier and who took over the silver-fox experiment after Belyaev's death in 1985. "And he believed that these complex interdependencies accounted for variations in behavior." Trut, a very small woman wearing very large eyeglasses, had been teaching for more than four decades, and she spoke like a lecturer: slowly, clearly, and using complete sentences that she could drop upon being interrupted and then pick up again. "So that when animals were selected for behavior—let's call it domesticated behavior—this actually led to changes in the regulatory mechanisms: the hormonal systems, the neuronal systems in the most general sense. And Dmitry Konstantinovich supposed that by changing the state of these regulatory molecules in the process of domestication, we changed the functional activity of a great many genes, causing the broadest spectrum of changes."

Dmitry Konstantinovich Belyaev turned out to be right. First, he proved that it is possible to breed for behavior—in other words, that behavioral traits are genetically determined. He began with a hundred female and thirty male foxes, whose reactions to humans were closely observed by researchers. Since the foxes came from a commercial fur farm, they had cleared the first hurdle on the path to domestication: They were capable of reproducing in captivity, and they were not afraid of humans. Fear, it could be said, had already been bred out of them. What differentiated the foxes from one another was their degree of friendliness toward humans.

The Siberian researchers eventually developed a 9-point friendliness scale, where +4 denotes an animal that eagerly jumps out of the cage to be cuddled by a human, while a −4 aggressively lunges at a human even when the cage is closed. A 0 reflects an animal that exhibits neither aggression nor friendliness: It does not attack but does not allow itself to be handled either. Out of the first group of foxes, Belyaev selected a very small percentage of the females and an even smaller percentage of the males—those deemed the most tame—and bred them. Nine years and nine generations of foxes later, Belyaev was in possession of a small population of silver foxes that were unlike any silver fox that had gone before. They loved people. They loved to be given attention by people. They loved to be handled by people. They were, from all appearances, ready to leave their cages and go live with humans in their homes, becoming fully domesticated.

The foxes stayed in their cages, though. The experiment was designed to limit human contact to avoid contaminating genetic influences with environmental ones. The farmworkers who fed the animals were not allowed to linger near the cages of friendly foxes. To ensure that workers maintained the same attitude while walking past the cages of tame and aggressive animals, the experimenters often mixed the animals, placing the friendly and unfriendly types in adjacent cages. There was no doubt: Animals could be bred for behavior, and dramatic changes in the behavior of a species took mere generations to achieve.

Once they proved that they could domesticate the wolf's nearest relative in nine generations flat, Belyaev's researchers moved on to other animals. They took on minks, which seemed to present a challenge. Unlike the fox and the wolf, minks are not universalists: While the canines can live just about anywhere, minks have strict location requirements, setting up home only near water. In addition, unlike wolves and foxes, they are loners. So the idea that a mink could be bred to the point of being domestication-ready seemed dubious. Still, the experiment worked, once again proving that it is possible to breed animals for behavior, which proved that behavioral traits are genetically determined.

Both the minks and the silver foxes were farm animals, accustomed to living in captivity and unafraid of people. Now the experiment had to be tried on wild animals. In the 1970s Belyaev's people went on a rat hunt in the forests around Novosibirsk. They looked not for rats that lived in building basements but for the certifiably wild sort. Many of the captured rats died. Many others refused to reproduce. The rest—roughly two hundred, including the first generation born in captivity—went on to form another historic experiment. The researchers divided them, at random, into three groups. One group would be bred for domestication: Only the friendliest of the rats would be allowed to reproduce while the rest were killed. Another group would be bred for aggression: Only the meanest ones would be allowed to have babies. The third was a control group: A small number of randomly chosen rats would be allowed to reproduce in each generation, presumably producing offspring that were little different from their wild forebears. The control group fell victim to budget cuts around the time the Soviet Union collapsed, but the tame and aggressive groups lived on, the latter maintaining an average aggression score of −3.

Irina Pliusnina, a small middle-aged woman with close-set eyes and a broad smile, led me into a large barn filled with a noxious smell. She handed me a scrubs-green robe and slippers to change into: If Anastasia had changed to keep her clothes clean when she handled the foxes, here the goal was to keep me from infecting the rats' environment with anything I might have dragged in from the outside. She opened a cage and gently picked up a rat by putting her gloved hand around the rodent's body.

"Here, my little girl," Pliusnina cooed. "Look, you could never handle a laboratory rat this way. They pick them up by the tail in labs. Our rats can't stand to be handled by the tail. They like it only when they are handled like this." She was basically embracing the rat. "What a good girl. She is one of those scary wild rats of Novosibirsk, you know, the stuff of legend for many years." The rat stretched its head toward Pliusnina, as though reaching for a kiss. The researcher leaned down to her subject, gushing: "So you are my mean wild rat. Mean mean mean rat. Mean mean mean mean."

The real mean rats were in cages in the room next door. Pliusnina carefully chose a cage to open: All of the rats were either pregnant or nursing, which made them especially aggressive (the tame rats in the other room were also in various stages of motherhood, but this did not deter them from cuddling with Pliusnina and even letting her handle their tiny offspring). Finally, she opened a metal door. The rat stood on her hind legs, her front paws lifted as though ready to strike.

"This one is a boxer," said Pliusnina. "If I reach for her, she will attack." This made her roughly a −3 on the tame-aggressive scale. "Quite a contrast, yes?" said Pliusnina. "That's sixty-seven generations of selection." Small change in evolution time, but further proof that breeding for behavior was possible, even in a population of animals caught in the wild.

Pliusnina continued to sing the praises of domesticated rats. They were, she explained, basically much nicer creatures. They were calmer, more self-confident, less nervous, and more inclined to try new things than their aggressive cousins—better-adjusted all around, in other words. In an "open-field" test, where rats that are normally caged are placed in a relatively large round arena and videotaped, the tame rats showed themselves to be much calmer than the aggressive ones: They ran around less spastically, and they defecated less. In another test, the rats were placed in a shallow tub filled with milky water, so they could not see the bottom of the tub. Their goal was to swim to a small platform on which they could sit: Otherwise, the tub was just deep enough to force them to keep swimming. The top of the platform was obscured by the milky water, but its location remained stable. With four swims a day, it took the domesticated rats just one day to learn to find the platform, while the aggressive ones took five days.

The tame rats reaped rewards for their tameness: They had more babies, and, like the tame foxes, they could have them more often than the aggressive group or than is normal in the wild. By the time the tame rats were in their thirteenth generation, tests showed that their pituitary-adrenal system had grown far less active in responding to emotional stress. By the twentieth generation, their basal levels of corticosterone, the rodents' primary stress hormone, were consistently lower than normal. The Siberian geneticists were breeding happy, well-adjusted, and highly fertile rats.

After a while, they got into breeding other kinds of rats as well: disturbed rats, cataplexic rats, and hypertensive rats. The last were laboratory chief Arkady Markel's pet project. Markel was a soft-spoken bespectacled white-haired man who had a large office with a corner cordoned off to form a sort of cubicle, in which he sat, partly obscured by wooden partitions, staring calmly at his small flat-screen computer monitor. He wore a gray soft-knit cardigan. One got the feeling that if Arkady Markel were a rat, he would not do well on the open-field test. He would, on the other hand, be tame to the point of depression, and would certainly not be hypertensive. But, being human, Arkady Markel had great compassion for his hypertensive research subjects.

Markel's rats, bred for hypertension, turned out to share a very specific psychological profile. In unfamiliar surroundings, they tended to have a higher-than-average interest in exploring. But while they shared this trait with Pliusnina's supertame rats, they were different on other scores: When they saw a stranger rat, for example, they expressed less fear and more aggression than might have been expected. They were, in other words, doers.

"Strictly speaking," explained Markel, "hypertension is not a disease but an aspect of an individual's psychological makeup. Its purpose is to ensure an active state. Hypertension is the price of success." Hypertension, in Markel's opinion, is a classic example of a trait that provides a selective advantage. First, hypertension—at least the sort he studied—is genetic, though probably extremely complex, caused by a large set of genes, each of which might play only a small role in shaping the condition. Second, the condition affects an extraordinarily high percentage of the human population: more than a quarter of all adults in developed countries. Third, if the effects of hypertension were solely negative, the trait would have been washed out of the population, or at least watered down. But Markel argued that the condition goes hand in hand with modernity, enabling the individual energy output that competitive contemporary societies demand— which was how the condition grew more and more common.

Markel had set out to breed rats that would have high reactive blood pressure—that is, their blood pressure would rise higher than other rats' in response to stress. In order to measure the rats' blood pressure, the researchers placed a tiny cuff on the animal's tail, thereby fulfilling both requirements of the experiment: The cuff provided stress to the rat and measurements for the scientist. To measure a rat's resting blood pressure, the researchers placed the creature in a jar saturated with enough ether to put the rat to sleep for about five minutes, which was enough time to place the cuff on its tail and take the measurements. It turned out that rats bred for high blood pressure under stress also had higher-than-average resting blood pressure. "Because all of life is stressful," explained Markel plaintively.

As responses to stress go, spiking blood pressure is perhaps the most constructive. In any case, it corresponds to our idea of the norm—unlike, say, going catatonic, which corresponds to our idea of mental illness. Markel and his colleagues created a line of rats that froze in response to stress, such as being scared or being pinched. Whatever gene predisposed rats to going catatonic, it was inherited in a dominant pattern—that is, if one parent was affected, half the offspring would be affected as well—which made it relatively easy to create a breed. They got a population of rats inclined to freezing in an uncomfortable vertical position, and proceeded to decapitate them and study their brains and blood. The catatonia-prone rats turned out to share a number of neurological and hormonal traits with human schizophrenics. At the same time, they also shared a number of traits with humans who are depressed: They had lower levels of certain hormones and neurotransmitters, and they had sleep patterns characteristic of depressives. They failed one of the standard rat-depression tests, though: They appeared to maintain their ability to enjoy sugar water, which may have meant they retained their ability to enjoy life generally. At the same time, they acted depressed in another common test: When they were placed in a tall glass jar partially filled with water, they gave up and started floating after only a short period of kicking their legs in a doomed attempt at escape. This, Markel's researchers suggested, may have marked their cataplexic rats as suitable models for studying both schizophrenia and depression—making them a good first stop for testing new anti-depressants and antipsychotic medication. At the same time, some other researchers at the institute were arguing that the quick-to-give-up rats were actually wiser rats, rational individuals who opted to save their energy. Pliusnina's tame rats also tended to give up faster in the glass jar.

"The problem with these tests is that we cannot ever tell what they are thinking," said Nina Popova. "Did the animal go still because it has no fear or because it is paralyzed by fear?" Popova, who founded the Laboratory of Behavioral Neurogenomics in 1971, received me in her office about a block from Markel's lab, in a dilapidated gray concrete building, the sort that had looked modern half a century earlier. The office was square, with glassed-in bookshelves of polished wood and a massive oak desk that took up most of the room. It looked like the study of a classics professor, someone who studied things so established they were virtually calcified. But Popova was studying something very new. She aimed to trace what she acknowledged was "the very long road from protein synthesis to behavior"—that is, she was trying to define the relationship between genes and the ways people, or at least mice, act.

Popova's lab had its own strain of cataplexic rodents: She and her staff studied mice. Her mice were quite like Markel's rats: They froze upon being pinched and stayed that way. They also did not move around much in the "open field," and they quickly gave up in the glass jar with water. Were they simply calm, rational, and well-adjusted? Unlikely. Just ten days on antidepressants turned them into normal-acting mice. Which may not have told the researchers much about what the mice were thinking, but it meant that the mice were good models for preliminary testing of new antidepressant medication.

Popova's lab had other weird mice, too. A group of French researchers had accidentally bred a line of mice with the monoamine oxidase A (MAOA) gene knocked out of them. The MAOA gene was probably the first so-called behavior gene on the map. In the early 1990s a Dutch geneticist identified a large family in which roughly half of the males suffered from mild mental retardation and made others suffer from their odd, often aggressive behavior. They engaged in arson, attempted rape, exhibitionism, and random impulsive violence. The condition was apparently hereditary, and, since it affected solely the males in the family, most likely linked to the X chromosome: The females in the family had two X chromosomes, one normal and one abnormal, and in half the cases they passed the abnormal one on to their offspring. The boys who got it would be affected; the girls would be healthy but could pass the gene on to their sons. In 1994 the Dutch researchers published their findings showing that the culprit was a badly damaged MAOA gene, which fast became known as "the aggression gene." A couple of years later Han Brunner, the lead researcher on the project, published an article arguing that the moniker was inaccurate, if only because "genes are essentially simple and behavior is by definition complex."

The relationship between genes and behavior is probably equally complex in simpler creatures, such as mice, but researchers who study rodents do not have to fear offending their subjects when describing their findings. What Popova observed was that the male mice with the MAOA gene knocked out were more likely to attack other male mice; that they responded less actively to stress than normal mice; and that they had an increased tolerance for alcohol. They were not merely aggressive but aggressively antisocial. "If we put strangers in a cage together," explained Popova, "we are collecting dead bodies every day. That is unusual for animals in general: Normally the purpose of a fight is to establish dominance, not to kill."

Human males with the less-active-than-normal MAOA gene tend to be both antisocial and alcohol-dependent. In the dozen years after the original Dutch discovery, the gene was studied extensively in different populations, including Germans, Brazilians, and Han Chinese. Most researchers found an association between the mutation, aggression, and alcoholism, and most leaned toward the theory that antisocial behavior, not substance abuse, was the primary factor. On the subject of how the mutation might influence or shape behavior and alcohol consumption, the scientists tried to tread most carefully.

Genes, it seemed clear, do not regulate behavior directly: The brain does. But genes may determine or influence the amounts of different neurotransmitters in the brain, thereby regulating both behavior and emotions. In the case of "the aggression gene," the neurotransmitter serotonin, implicated in regulating mood, sleep, sexuality, and appetite, among other things, emerged as the primary suspect. A genetic variant may interfere with serotonin processing in one of three ways: by lowering serotonin synthesis or making its breakdown less efficient; by disabling the reuptake system, through which we recycle neurotransmitters; and by reducing the sensitivity or the density of serotonin receptors. The MAOA gene appeared to regulate the amount of enzyme required to process serotonin, and that, the researchers believed, was why serotonin levels in both mice and men were higher than normal, which may have made them act similarly antisocial, impulsive, and aggressive.

After three decades of researching mice, Popova decided to venture into studying the brains and genes of humans. Her staff traveled to regions of the Russian North where various indigenous populations continued to enjoy a fair amount of isolation and looked for inroads into studying their serotonin systems. Preliminary research on a Khant family that seemed to have suicide running through it suggested that its members might have a gene variant that lowered the level of serotonin transporters, the ones responsible for sending the neurotransmitters on their way again once they had reached a receptor.

Popova was both eager and reluctant to talk about her human research. Her results were strictly preliminary, she stressed, and we had so much more to learn still about the ways humans work. And then maybe one day we would learn not only why men in a particular Khant kindred often committed suicide but also why Russians—regardless, it seemed, of the ethnic group to which they belonged—drank so much.

~

On June 2, 2006, I looked out the window of my Moscow apartment as Vova, my eight-year-old son, went out to walk the dog. I watched as he sauntered past a playground, empty at the early hour, and stopped in a small grove at the far end of the building's courtyard. A man was standing in the grove already, watching over his own dog. Vova stood, very adultlike, and chatted with his fellow dog-walker. I took pleasure in watching them: Vova, very small for his age (I bought him clothes for six-year-olds), proudly holding our medium-sized mutt, and the man, tall and scrawny, walking a dachshund, of all things. Vova said something to the man, the man extended his arm, with a beer bottle in it, and I watched Vova take a sip.

Russia is a nation of drinkers. Sipping a morning-after beer out in the courtyard while walking his nice purebred dog marked the man as perhaps a well-behaved, domesticated alcoholic, or even just a man who had hosted a party the night before. But what was my eight-year-old doing? I asked him when he came back in.

"I was just socializing," he said. My son was giving me a preview of himself as a teenager.

"That's nice," I said. "What was with the bottle?"

"He had some nonalcoholic beer with him, and he offered me a sip."

I decided to ignore the doubtful alcohol-content assertion and focus instead on the germs that Vova had clearly risked getting by engaging in risky behavior. I chose my words carefully, and I had the distinct sense that I was speaking to an alcoholic in training. I had some experience in speaking to alcoholics about drinking: My partner, Svenya, and I had just separated as she entered rehab for drugs and alcohol. Svenya had been using mind-altering substances for sixteen years—half of her life. Now I wondered whether I was seeing the consequences of her drinking: Our son had emulated her by drinking nonalcoholic beer (a popular beverage in Russia, and apparently harmless from a purely biochemical standpoint). Then again, I could be seeing the evidence of Vova's genetic heritage.

Vova was born in Kaliningrad, Russia, to a twenty-year-old woman who was HIV-positive. Statistically speaking, that meant she was probably a drug user: At that point nearly all HIV-positive people in Russia were injection drug users. My memory helpfully produced episodes demonstrating that Vova had inherited a predilection to chemical dependency. At the age of four he had swallowed an infant's daily dose of phenobarbital, which had been prescribed for my two-month-old daughter, Yael, who had been having non-fever-related seizures. Two years later, at Svenya's and my wedding, a Jewish ceremony in Massachusetts, each of our children was given a sip of white wine under the chuppah. Both scrunched up their faces in disgust, but then Vova said, "Can I have some more?"

So if Vova went on to develop a drug or alcohol problem, I would have my pick of people to blame: Vova's biological mother, for having saddled him with a genetic predisposition to addiction, or Svenya, for having taught him the behavior by example. The question was, with this frightening confluence of genes and environment, was there anything I could do about it now that he was almost nine and sharing a beer with a stranger in the courtyard?

~

A particular problem in our family was that I myself could not understand addiction. I was the sort of infuriating person who could—and did—smoke two to three cigarettes a day for nearly twenty years, and then stop without a second thought. When Svenya and I were first together, I drank heavily alongside her: We were young, childless, and in love, and I could still get up for work the next morning. Then the children came, and I stopped drinking heavily, expecting Svenya to do the same. It took me years to understand that she could not stop. But even now, as I write this, I cannot imagine what it is like to be unable to stop doing something that you know is ruining your health and your family.

A few months into our separation and her sobriety, Svenya became an exemplary recovering substance abuser. She talked to me about codependency, family dynamics, and vicious cycles. I talked to her about genes and neurotransmitters and the substance-abusing brain. Sometimes, our understandings collided. It was easier for me to come around to her view of the forces that shaped us: There was a huge arsenal of tools on her side, created over decades of understanding human behavior through psychology and, at least in the

United States, through psychoanalysis. Psychobabble was so common, it felt like plain English.

But this was about to change. In fact, our understanding of human behavior was already changing, in ways perhaps more dynamic and more profound than our understanding of race and ethnicity. Physiology was starting to make common sense. The names of various neurotransmitters could be dropped into conversation without explanation. And everyone knew that the genome offered the ultimate map of the human being. I could imagine that a new language—let us call it biobabble—would start insinuating itself into the meetings of Alcoholics Anonymous, where people would talk about their family's drinking in terms of heredity, and into, say, the dating scene, where well-educated midcareer professionals would start explaining to potential partners that they had never formed an intimate relationship because of this or that polymorphism.

By the time this book comes out, one or more U.S. companies will likely start offering commercial testing for genes associated with behavior. These are likely to be the same companies that already offer commercial ethnicity and paternity testing—they are the ones with the technology and the expertise—but they will likely do it in the guise of new companies, to separate their behavior-gene testing from their established ethnicity-gene testing. They will be aware they are treading on thin ice. They will risk looking like charlatans. But they will do it anyway: They will offer the testing, because we really want them to. In the era of the genome, we all want a printout of who we really are.

~

I certainly did. Which is why a late-spring afternoon found me in the basement of a mental health facility in Jerusalem, gargling. The geneticist Richard Ebstein and his team used this ingenious method for collecting DNA: Instead of finger pricks or cheek swabs, they gave their patients a cap of regular mouthwash. One swished it around in one's mouth, then spat into a plastic tube. Ebstein was not happy with my first tube: The liquid was too clear. I had gargled instead of swishing. My second plastic tube was half filled with satisfyingly murky liquid. This meant my DNA sample had been collected.

Brooklyn-born and Yale-educated, Ebstein had moved to Israel with his American wife in the 1960s, for fear of the American draft. He chuckled when he told me this. I wanted to ask him more about this oddly courageous choice of haven, but he was not particularly good at answering questions about himself and his motivations: He was, he explained to me, fairly autistic, in the way of many scientists. And he had made a reputation out of such well-founded generalizations.

A biochemist by training, he made his first genetics headlines in the early 1990s, when he claimed to have identified a "novelty-seeking gene." A particular polymorphism of a dopamine receptor gene, found in a sizable minority of people, seemed to make them more likely to try new things and take concomitant risks. Using a captive population of Hebrew University students in Jerusalem, over the years Ebstein continued to administer psychological questionnaires and look for correlations in the genome. He got a fair amount of mileage out of the dopamine receptor gene: The same polymorphism, it seemed, was responsible for conferring on a minority of people a highly active sex drive, but worked against one of the most mysterious human traits—altruism. The propensity to help others at one's own expense, and sometimes even at one's own risk and peril, had long fascinated researchers. Some have argued that altruism as nature intended it extends only to kin. Others have argued that altruism does not contradict what we know about evolution because it is geared toward the survival of the species, if not the individual. Ebstein contributed to the debate by suggesting that altruism is hardwired: The most common polymorphism of exon 3 of the DRD4 dopamine receptor gene, a combination of two alleles of equal length—four base pairs—seemed correlated with a propensity for helping others. It also worked against both novelty-seeking and heightened sex drive: The sexy, risk-taking few were marked with one seven-pair allele. The purported altruism polymorphism also looked like it might raise one's risk of depression—although, being very common, it seemed an unlikely suspect for causing a disorder. Then again, depression is very common, too.

I was fairly certain of my pending test results. I was a novelty-seeking, sex-crazed, egomaniacal optimist, a preternaturally hard worker who was always basically happy. Most of my friends thought this too. Most of the people who could say this sort of thing to me would claim they tolerated my inattention to others in order to feed off my boundless energy. So I was caught off guard when one of Ebstein's coauthors, a talkative Russian Israeli named Inga Gritsenko, informed me I was a common four-four. She refused to interpret my test results for me, but I knew what the implications were: The survey of my genes showed that I had a lowered sex drive and did not possess the novelty-seeking gene but was an altruist at risk for depression. This was pure silliness. I had the evidence: the time I was sentenced to death by Kosovar guerrilla fighters after I dragged a photographer and two interpreters into the mountains to look for action (if that was not novelty-seeking, what was?); the rather significant number of sexual partners and romantic involvements of both sexes accumulated over the years; my ability to bounce back from all sorts of trying experiences while maintaining an even demeanor. I had indeed had two bouts of depression at the classic junctures—as a college freshman and postpartum—but I hardly thought of these miserable periods as my defining moments. I had never been treated for depression.

Feeling slightly offended, I continued combing through Ebstein's published papers. He had studied the dopamine receptor gene in women with fibromyalgia, a common but mysterious illness that hits people—many more women than men—in midlife, causing sleep disturbance and painful tenderness all over. Women with fibromyalgia were said to have a particular personality, to be more prone to anxiety and depression. They also turned out to have a disproportionate number of four-four repeats, the kind I have. Blogs and alternative medicine sites told me that fibromyalgia sufferers tended to be hard-driven overachievers. Psychobabble would tell us their bodies were telling them to stop. Biobabble would tell us what Ebstein wrote: that with the unusual personality profile, it was no surprise that there seemed to be a link between the disease and a gene associated with behavior.

Inga also informed me that I was homozygous for the short form of a serotonin transporter gene, which itself is linked to harm avoidance and neuroticism and anxiety. But not all studies confirmed these findings, so Ebstein and his team decided to study neonates—eighty-one two-week-old babies, male and female, healthy babies of healthy mothers. Babies with my variant of DRD4 scored lowest on orientation, motor organization, range of state (just what it sounds like—a measure of how aroused a baby can become) and second lowest on regulation of state (the baby's ability to calm down, essentially). Babies with my variant of the serotonin transporter scored lowest on regulation of state. I actually felt embarrassed: All these people in Ebstein's lab had seen my results. Of course, I was no longer two weeks old, but I did have certain lifelong problems with regulation of state, and anyway, I *would* feel embarrassed, with my tendency toward anxiety and neuroticism. They observed that, overall, babies tended to want to engage in interactive behavior—with the exception of those who, like me, had both short DRD4 alleles and were homozygous for the short serotonin-transporter allele. In other words, I thought they might note, the fact that I can pick up a phone and call someone—even if it did take me years to learn to—was an impressive achievement. At the end of this paper, there was a long passage outlining all the problems with corroborating the findings in other ethnic populations, but this did not help me: All the babies were Jewish, twenty of them Ashkenazi.

Another experiment concerned second graders. The researchers looked at shyness. I know shyness. And I know that shyness is inherited. I got mine from my mother. My mother's shyness was what scientists would have called "totally inhibiting social phobia." She was terrified of strangers, even in her own home. When I was a child, what should have been routine socializing had the spirit, if not the trappings, of a major event: a function, I realized later, of my mother's fear and a reflection of the effort she had to make to overcome it. Still, her social circle was largely limited to members of our extended family and others she had known for a very long time. When I was in my late teens and early twenties, my own brand of shyness would render me essentially speechless. Entire dinner parties would go by without my uttering a word; or I might be producing words, but I was basically inaudible. When I was able to overcome this, which was not always, I might find myself yelling, which embarrassed me further.

When I was twenty-four, at the end of my first major relationship—which suffered unduly from my shyness; I basked in the luxury of human contact while making others uncomfortable—I was offered a job in California, a place where no one knew how shy I was. I decided to pretend I was not shy. For the most part, it worked. In the fifteen years since, I have made friends, had relationships, changed jobs, interviewed hundreds of people—all with only occasional cold sweats and mere minutes spent paralyzed, staring at the telephone pad. Here was a clear example of environment—one I created myself—beating genetics. But I would find no affirmation in Ebstein's study: He found a significant correlation between shyness and the long serotonin-transporter allele, which I did not have.

They seemed to find that the long allele of the serotonin transporter was far more common among smokers: ever-smokers (those who have ever smoked in their lives), past smokers, and current smokers. I was never really a smoker, which is why I quit so easily. "You are what the literature calls 'a tripper,'" said Ebstein, sounding slightly amused. "There are some people who never go over two or three cigarettes a day, or a few cigarettes a week. Very unusual." I had a slight suspicion that he was trying to comfort me for having the most common dopamine-receptor gene variant: at least I had an unusual relationship to smoking. He acknowledged that my short-short serotonin may have had something to do with my failure to become a smoker—and the lack of the "novelty-seeking" dopamine-receptor variant helped to avoid pushing me in that direction: Novelty-seeking and risk-taking seemed to be associated with smoking, at least in the Israeli population. The "novelty-seeking" allele was also found to be a risk factor for opioid dependence, when Ebstein looked at Israeli heroin addicts. My own relationship with opiates was marked by revulsion: I had undergone enough surgeries to try a variety of opiate painkillers, and I consistently chose pain over the loss of control and brain fog. Now here was an explanation: I did not have the dopamine receptor-related polymorphism that would predispose me to opioid addiction, and I had the serotonin-transporter polymorphism that predisposed me to anxiety, which made me a downer-averse control freak. This, in turn, created significant stores of Percocet and Tylenol-3 with codeine in our medicine cabinet, laying the environmental foundation of Svenya's easily triggered heroin addiction.

Ebstein studied professional dancers, using professional athletes as a control group—to weed out genes that accounted for the athletic ability both fields would require. He found that there appeared to be a gene variant connected with creative dance ability, the same gene variant that had earlier been linked to spirituality. I was not tested for dancing ability. He studied hypnotizability and seemed to find genes related to one's response to hypnosis. I was not tested for this, either. Ebstein also looked at genes connected with concern for appropriate self-presentation and the propensity for forming social relations (which they measured by looking at sibling relationships). I was not tested. I was, however, tested for ten different markers on a gene related to Prodynorphin, the body's natural opioid, a building block for endorphins. I had my results, but Ebstein could not yet tell me whether they meant anything: It was an ongoing project. I had a feeling, though, that it would tell me something interesting.

I asked Ebstein whether he thought that we were entering an era when people would look to genes for the truth about themselves. He mumbled something, then asked me to rephrase the question. I asked whether behavioral genetics was on its way to becoming the source of self-knowledge, whether people would be going for genetic testing the way they might now go for psychotherapy. Ebstein looked absent for a moment. "Oh." He finally smiled. "Self-knowledge! Yeah, my daughter is really interested in that. I am not. I tend to be autistic, like most scientists," he reminded me.

On the subject of my own genes, Ebstein cautioned me against putting too much credence in my results. Many of my variants were so common as to have no statistically significant correlations. "Of your results, the short-short serotonin transporter is most believable," he said, meaning that the variant was relatively uncommon and the evidence from various studies reasonably solid. "It's the tendency toward neuroticism."

I knew it, really. The entire personality I had constructed was a function of my hypercompensation. Deep inside, I had always known myself to be a passive, cowardly person, unlikely to journey anywhere outside the realm of the very well-known. Had I been born in a middle-class U.S. suburb, I would probably still be living there. Knowing this about myself, and not liking what I knew, I tried to prove I could be my own hero: worldly and adventurous, self-absorbed and even seemingly callous, hypercompetent and overconfident. I had managed to convince many people, including myself. But now my true personality had been exposed by a genetic test.

I was only partly joking.

~

Of all the results that Inga had dictated to me over the phone, the one that struck me did not have to do with novelty-seeking or neuroticism. It was the MAOA gene, the "aggression gene"—the gene that landed all those Dutch men in trouble and the gene whose absence made Nina Popova's mice eat their own. I turned out to have one short and one long allele. That meant I had a 50 percent chance of passing on either variant of the gene to any child I had. If the child was a boy, he would inherit the gene from me only: It was located on the X chromosome, of which he would have only one copy. So any biological son of mine would stand a 50 percent chance of inheriting a less-active MAOA gene, putting him at risk for aggressive and impulsive behavior and alcohol dependence. This described my adopted son perfectly.

I not only loved Vova but admired him. He was smart, talented, and fantastically sensitive to the needs of others—even if one did not take into account his age and gender. He was at an age when I felt him becoming my friend. But sometimes, when I came up against traits that seemed foreign to me—like when I saw him swigging beer in the courtyard—I flashed an image of his birth mother, who I assumed had saddled him with an addiction gene.

Years earlier, long before we adopted Vova, I had a conversation with a man who had adopted several children. I do not remember what I asked him about it, but I remember his answer: "I just don't think that my genes are better than my neighbor's genes." He had been right about me too, and I had the evidence before me, scribbled in my notepad in blue ink: "MAOA = 3–4."

~

If there is one thing behavioral geneticists can agree on, it is that all of their findings are nothing but a reason to do further studies. These studies often fail to corroborate the earlier findings. Human behavioral genetics is a very young science: It is only now constructing its own foundation by sketching out an approximate map of the genes that may be linked to behavior. But, in addition to the need for caution, two other facts about behavioral genetics seem certain: It is a science of the future, and its findings will be staggering, transforming our understanding of ourselves in ways we may be unable now to imagine.

Consider, again, the foxes and the rats in Siberia. The original purpose of Dmitry Belyaev's study was to try to understand how the process of domestication may have turned the fairly homogeneous wolf into the wildly heterogeneous dog. He hypothesized that selection for behavior may have thrown the wolf's genome out of whack, causing new mutations to appear faster than normal and activating or deactivating various genes. The hypothesis was confirmed: Soon enough, Belyaev's tame silver foxes started producing offspring that had certain doglike traits: Some had floppy ears; some had tails that curled upward; some had shortened snouts; and some had characteristic changes in their coloring. These traits appeared at a rate several times higher than might have been explained by spontaneous mutation (the researchers were extremely careful to avoid inbreeding, so a founder effect could be ruled out). They appeared in some of the aggressive foxes, too—though not as often as in the domesticated ones—suggesting that it was the fact of selection for behavior, not the type of behavior chosen for selection, that wreaked havoc with the foxes' genes.

The Siberian researchers noticed, too, that all animals subjected to domestication, either in the lab or out in the world, seemed to exhibit similar traits—most often, changes in coat coloring: Dogs, horses, cows, foxes, and rats seemed to acquire a similar look, with discoloration on the chest. In the Novosibirsk lab, the rats changed in the most curious—and most stereotypical—ways. First, some of the tame rats showed the characteristic chest spots: These rats, the researchers determined, were heterozygous for the gene that seemed to determine coat coloring. The homozygotes, who appeared a few generations later, had a very specific "hooded" appearance, which was just what it sounds like: They looked like they were wearing light hoods. No "hooded" rats appeared on the aggressive side, where, instead, some all-black rats started to show up. It looked curiously appropriate: The "hooded" rats had the cuddly appearance of pets; the black rats looked especially mean—and they were.

Extrapolating the results of such research to humans is, of course, very dangerous business. The most obvious argument against such extrapolation is that behavior traits such as tameness or aggressiveness are so crude that this sort of selection could never occur among modern human beings. At the same time, human beings are probably more complex and possibly more easily thrown out of genetic balance than rats or wolves—and certainly less genetically straightforward than Mendel's peas. I asked Arkady Markel, the shy man who bred hypertensive rats, whether he thought his lab experiments pointed to the possibility that in small populations of humans, selected or self-selected for certain behavioral traits—like, say, the Amish—the rate of spontaneous genetic mutations might go up. He looked pleased with the question. "It's possible," he said, smiling conspiratorially. I thought of the paper on Ashkenazi intelligence I described in chapter 2: Perhaps the idea that Jews were so sickly because we were so smart—or, rather, selected for particular behavioral traits—was not so far-fetched.

∼

It will be years before scientists reach true consensus on matters such as genetic drift versus selection. It will be a while before a canon of scientific literature on behavioral genetics takes shape. A cultural prediction is easier to make: Biobabble is not just around the corner, it is already here.

In a single month when I was going over the edits of this book, between July 15 and August 15, 2007, the *New York Times* published at least a dozen stories that mixed genetics and lifestyle. There was, for example, a story on the causes of autism; a story on prophylactic surgery for hereditary pancreatic cancer; a story on sweating that mentioned genetics almost in passing, as though it were fairly obvious; a story suggesting that the study of oddly genetically programmed bees might yield insight into human family dynamics; a story on one of those perennial surveys that show men have more sexual partners than do women, mentioning that men are compelled to spread their genes around; a story that suggested the pharmaceutical industry will be the first beneficiary of genetic engineering of livestock; a short piece debunking the myth that eating garlic helps repel mosquitoes, mentioning that only one's genes could offer protection; an oddly incoherent op-ed about a cognitive theory of the self that mentioned genetic influence on who we are; a large magazine piece about children conceived with the help of a donor egg, noting offhandedly that genes do not families make; and a story on the identification of mutations linked with restless legs syndrome that suggested sufferers may now command more compassion and respect. But most interesting, there were at least six stories in that one month that had nothing to do with health or genetics but mentioned genes anyway. A story on the physicist and writer Gino Segre asserted that physics was in the protagonist's genes; the review of a novel claimed casually that a character's motivation was in his genes; and a long piece on the theories of industrial revolution claimed that "middle-class values needed for productivity could have been transmitted either culturally or genetically"; an op-ed contributor who suffered from excessive sweating dreamed up an antiperspirant that could "alter your gene structure"; a review of a store claimed that "Art Deco is the dominant gene" in the establishment; a sportswriter referred to a "cheating gene."

When I did a similar analysis of articles published in the *New York Times* a year earlier, I found that the number of references to genetics—both real and imagined—had roughly doubled. When I looked at the *New York Times* archives for the same period fifteen years earlier, I found articles on gene cloning and gene splicing, as well as Gene Hackman and Gene Tierney, but no casual use of the word *gene* in nonscience pieces, with one small exception: an amusing reference to "a gene for humor" in a review of a book by a geneticist. The reviewer was presciently ahead of his times.

Language, it seems, is running ahead of knowledge: We have started speaking of genes that determine behavior, personality traits, and even lifestyle preferences before we have had a chance to learn what these genes might be. Industry is rushing to catch up: While I was writing this chapter, a company called Consumer Genetics was formed in Sunnyvale, California, and began advertising genetics tests that would show whether one's rate of metabolizing caffeine or alcohol was fast or slow. By late summer of 2007 the caffeine test had become available, at the cost of $139. The alcohol-metabolizing test, which promised to tell you whether moderate drinking might lower your cholesterol level, was not available yet.

Chapter 12

WHAT WE FEAR MOST

All our fears about the future of genetics are housed neatly, and unglamorously, in a two-story brick office building in Chicago's Boystown. The reception area on the second floor, outfitted with the same gray carpet and months-old magazines as every medical waiting room in America, belongs to the Reproductive Genetics Institute, which offers in vitro fertilization services and prenatal testing. This was the first place in the United States, and one of the first in the world, where chorionic villus sampling (CVS)—an alternative to amniocentesis that can be offered at an earlier time in the pregnancy—was performed. This was the first place in the world where preimplantation genetic diagnosis, or PGD, was offered, allowing future parents to weed out unsuitable embryos before implantation. This was where the first "spare parts baby" was conceived—a child selected as an embryo because he would be a suitable donor for his ill older sibling. This is also the place that claims to house the largest number of stem-cell lines in the world. For talk show hosts and science reporters worldwide, the men who run this institute are invaluable experts for commenting on things some of us are afraid to think about. For opponents of stem-cell research, these are the monster scientists of the future. The scientists themselves refuse to recognize the controversies.

"Oh, this ethics!" barked Anver Kuliev, director of the institute, when I brought up the topic. "I personally wrote the section about ethics in our book." He and Yury Verlinsky, president of the institute, have coauthored several books on preimplantation genetics. "Usually people say 'slippery slope,' people talk a lot, they will never understand much—it's just talk and talk and talk. And actually I did this in three or four pages. Because we have not much time for talk, we have to do the actual work. And also, the place for PGD is just in general practice. Because it's already there. It's primary prevention." I found Kuliev and Verlinsky's book, and I read the section on ethics. It was four and a half pages of dense and cogent argument, which testified to the fact that Kuliev took this area more seriously than he let on. But the main argument, both for his impatience and his institute's practice, was singular: There is no stopping the genetics future you say is so scary, because that future is already here.

The history of the Reproductive Genetics Institute spans two continents and a couple of decades and includes the stories of scientific quests and inspired discoveries, mixed with a dash of international intrigue. Back in the 1970s, Yury Verlinsky, born in Ukraine and educated there in cancer cytogenetics, and Anver Kuliev, born in Azerbaijan and trained in medicine and genetics, teamed up at a research institute in Moscow. Their goal was to develop a prenatal test that could be administered earlier than amniocentesis, allowing women to avoid second-trimester abortions. To do this, they were trying to test tiny samples of placental tissue rather than amniotic fluid, as with amniocentesis. Various researchers had tried this since the 1960s, but no one had succeeded in getting an accurate sample safely. In the midseventies, a group of Chinese researchers reported limited success (four out of a hundred women in their study lost their pregnancies, and six of the diagnoses were erroneous). In Moscow, Kuliev, Verlinsky, and a Hungarian postdoc named Zoltán Kazy were trying to develop a technique for performing a chromosomal analysis of the tissue. They failed.

In 1979, along with about fifty thousand other Soviet Jews, Verlinsky, his wife, and their nine-year-old son emigrated. They settled in Chicago, where Verlinsky was hired to run the cytogenetics laboratory at the Michael Reese Hospital—because, he explained to me, "of my previous experience and because a chromosome in any language is a chromosome and a microscope is a microscope." Verlinsky was working on identifying chromosomal polymorphisms in Down syndrome families and dabbling in chorionic villus sampling on the side: He was using tissue obtained from abortions to try to develop a technique for performing a chromosomal analysis. This time, he succeeded.

Like all Soviet émigrés, Verlinsky could have no contact with colleagues back home. Zoltán Kazy was publishing papers with Russian researchers, reporting some success with chorionic villus sampling. There was no trace of Anver Kuliev in international journals, however. Then his name came up during a scientific meeting in London, where Verlinsky was reporting on his success with chromosomal analysis and a British researcher named Bruno Branbatti was boasting that he had learned to perform a molecular analysis of chorionic villi tissue. They were talking about marrying the techniques. Someone mentioned that all of this should go to a Dr. Kuliev, who was now head of genetics at the World Health Organization. While Verlinsky the émigré had been stripped of his Soviet citizenship, his former colleague had made a brilliant career as a Soviet medical bureaucrat: He was serving the WHO as a representative of the USSR.

"I said, 'Dr. Kuliev is my close friend,'" said Verlinsky, who after nearly thirty years in the United States spoke fluent and idiomatic English with one of the most stereotypically Russian accents I had heard. "They thought I was bluffing. So we called Geneva, and I said, 'Anver, this is Yury, I know how to do what we were trying to do in Moscow.' And he said, 'Oh, come over and let's make a meeting in Geneva.' And in '82 we started to do—the first in the United States—CVS. So when I was in the meeting in Geneva, I was already looking forward and I was saying, 'Let's do preimplantation.'"

That same year, Verlinsky began his research into in vitro fertilization and the possibility of performing chromosomal analysis on embryos. Now that his old research partner was again part of the process, every couple of years, international scientists working in this area would meet somewhere under the auspices of the World Health Organization. In 1987 the group was meeting in Israel. Verlinsky wandered into an art gallery and found himself staring at a painting by Joan Miró. "The circle was red and yellow and collapsed at the end," remembered Verlinsky. "And I thought, 'This is it, this looks like an oocyte. And if this is yellow, we know exactly where will be red. And if it's red, we know where will be yellow.' I marked this idea on this business card from this art gallery."

The idea was this. A human egg cell contains forty-six chromosomes, twenty-three of which it expels on its way to fertilization. So if a woman is heterozygous for a gene, then by looking at the discarded part of the egg researchers would know whether the embryo contained the gene with the mutation. If, for example, the woman was a BRCA carrier and the discarded part of the egg contained the "bad" copy of the gene, then the embryo would not be affected: "If this is yellow, we know exactly where will be red." This worked for recessive conditions, too—to an extent: If both parents were known to be carriers—that is, if both carried just one "bad" copy of a gene—and the "bad" gene was expelled by the egg, the embryo would not be affected; if, however, it was not expelled, there was still a 50 percent chance the embryo would be healthy, since it may inherit a "good" copy of the gene from the father as well.

Verlinsky started running experiments. By the late 1980s he had performed this kind of test for women going through in vitro fertilization. By the early 1990s Verlinsky developed a technique for removing one of the cells from a six-to-eight-cell embryo and testing it for chromosomal and genetic abnormalities. He had also left Michael Reese Hospital and started the Reproductive Genetics Institute. Then the Soviet Union collapsed, and Anver Kuliev accepted his old friend's long-standing offer and came to join him in running the Chicago institute—which, by the time I visited it in the summer of 2006, was offering prenatal testing through chorionic villus sampling and in vitro fertilization with preimplantation genetic testing for more than 120 conditions, as well as testing that would allow couples to have babies who would make suitable donors for their sick siblings.

~

In his brief on ethics Anver Kuliev compared preimplantation genetic diagnosis to such primary prevention measures as adding folic acid to the diet of women of childbearing age—a step that can vastly reduce the risk of giving birth to children with congenital abnormalities. The beauty of such measures, he argued, was that "they provide the actual gain in infants free of congenital malformations rather than the avoidance of birth of affected children." In other words, if a woman discovers she is carrying a fetus affected with chromosomal or genetic abnormalities, she might have an abortion. If a woman discovers that one of her embryos obtained through in vitro fertilization is affected, she will opt to have another, unaffected embryo implanted and go on to have a healthy baby. This is an overargued way to express a simple truth: It is far easier for humans, whatever their ethical and religious beliefs, to reject embryos than to abort fetuses.

The clinical sterility of such discussions is a luxury compared with the human torment geneticists witnessed, sometimes experienced, and often tried to alleviate back in the days of prenatal testing. "I will never forget the spinal muscular dystrophy case," said Barbara Handelin, who ran the first commercial genetic testing laboratory in the

Boston area in the 1980s. "One of the first cases we did was this young couple whose daughter had died about six months before they knew, before we knew that we would be offering prenatal testing. They were our first prenatal case. And we said, 'You know, the thing is, we have to have tissue from your daughter, so if there is a muscle biopsy, we'll try that,' which we did and we were not able to get DNA out of it: It was a tiny section that was left and it had been formalin-fixed in paraffin, and we were not able to get enough useful material out of it. So I had to call her and say, 'Can't do it.' And she said, 'Well, what about hair?' And I said, 'You mean, like a lock?' And she said, 'Well, I have her neonatal baby cap that I kept. And there is quite a bit of hair in it.' One of the first signs is hair loss. And she said, 'If I take the hair out, would you be willing to try that?' I said, 'Yeah, we will, but I'd hate for you to give up the only thing that you have.'" Barbara had tears in her eyes when she told me this story twenty years later.

"So the day we got DNA out of that hair sample—oh! People were running through the hallway: 'Got to call her and tell her!' And then her CVS sample arrived, and people held their breath for a couple of weeks. So when we got those results—she was carrying a carrier, not affected—that was a very happy day. And she sent a postcard with a photograph after the baby was born. That—I still have that."

There was another case Handelin remembered in detail two decades later. "There was a woman who had polycystic kidney disease—one of her siblings had died of renal failure. She was actually doing very well herself. But she didn't want to pass it on." The condition is what it sounds like: a disease that causes fluid-filled cysts to form on the kidneys, and sometimes on other organs, reducing kidney function and ultimately leading to kidney failure. Hereditary polycystic kidney disease can be caused by either a dominant or a recessive gene. The woman who came to Handelin's lab had the more common dominant form.

"The first time we did her case—I mean, it was depressing: depressing for the tech who gets the results, and they bring it to you, and we go over it in a lab meeting and say, 'Well, you know, bad news, bad news for this lady.' And then, a year later, her case comes in again, and bad news again. And you think, 'Okay, this will be it, she is not going to be able to take this anymore. Or maybe she'll just go ahead with this pregnancy.' And the third time she got pregnant, I just felt like, really, *we* can't take it!"

The genetic counselor working with the woman asked Handelin to meet with her: The counselor wanted to tell the woman to stop, but the rules governing genetic counseling forbade that. "So I did talk with her, and she was just so incredibly determined, and I said, 'We are just feeling for you.' And she said, 'I'm okay. Do your job. I cannot bring children into the world with this hanging over their head.'" It took five years and four pregnancies.

These days, either woman would have been able to seek in vitro fertilization with preimplantation genetic diagnosis either through polar body biopsy, as the Miró-inspired technique is called, or embryo biopsy—or both, as the Reproductive Genetics Institute usually does it to get the most reliable results—then picked a healthy embryo or embryos, and gone on to have the healthy baby she sought without the torment to which both women were subjected in the 1980s. This is very much how Verlinsky and his colleagues saw preimplantation genetics: If the technical opportunity was there, it had to be used, for its benefits were tangible while its risks were really just other people's fears.

～

Preimplantation genetic diagnostic procedures are banned in Austria, Germany, Italy, and Switzerland. "I think Germans collectively are unsound," wrote the journalist Martha Gellhorn in 1990. "I think they have a gene loose, though I don't know what the gene is." Germans clearly hold a similar opinion of themselves, as do other countries that either chose fascism or know themselves to have aided it. "It's really more about a fear," Karen Avraham, a geneticist at Tel Aviv University, explained to me. "It's because of what the Nazis did, and eugenics. And there should be fear. Because if any of those tools get into the wrong hands..." She trailed off.

Israel, being the opposite of Germany, Austria, and Italy, not only allows preimplantation genetics but, in some cultural way, encourages the practice. All Israeli general hospitals have their own fertility units; public health funds cover the cost of producing two babies for any infertile woman under the age of forty-five, and the standards of what constitutes infertility tend to be very liberal. "In Israel people want a child—they have fertility done," said Avraham, who was born and raised in the United States. "People don't wait a year to see if they can get pregnant. And you see a lot of twins." Prenatal testing in Israel is performed routinely, regardless of the pregnant woman's age: The quest for perfection, or at least for health, overrides the economic considerations that, in other countries, limit prenatal testing to pregnant women over a certain age. Nor are there any religious obstacles to preimplantation genetics: A fetus is not considered a human life before forty days' gestation—and this is also one of the reasons Israel has been able to make great strides in stem-cell research.

Avraham herself studies the genetics of deafness, one of the most ethically fraught issues even in Israel's permissive climate. One out of twenty-one Ashkenazi Jews carries a mutation in the connexin 26 gene, a gene that codes for one of the proteins known as connexins, which are essential to hearing. That would mean that roughly one in eighteen hundred Ashkenazi babies would be born deaf. That makes what is called "nonsyndromic deafness"—hearing loss that is not associated with any other symptoms—one of the most common genetic disorders among Ashkenazi Jews. But screening has been a matter of some controversy.

Dor Yeshorim, the wildly successful Jewish premarital genetic screening program, holds to a principle of testing only for conditions that are life-threatening. That standard was relaxed slightly when the program began testing for Gaucher's disease, which can be mild but can also be debilitating. Deafness is isolating, limiting, and difficult to overcome, but it is neither life-threatening nor physically painful. Some people argue that deafness provides for an alternative form of communication that should be protected. The active deaf community in Israel has been consistently opposed to developing genetic therapy for deafness, but there are also deaf and hearing-impaired scientists at Avraham's laboratory working on this research. Premarital, preimplantation, and prenatal genetic testing for deafness, though, goes to the question not of whether deafness should be treated but whether deaf people should be born.

Michal Sagi, who ran the genetics unit at Hadassah, told me in 2005 that her program had solved the dilemma the ostrich way: by omitting deafness mutations from the list of available prenatal genetic tests lest it look like a recommendation for testing—and possibly for the termination of pregnancy. So much for full disclosure, one of the basic tenets of genetic counseling.

Genetic tests for deafness mutations are, however, available to pregnant Israelis who ask for them. If they ask, it is usually because they already have deaf children. "There is a physician who had two deaf children, and he decided they were not going to have another deaf child," Avraham told me. The physician had apparently been very public about his story, even using it to aid in fund-raising efforts for Avraham's work. "So they were going to have prenatal diagnosis. They were not negating their first children, but they felt they couldn't provide for their first two if they had another deaf child. Such a burden. We discovered the mutation, we worked with the clinic, they did the diagnosis: The child was hearing. The fetus did not have the mutation. They carried the pregnancy, the child was fine. But they do say that if they found the fetus had the mutation, they don't know that they could have had an abortion. My guess is they would have done it, because it would have been just too much of a burden." Deaf children whose parents want to integrate them into hearing society require time-consuming, difficult, and expensive training: Mothers quit jobs, families go into debt and sometimes run out of resources.

What makes us uncomfortable about stories like these is the question of selection. A woman or a couple who choose an abortion because they do not want a child at all are somehow less ethically suspect than people who do not want a child with particular characteristics. For decades, Soviet women used abortion as their means of birth control; Russian women still average 2.43 abortions in a lifetime. Yet distasteful as this is, it is not frightening in the same way as abortion resulting from prenatal testing—or even embryo rejection resulting from preimplantation diagnosis—because it does not raise the question of our criteria for the suitability or the worthiness of life.

~

Most people with whom I talked as I did research for this book claimed to have specific ideas about the dividing line between suitable and unsuitable criteria in screening potential offspring. Sex selection is one of the usual border posts, and a mainstay of prenatal counselors everywhere: If a woman's or a couple's apparent primary reason for seeking a prenatal test such as amniocentesis is to learn the sex of the fetus, the test is normally denied. "I'm not a criminal investigator to try to find out what reason they have to transfer this embryo or another one," bristled Yury Verlinsky: His clients learned all sorts of things about their embryos, including their sex, and they made choices to which, he felt, they were entitled—like to transfer a healthy embryo that was male or female.

Not only that, some of his clients had sound medical reasons for seeking sex selection: A woman who carries an X-chromosome-linked mutation, such as the one associated with hemophilia or "bubble boy" syndrome, or a variety of others, could reasonably choose to give birth only to daughters, who would not be affected by the genetic disease. One of Verlinsky's earliest successful experiences with preimplantation diagnostics involved identifying a female embryo for a couple in which the woman was a carrier for hemophilia.

. In some societies, preimplantation sex selection for social reasons has gained acceptance. In India the practice is called "family balancing," although it may more reasonably be called "boy selection"—a service for families already burdened with girl children in a country where boy babies are prized. Israel's first approved case of preimplantation sex selection was a perfect Talmudic case study. It involved an Orthodox couple in which the husband was infertile. The couple wanted to have a child by using donated sperm, but the situation was complicated by the fact that the future father was a Cohen, whose son would be expected to play a certain role during synagogue services. But the Cohen heritage is passed on to biological children only: One cannot become a Cohen by dint of adoption, marriage, or any other nonbiological relationship. If a boy was born, therefore, the father would face a dilemma: Either allow him to be treated as a Cohen, thereby perpetrating a lie, or tell the community that the child was not his biologically, which was not something he wanted to do. The obvious solution, the rabbinical authorities agreed, was to have a girl. Since then, authorities have also allowed couples who have four or more children of the same sex to seek preimplantation sex selection.

"Some people do it the natural way." Verlinsky laughed. "Trish, our secretary, she has seven children: She did six boys until she got a girl. One of the ways to do sex selection." He had a point: People have always engaged in sex selection of some sort or another. Every culture has its superstitions regarding when and how boys or girls are most efficiently conceived. To argue that just as humankind has finally acquired something it has sought all along—a truly effective way to choose boy or girl babies—it should be banned, seems hardly logical and entirely unnatural.

Selection for intelligence raises another universal objection. The obvious objection to this objection is that the most common understanding of the most common prenatal test—amniocentesis—is to avoid giving birth to a child with Down syndrome, at least in part, and possibly primarily, because these children have mental retardation. The common amendment to the selection-for-intelligence objection aims to exclude children who would be mentally retarded: Giving birth to a child like that, the logic goes, requires a different sort of parent-child relationship, one that is far more demanding, much less rewarding, and largely devoid of development.

But for the past twenty years and more, potential parents in the United States have been routinely aborting fetuses that would develop into children of below-average intelligence. These are fetuses identified either through amniocentesis or through chorionic villus sampling as having an abnormal set of sex chromosomes: a lone X; three Xs; two Xs and a Y; and one X and two Ys. The abnormalities carry different prognoses: XXYs are males who are usually infertile; single-X females (or females with a partial second X chromosome) tend to be infertile, often fail to develop secondary sex characteristics, are usually of short stature, and often have heart problems; triple-X females tend to be tall and to have behavioral problems; XYY males are just what you would expect: very tall and exceedingly active. Two things unite all the sex-chromosome abnormalities: The affected individuals tend to have learning problems and intelligence just below average, albeit in the normal range; and fetuses identified as having a sex-chromosome abnormality are usually aborted—a twenty-year study of prenatal testing in the United States showed that between 1983 and 2003, 60 percent of such pregnancies ended in termination. Presumably, at least some of the decisions are attributable to a general discomfort with abnormalities, especially ones that have some relationship to sex and gender. But the weeding out of these fetuses is most certainly a form of selecting for intelligence.

"Nobody knows," objected Verlinsky. "There is definitely the biggest percentage of the triple X in the psychiatric institutions." But selection for mental health, with its perpetually shifting definition, seems an even more precarious proposition than selection for intelligence.

Preimplantation genetics skirts the issue, in a sense. Chromosomal abnormalities are usually spontaneous, not hereditary, which means that a set of half a dozen to a dozen embryos might contain several normal ones and one or a few with chromosomal abnormalities. Most of the abnormalities would be incompatible with a normal pregnancy—these would be embryos that, under natural conditions, either would not have implanted or would have been spontaneously aborted—but in any case, a couple going through in vitro fertilization would never have a reason to choose to transfer, say, an XXY embryo, which might have occurred naturally and would, if it successfully implanted, develop into an overweight, infertile, and intellectually rather slow male, over a normal healthy one.

An organization like Dor Yeshorim sets its limits at life-threatening or extremely debilitating early-onset diseases that are inherited through a recessive gene. The logic is clear: One needs grave reasons to interfere with a marriage, and a disease that is not particularly severe, or one that sets in later in life—by which time treatment may be available—is just not grave enough. Testing for dominant disease-causing genes would be unfair because carriers, under the Dor Yeshorim system, would be marked as unmarriageable.

Preimplantation testing demands a different set of rules. There is no reason to exclude dominant diseases. Is there reason to exclude late-onset diseases such as hereditary breast cancer? I asked Verlinsky whether he would undertake in vitro fertilization for the purpose of weeding out embryos affected with my mutation. "Sure," he answered. "It's a predisposition, it's high risk, and why do you have to give your child the same worries you have?" Seriously, why would I consider saddling a daughter of mine with the nighttime fears, the obsessive medical examinations, or, worse, the surgeries, or, worse yet, the torturous treatment for breast or ovarian cancer? Any parent wants her children to grow up as free of worry as possible—and especially to be free of the fears that plague the parent. The thing with in vitro fertilization is, the child or children are theoretical—few of us are capable of thinking of eight cells in a dish as a human life—while the genes, and the risks they carry, have names, identities, and tangible consequences. Weeding out any embryo with impaired chances for turning into a healthy, long, and successful human life is easy and painless by most people's ethical standards.

For those whose beliefs hold an embryo to be a human life, there is Verlinsky's method of testing the chromosomes shed by the maturing oocyte. Even German law allows this method. It is not foolproof—in particular, it cannot weed out any genes on the father's side, which makes it possible to pass on dominant disease-causing genes, and it cannot catch chromosomal errors that occur at conception—but it is effective in preventing most conditions.

The more stringent one's ethical demands, the more expensive it is to meet them. The economic hierarchy is clear: Amniocentesis, which can lead to a second-term abortion, is cheaper than chorionic villus sampling (roughly a thousand versus fifteen hundred dollars), which forces the decision to terminate a pregnancy into the first trimester and is in turn cheaper than preimplantation genetics (roughly ten thousand dollars for the entire fertilization procedure), the cost of which is raised further by testing the polar bodies discarded by the oocyte. For more than twenty years of its existence, chorionic villus sampling has remained an expensive procedure limited to the moneyed and the informed, but in vitro fertilization, at least in countries like the United States and Israel, is fast trickling down to the insured masses. Most of Verlinsky's early diagnostic work was supported by research grants; now his research is supported by insurance fees paid for in vitro fertilization. With parents getting older and increasingly impatient, in vitro fertilization is on its way to becoming the standard of care. That means only the uninsured will have abortions—or children who carry genetic abnormalities.

~

The Reproductive Genetics Institute's Web site has a list of single-gene disorders for which an embryo can be screened. The list is alphabetical, and the first condition on it is achondroplasia, a bone-growth disorder that results in very short stature and disproportionately short limbs and large head. There are sound reasons for wanting to avoid passing on achondroplasia: The pregnancy and especially the birth can be very complicated, and life for extremely short people can be difficult. Each disorder on the list is a hyperlink. I clicked on achondroplasia and landed on the Web page for Little People of America. It opened with what looked like a family picture: a man, a woman, and three children with achondroplasia, all hugging and looking very happy.

I remembered a conversation I had almost twenty-five years ago. It concerned a news article about a group of scientists who claimed to have determined that homosexuality is genetic. I thought it was interesting, and vaguely affirming. My first girlfriend, who read the article with me, was appalled. "There is nothing like being the last of your kind," she said. I imagined Little People of America knew that feeling.

Every now and then a couple will ask Verlinsky and his associates to help them have a baby with a congenital abnormality. Several couples, said Verlinsky, had come in wanting to have a child with Down syndrome to provide companionship for an older affected sibling. Verlinsky had refused. Others had asked for congenital deafness syndrome or achondroplasia. Verlinsky had always refused: "Our role is to diagnose disease and avoid it, not to create disease."

Verlinsky could not have taken a different position—both because he runs a medical institution (though he is not a medical doctor) and because of the legal climate that is taking shape in the United States and some other Western countries. Since the late 1970s American courts have usually awarded financial damages in so-called wrongful-birth lawsuits, directing doctors to pay the parents of children born with birth defects that could have been identified or predicted during the pregnancy. Some people, the legal logic goes, should never have been born. The courts have stopped short of saying that some people should not be alive—they have generally rejected so-called wrongful-life lawsuits, where the plaintiffs are disabled children themselves—essentially creating a grandfather clause for the existence of people with congenital defects.

But if it is wrong to enable the birth of a child with abnormalities, is it always right to enable the birth of a healthy child? In the whole field of preimplantation diagnostics, there is probably nothing that scares people as much as so-called spare parts babies, whom professionals in the field prefer to call "designer babies"—children born to serve as donors for an ill sibling. The first such child, Adam Nash, was born in August 2000 with the help of Reproductive Genetics Institute. Stem cells from his umbilical cord were transplanted to his six-year-old sister Molly, effectively curing her Fanconi's anemia, a fatal bone-marrow disorder. Over the course of the following six years, about fifty more designer babies were born, about a dozen of them conceived at the Reproductive Genetics Institute, but the controversy over the commodification of human life did not quiet down.

Neither Verlinsky nor Kuliev had much patience for discussing the ethics of creating designer babies. "Usually families spend years looking for an identical donor," said Kuliev. "Often the child is dead before they can find the donor. Maybe somebody is lucky, but they can never be 100 percent identical: It's more or less close, and it gives some side effects. And this is 100 percent identical. There was a spectacular case, diamond-blackfan anemia—you can imagine the misery. Now that child no longer needs any supplementation. And that Molly, she is going to school. And Adam is five, and he is fine."

~

Of all the diseases in the world, Kuliev had to mention diamond-blackfan anemia. Five years earlier, distant relatives of mine came to Moscow from their far northern town because they had been referred to the big children's hospital here. Their daughter, Diana, was a month older than my daughter, Yael. They stayed with us most of the summer: The girls played together, had babbling conversations of sorts sitting in their high chairs, and sometimes struggled for control over toys. Diana was a bit thinner and less than an inch shorter than Yael, but, on the whole, she did not seem unhealthy. But then every couple of weeks she would spike a fever and grow weak and helpless. Then she would get a blood transfusion that brought her back to life. It was a hellish roller coaster, but it did not prepare her parents or anyone else for hearing that the child had a fatal disease. She was diagnosed with diamond-blackfan anemia, a disorder in which the bone marrow fails to make red blood cells. Only about six or seven hundred people worldwide were known to be living with it. They lived from transfusion to transfusion. About half of them could get some help from corticosteroids, which stimulated the bone marrow, but at a great cost: The steroids made their bones brittle, ultimately contributing to the body's overall weakness. Some people lived into their twenties; most died earlier.

After Diana's parents had the diagnosis confirmed, they returned to their hometown. Diana was lucky: Corticosteroids worked for her. They made her bloated and stunted her growth—by the age of five she was almost a head shorter than Yael—but they reduced her need for transfusions. She was a happy, smart girl, a born dancer. Her parents, who were in their early twenties when she was born, sometimes tried to talk to doctors about the possibility of having another child, and of perhaps obtaining a bone-marrow transplant for Diana. The doctors tried to discourage them: An outside donor would be hard, probably impossible to find, and a second child might also be born with the disorder. How would they take care of two severely ill children?

After I met with Kuliev and Verlinsky, I wrote Diana's mother a short, dry e-mail message, explaining that I had visited a place where designer siblings for children with Diana's disorder had been conceived. She wrote back the next day. There was no anguish, no hand-wringing over the ethics and the difficulties of having a designer baby. In fact, there was no time to waste. There was only a chance to save her daughter.

Acknowledgments

As I worked on this book, I posted chapters to a blog that was read by a small group of people, some of whom are my friends while others I had never met. They caught errors (some of the readers were, fortunately, trained in the fields into which I had carelessly ventured), pointed out logical inconsistencies, suggested edits, and generally kept me company while I worked. I owe many thanks to Ellen Todres Gelfand, Nikolai Klimeniouk, Ilya Kolmanovsky, Mark Schoofs, who actually took the time to edit entire chapters, and Lena Shagina. Katya Krongauz on several occasions saved me from collapsing under the anxiety of producing this book. As always, my father and brother, Alexander and Keith Gessen, were among my first and most important readers. And many thanks to my amazing agent, Elyse Cheney, and Becky Saletan, the most thoughtful of editors.

Allele. A sequence of DNA code that occupies a given position in a chromosome.

Autosome. A non-sex chromosome. People normally have twenty-two pairs of autosomal chromosomes. Each pair contains two copies of every gene, one inherited from the mother and one from the father.

Carrier. Someone who has a single copy of a gene that determines a recessive trait. A child of two carriers has a 25 percent chance of inheriting the gene from both parents and therefore exhibiting the trait or being affected with the disorder.

Chromosome. The building block of the genome, a single DNA molecule that contains many genes, regulatory elements, and other nucleotide sequences. People normally have one pair of sex chromosomes (XX in women and XY in men) and twenty-two pairs of autosomal chromosomes.

Complex trait. A trait whose expression is dependent on the presence of more than one gene.

Consanguineous marriage. Unions between people who are second cousins or more closely related. The practice, common in a number of closed populations around the world, increases the risk of children who are affected with recessive conditions because the children are certain to be homozygous in a significant part of the genome.

Dominant trait. A trait caused by a single copy of a gene. For example, brown eye color is a dominant trait: A person who inherits a copy of the gene from either parent will have brown eyes. The genes that are linked to Huntington's disease or hereditary cancers are also dominant.

Endogamy. The practice of marrying exclusively within a social, religious, or ethnic group.

Eugenics. A social philosophy pioneered by Sir Francis Galton in the late nineteenth century. Its adherents advocate aiding the human race through improving its heredity. The eugenics movement was very strong in the United States in the first half of the twentieth century, leading to the adoption of sterilization laws in a number of states and the implementation of immigration policies aimed at selecting migrants with the best hereditary characteristics. The Nazi regime in Germany relied on eugenics to justify many of its policies, rendering the very concept of eugenics suspect. Many modern practices, however, follow the spirit of Galton's philosophy. These include premarital genetic testing and prenatal and preimplantation testing.

Exon. Any part of a gene that is actually transcribed to the final messenger RNA.

Gene therapy. Treatment based on inserting genes into a human organism to replace a deleterious allele with a functional one. Several techniques of gene therapy have been tried since 1990, some of them successfully, but all methods are still considered highly experimental.

Genetic drift. The idea that chance is primarily responsible for a given allele's becoming more or less common in a population. The smaller the population, the greater the potential influence of genetic drift. Many geneticists believe genetic drift is the primary reason small closed populations may have a high prevalence of a genetic condition rarely found elsewhere or may entirely avoid a genetic condition that is considered common in other populations. For example, the Amish of western Pennsylvania have a high prevalence of the rare disease glutaric aciduria Type 1 but have zero incidence of cystic fibrosis, the most common genetic disorder in the world. A corollary to the concept of genetic drift is the concept of selective advantage.

Genome. All the hereditary information of an organism as encoded in DNA.

Haplotype or haploid genotype. A set of alleles that are transmitted together. Certain haplotypes are believed to pinpoint a person's geographic and/or ethnic origins.

Heterozygous. Possessed of two different copies of an allele.

Homozygous. Possessed of two identical copies of an allele.

Mendelian inheritance. The mechanism of genetic inheritance described by the nineteenth-century Moravian monk Gregor Mendel. His two laws are the cornerstone of the modern understanding of genetics. Mendel's first law holds that alternative versions of genes explain variations in inherited characteristics, that an organism inherits two alleles for each characteristic, and that if the alleles differ the one accounting for the dominant trait will be fully expressed. Mendel's second law holds that different traits are inherited independently from one another.

Mitochondrial DNA. DNA that resides outside the cell nucleus and is inherited from the mother.

Polymorphism. A mutation—a distinct form of an allele—that is present in at least 1 percent of a given population.

Recessive trait. A trait expressed only in the presence of two copies of a gene. Blue eye color is an example, as are a variety of disorders that occur in children born to two carriers.

Selective advantage. The concept at the root of the theory that natural selection, not genetic drift, is the primary reason an allele becomes common in a population or gets entirely washed out. Mutations that are deleterious in homozygotes, the theory holds, are beneficial to heterozygotes—confer a selective advantage on them—hence the allele is carried forward through the generations. An example is the mutation that causes sickle-cell anemia in those who inherit two copies of the mutant gene but, it is believed, helps protect carriers (those with just one copy) from malaria.

Stem cell. A sort of cell that has the ability to renew itself and to differentiate into any number of kinds of cells.

X-linked. Mutations that reside on the X chromosome. Females have two X chromosomes, inherited in equal part from the mother and the father, and males have one X chromosome, inherited from the mother, and one Y chromosome, inherited from the father. If a woman carries an X-chromosome allele with a deleterious mutation, she herself will not be affected, nor will her female children: They all have another, healthy copy of the allele, inherited from the father. Male children, however, will invariably be affected, since they are certain to inherit the X chromosome from the mother and will not have a healthy copy of the allele. Examples of X-linked disorders include a form of hemophilia and "bubble boy" syndrome, a severe congenital immune deficiency.

CHAPTER I: MY MOTHER'S FATAL FLAW

The search for the breast cancer gene is described in Kevin Davies, Michael White, *Breakthrough: The Race to Find the Breast Cancer Gene* (John Wiley & Sons, 1996).

On lifetime risk for mutation carriers and on earlier age of onset for successive generations: Mary-Claire King, Joan H. Marks, Jessica B. Mandell, The New York Breast Cancer Study Group, "Breast and ovarian cancer risks due to inherited mutations in BRCA1 and BRCA2," *Science* 5645 (October 24, 2003): 643–646.

The function of the BRCA genes: K. Yoshida, Y. Miki, "Role of BRCA1 and BRCA2 as regulators of DNA repair, transcription, and cell cycle in response to DNA damage," *Cancer Science* 11 (November 2004): 866–871.

Nonviability of BRCA homozygotes: Srdjan Denic, Lihad Al-Gazali, "Breast cancer, consanguinity, and lethal tumor genes: Simulation of BRCA1/2 prevalence over 40 generations," *International Journal of Molecular Medicine* 10 (2002): 713–719.

Some cancers in mutation carriers are more aggressive than in noncarriers: M. O. Nicoletto, M. Donach, A. De Nicolo, G. Artioli, G. Banna, S. Monfardini, "BRCA-1 and BRCA-2 mutations as prognostic factors in clinical practice and genetic counseling," *Cancer Treatment Review* 5 (October 2001): 295–304.

Male breast cancer in mutation carriers: N. Wolpert, E. Warner, M. F. Seminsky, A. Futreal, S. A. Narod, "Prevalence of BRCA1 and BRCA2 mutations in male breast cancer patients in Canada," *Clinical Breast Cancer* 1 (April 2000): 57–63.

On the risk of developing cancer for mutation carriers: A. Antoniou, P. D. P. Pharoah, S. Narod, H. A. Risch, J. E. Eyfjord, J. L. Hopper, N. Loman, H. Olsson, O. Johannsson, Å. Borg, B. Pasini, P. Radice, S. Manoukian, D. M. Eccles, N. Tang, E. Olah, H. Anton-Culver, E. Warner, J. Lubinski, J. Gronwald, B. Gorski, H. Tulinius, S. Thorlacius, H. Eerola, H. Nevanlinna, K. Syrjäkoski, O.-P. Kallioniemi, D. Thompson, C. Evans, J. Peto, F. Lalloo, D. G. Evans, and D. F. Easton, "Average risks of breast and ovarian cancer associated with BRCA1 or BRCA2 mutations detected in case series unselected for family history: A combined analysis of 22 studies," *American Journal of Human Genetics* 72 (2003): 1117–1130.

For an elucidation on Judaism's position on genetics, see Fred Rosner, "Judaism, genetic screening and genetic therapy," available online through the Jewish Virtual Library, http://www.jewishvirtuallibrary.org/jsource/Judaism/genetic.html (accessed September 13, 2007).

The prevalence of my mutation among non-Ashkenazi Jews and the dating of the founder: Revital Bruchim Bar-Sade, Anna Kruglikova, Baruch Modan, Eva Gak, Galit Hirsh-Yechezkel, Livia Theodor, Ilya Novikov, Ruth Gershoni-Baruch, Shulamit Risel, Moshe Z. Papa, Gilad Ben-Baruch, Eitan Friedman, "The 185delAG BRCA1 mutation originated before the dispersion of Jews in the Diaspora and is not limited to Ashkenazim," *Human Molecular Genetics* 5 (May 1998): 801–805.

On the prevalence of a BRCA1 mutation in U.S. Hispanics and the fact that it shares a founder haplotype with the Ashkenazi mutation: J. N. Weitzel, V. Lagos, K. R. Blazer, R. Nelson, C. Ricker, J. Herzog, C. McGuire, S. Neuhausen, "Prevalence of BRCA mutations and founder effect in high-risk Hispanic families," *Cancer Epidemiology Biomarkers & Prevention* 7 (July 2005): 1666–1671. This study reports that not all carriers of this mutation have the "Ashkenazi" haplotype: D. B. Berman, J. Wagner-Costalas, D. C. Schultz, H. T. Lynch, M. Daly, A. K. Godwin, "Two distinct origins of a common BRCA1 mutation in breast-ovarian cancer families: A genetic study of 15 185delAG-mutation kindreds," *American Journal of Human Genetics* 6 (June 1996): 1166–1176.

CHAPTER 2: THE FOUR MOTHERS OF JEWS

This paper explains why my risk of developing cancer would have gone down if I had discovered that my mutation came from the side of the family unaffected by cancer: J. M. Satagopan, K. Offit, W. Foulkes, M. E. Robson, S. Wacholder, C. M. Eng, S. E. Karp, C. B. Begg, "The lifetime risks of breast cancer in Ashkenazi Jewish carriers of BRCA1 and BRCA2 mutations," *Cancer Epidemiology Biomarkers & Prevention* 5 (May 2001): 467–473.

Carrier frequency for "my" mutation among Ashkenazi Jews: B. B. Roa, A. A. Boyd, K. Volcik, C. S. Richards, "Ashkenazi Jewish population frequencies for common mutations in BRCA1 and BRCA2," *Nature Genetics* 2 (October 1996): 185–187.

Studies showing that exercise and weight modify the risk of cancer in mutation carriers: M. C. King, J. H. Marks, J. B. Mandell, The New York Breast Cancer Study Group, "Breast and ovarian cancer risks due to inherited mutations in BRCA1 and BRCA2," *Science,* October 24, 2003, 643–646; Joanne Kotsopoulos, Olufunmilayo I. Olopade, Parviz Ghadirian, Jan Lubinski, Henry T. Lynch, Claudine Isaacs, Barbara Weber, Charmaine Kim-Sing, Peter Ainsworth, William D. Foulkes, Andrea Eisen, Ping Sun, and Steven A. Narod, "Changes in body weight and the risk of breast cancer in BRCA1 and BRCA2 mutation carriers," *Breast Cancer Research* 5 (2005): R833–843.

This paper discusses increased risk of other cancers in BRCA mutation carriers: B. Friedenson, "BRCA1 and BRCA2 pathways and the risk of cancers other than breast or ovarian," *Medscape General Medicine* 2 (June 2005).

On Judaism's position on marrying epileptics, see Rosner, "Judaism, genetic screening and genetic therapy." My great-uncle's decision not to get married may have been unjustified because it seems unlikely that his form of epilepsy was heritable: No one else in the family had similar symptoms. Epilepsy is a condition that can be caused by a great many factors, including injuries, infections, and genetics.

For definitions of "Jewish diseases," see, for example, the Web site of the Center for Jewish Genetic Diseases at the Mount Sinai School of Medicine, http://www.mssm .edu/jewish_genetics/genetic_diseases.shtml (accessed September 13, 2007).

The study suggesting that Jews are so sickly because they are so smart: Gregory Cochran, Jason Hardy, Henry Harpending, "Natural history of Ashkenazi intelligence," *Journal of Biosocial Science* 5 (September 2006): 659–693. Steven Pinker commented on the paper in "Grops and Genes," *New Republic*, June 26, 2006. For a review of work on the genetic basis for intelligence see Ian J. Deary, Frank M. Spinath, and Timothy C. Bates, "Genetics of intelligence," *European Journal of Human Genetics* 14 (June 2006): 690–700. Joseph Jacobs wrote on "germ-plasm" in Joseph Jacobs, *Jewish Contributions to Civilization: An Estimate* (Philadelphia: Jewish Publication Society in America, 1919), quoted according to Yuri Slezkine, *The Jewish Century* (Princeton and Oxford: Princeton University Press, 2004).

The Cohanim studies: Michael F. Hammer, Karl Skorecki, Sara Selig, Shraga Blazer, Bruce Rappaport, Robert Bradman, Neil Bradman, P. J. Waburton, Monic Ismajlowicz, "Y chromosomes of Jewish priests," *Nature* 385 (January 2, 1997): 32; Mark Thomas, Karl Skorecki, Haim Ben-Amid, Tudor Parfitt, Neil Bradman, David Goldstein, "Origins of Old Testament priests," *Nature* 394 (July 9, 1998): 138–140.

On the Lemba: Mark G. Thomas, Tudor Parfitt, Deborah A. Weiss, Karl Skorecki, James F. Wilson, Magdel le Roux, Neil Bradman, and David B. Goldstein, "Y chromosomes traveling south: The Cohen modal haplotype and the origins of the Lemba, the 'Black Jews of Southern Africa,'" *American Journal of Human Genetics* 66 (February 2000): 674–686; Paul Brodwin, "Genetics, identity, and the anthropology of essentialism," *Anthropological Quarterly* 2 (Spring 2002): 323–330; Tudor Parfitt, Yulia Egorova, "Genetics, history, and identity: The case of the Bene Israel and the Lemba," *Culture, Medicine and Psychiatry* 2 (June 2005): 193–224.

Marina Faerman, Gila Kahila Bar-Gal, Patricia Smith, Charles Greenblatt, Lawrence Stager, Dvora Filon, Ariella Oppenheim, "DNA analysis reveals the sex of infanticide victims," *Nature*, January 16, 1997, 212–213; Marina Faerman, Gila Kahila Bar-Gal, Dvora Filon, Charles Greenblatt, Lawrence Stager, Ariella Oppenheim, Patricia Smith, "Determining the sex of infanticide victims from the late Roman Era through ancient DNA analysis," *Journal of Archaeological Science* 25 (September 1998): 861–865.

Marina Faerman, Almut Nebel, Dvora Filon, Mark Thomas, Neil Bradman, Bruce Ragsdale, Michael Schultz, Ariella Oppenheim, "From a dry bone to a genetic portrait: A case study of sickle cell anemia," *American Journal of Physical Anthropology* 2 (February 2000): 153–163.

On beta-thalassemia: Dvora Filon, Marina Faerman, Patricia Smith, Ariella Oppenheim, "Sequence analysis reveals a beta-thalassemia mutation in the DNA of skeletal remains from the archaelogical site of Akhziv, Israel," *Nature Genetics* 9 (April 1995): 365–368; Tina Martino, Feige Kaplan, Stanley Diamond, Ariella Oppenheim, Charles Shriver, "Probable identity by descent and discovery of familial relationships by means of a rare beta-thalassemia haplotype," *Human Mutation* 1 (September 1997): 86–87; Deborah Rund, Tirza Cohen, Dvora Filon, Carol Dowling, Tina Warren, Igal Barak, Eliezer Rachmilewitz, Haig Kazazian, Ariella Oppenheim, "Evolution of a genetic disease in an ethnic isolate: Beta-thalassemia in the Jews of Kurdistan," *Proceedings of the National Academy of Sciences, USA,* January 1991: 310; Dvora Filon, Varda Oron, Svetlana Krichevski, Avraham Shaag, Yechezkel Shaag, Tina Warren, Ada Goldfarb, Yona Shneor, Ariel Koren, Mehmet Aker, Ayala Abramov, Eliezer Rachmilewitz, Deborah Rund, Haig Kazazian, Jr., Ariella Oppenheim, "Diversity of beta-globin mutations in Israeli ethnic groups reflects recent historic events," *American Journal of Human Genetics* 5 (May 1994): 836–843; A. Koren, L. Zalman, H. Palmor, E. Ekstein, Y. Schneour, A. Schneour, S. Shalev, E. A. Rachmilewitz, D. Filon, A. Oppenheim, "The prevention programs for beta thalassemia in the Jezreel and Eiron valleys: Results of fifteen years experience," *Harefuah* 11 (November 2002): 938–943.

Almut Nebel, Dvora Filon, Bernd Brinkmann, Partha Majumder, Marina Faerman, Ariella Oppenheim, "The Y chromosome pool of Jews as part of the genetic landscape of the Middle East," *American Journal of Human Genetics* 69 (November 2001): 1095–1112.

Doron Behar, Mark Thomas, Karl Skorecki, Michael Hammer, Ekaterina Bulygina, Dror Rosengarten, Abigail Jones, Karen Held, Vivian Moses, David Goldstein, Neil Bradman, Michael Weale, "Multiple origins of Ashkenazi Levites: Y chromosome evidence for both Near Eastern and European ancestries," *American Journal of Human Genetics* 73 (March 2003): 768–779.

D. M. Behar, E. Metspalu, T. Kivisild, A. Achilli, Y. Hadid, S. Tzur, L. Pereira, A. Amorim, L. Quintana-Murci, K. Majamaa, C. Herrnstadt, N. Howell, O. Balanovsky, I. Kutuev, A. Pshenichnov, D. Gurwitz, B. Bonne-Tamir, A. Torroni, R. Villems, K. Skorecki, "The matrilineal ancestry of Ashkenazi Jewry: portrait of a recent founder event," *American Journal of Human Genetics* 78 (March 2006): 487–497.

Amy Harmon, "Love You, K2a2a, Whoever You Are," *New York Times,* January 22, 2006.

Bryan Sykes, *The Seven Daughters of Eve: The Science that Reveals Our Genetic Identity* (New York and London: W. W. Norton and Co., 2001).

The estimate of the number of clients of commercial DNA ancestry tests: Harmon

DNAPrint Genomics, Inc., annual report (Form 10-KSB/A) filed with the U.S. Securities and Exchange Commission for 2004; Form 10QSB filed by DNAPrint Genomics for the first quarter of 2006.

Tony Frudakis, "Powerful but requiring caution: Genetic tests of ancestral origins," *National Genealogical Society Quarterly* 93 (December 2005): 260–268.

On drug reactions: L. Slordal, O. Spigset, "Heart failure induced by non-cardiac drugs," *Drug Safety,* July 2006, 567–586; A. Holdcroft, "UK drug analysis prints and anaesthetic adverse drug reactions," *Pharmacoepidemiology and Drug Safety* 16 (March 2007): 316–328.

M. Gregg Bloche, "Race-based therapeutics," *New England Journal of Medicine* 351 (November 11, 2004): 2035–2037; Anne L. Taylor, MD, Jay N. Cohn, MD, "A-HeFT: African-American heart failure trial," presented at the American Heart Association 2004 Annual Scientific Sessions, http://www.medscape.com/viewarticle/ 494186 (accessed September 14, 2007).

CHAPTER 3: THE POST-NAZI ERA

Materials of the museum at Spiegelgrund available at http://www.spiegelgrund.at (accessed September 14, 2007). Additional information comes from: Herwig Czech, "Research without scruples: The scientific utilization of patients of the Nazi Psychiatric Murders in Vienna," and Herwig Czech, "From welfare to selection: Vienna's public health office and the implementation of racial hygiene policies under the Nazi regime," both papers supplied by the author in manuscript form.

Brigitte Hamann, Thomas Thornton (translator), *Hitler's Vienna: A Dictator's Apprenticeship* (Oxford: Oxford University Press, 2000).

Robert N. Proctor, *Racial Hygiene: Medicine Under the Nazis* (Cambridge and London: Harvard University Press, 1998).

Johann Gross tells the story of his childhood and internment in Johann Gross, *Spiegelgrund: Leben in NS-Erziehungsanstalten* (Vienna: Verlag Carl Ueberreuter, 2000).

Robert N. Proctor, *The Nazi War on Cancer* (Princeton and Oxford: Princeton University Press, 1999).

Edwin Black, *War Against the Weak: Eugenics and America's Campaign to Create a Master Race* (New York and London: Four Walls Eight Windows, 2003).

Robert Jay Lifton, *The Nazi Doctors: Medical Killing and the Psychology of Genocide* (New York: BasicBooks, 1986).

Daniel J. Kevles, *In the Name of Eugenics: Genetics and the Uses of Human Heredity* (Cambridge and London: Harvard University Press, 1985).

Calvin Coolidge, "Whose Country is This?" *Good Housekeeping* 72 (February 1921): 14.

The ban on genetics in the Soviet Union is described in Valery Soyfer, Leo Gruliow, Rebecca Gruliow, *Lysenko and the Tragedy of Soviet Science* (Piscataway, NJ: Rutgers University Press, 1994).

The level of popular support for the idea of "Russia for the Russians" was measured by the Levada Center between 1998 and 2006. In that period the percentage of Russian residents supporting the idea grew from 46 to 54 percent. http://www.levada .ru/press/2006082500.html (accessed October 10, 2007).

On the correlation of the number of cigarettes smoked and lung cancer: N. S. Godtfredsen, E. Prescott, M. Osler, "Effect of smoking reduction on lung cancer risk," *Journal of the American Medical Association* 294 (September 2005): 1505–1510.

CHAPTER 4: INDECISION

Pakistani study: Alexander Liede, Imtiaz A. Malik, Zeba Aziz, Patricia de los Rios, Elaine Kwan, and Steven A. Narod, "Contribution of BRCA1 and BRCA2 mutations to breast and ovarian cancer in Pakistan," *American Journal of Human Genetics* 71 (September 2002): 595–606.

Risk of breast cancer: Lynn C. Hartmann, MD, Thomas A. Sellers, Ph.D., Marlene H. Frost, Ph.D., Wilma L. Lingle, Ph.D., Amy C. Degnim, MD, Karthik Ghosh, MD, Robert A. Vierkant, MAS, Shaun D. Maloney, BA, V. Shane Pankratz, Ph.D., David W. Hillman, MS, Vera J. Suman, Ph.D., Jo Johnson, RN, Cassann Blake, MD, Thea Tlsty, Ph.D., Celine M. Vachon, Ph.D., L. Joseph Melton III, MD, and Daniel W. Visscher, MD, "Benign breast disease and the risk of breast cancer," *New England Journal of Medicine* 353 (July 21, 2005): 229–237.

The incidence of breast cancer in the United States rose by 4.1 percent between 1975 and 1992 while mortality from breast cancer dropped by 2.3 percent (the statistics are even more dramatic for white people: a rise of 4.3 percent and a drop of 2.4 percent). For ovarian cancer, however, both incidence and mortality fluctuated very little, which is to say, nothing really changed. L. A. G. Ries, M. P. Eisner, C. L. Kosary, B. F. Hankey, B. A. Miller, L. Clegg, A. Mariotto, E. J. Feuer, B. K. Edwards (eds). "SEER Cancer Statistics Review, 1975–2002," National Cancer Institute, Bethesda, MD, http://seer.cancer.gov/csr/1975_2002/ (accessed September 13, 2007).

Deaths from ovarian cancer: G. Tortolero-Luna, M. F. Mitchell, "The epidemiology of ovarian cancer," *Journal of Cellular Biochemistry–Supplement* S23 (1995): 200–207.

Janet Hobhouse, *The Furies* (New York: NYRB Classics, 2004).

CHAPTER 5: A DECISION AT ANY COST

Daniel Kahneman's Nobel lecture, "Maps of Bounded Rationality: A Perspective on Intuitive Judgment and Choice," given December 8, 2002, http://nobelprize.org/economics/laureates/2002/kahnemann-lecture.pdf (accessed September 13, 2007).

The eBay.com experiment is described in Tanjim Hossain and John Morgan, "A test of the revenue equivalence theorem using field experiments on eBay," a working paper, http://emlab.berkeley.edu/users/webfac/dellavigna/e218_f03/morgan.pdf (accessed September 13, 2007).

There are two studies by Andrew Caplin and John Leahy. The conclusion that doctors mishandle information comes from "The supply of information by a concerned expert," *Economic Journal*, July 2004, 487–505. They focus on instances of choosing not to know and propose a model for measuring the usefulness of obtaining information about a possible stressful outcome in "Psychological expected utility theory and anticipatory feelings," *The Quarterly Journal of Economics* 116 (February 2001): 55–79.

The Berkeley study is Botond Köszegi, "Health anxiety and patient behavior," *Journal of Health Economics*, November 2003, 1073–1084.

The terms "monitors" and "blunters" are introduced in S. M. Miller, C. E. Mangan, "Interacting effects of information and coping style in adapting to gynecologic stress:

Should the doctor tell all?" *Journal of Personality and Social Psychology* 45 (July 1983): 223–236.

The two studies that conclude that women are likely to want to know whether they carry one of the BRCA mutations are: Lisa Soleymani Lehmann, Jane C. Weeks, Neil Klar, Judy Garber, "A population-based study of Ashkenazi Jewish women's attitudes toward genetic discrimination and BRCA1/2 testing," *Genetics in Medicine* 4 (September/October 2002): 346–352; and Ilan Yaniv, Deborah Benador, Michal Sagi, "On not wanting to know and not wanting to inform others: Choices regarding predictive genetic testing," *Risk Decision and Policy* 9 (September 2004): 317–336.

The Georgetown study: C. Lerman, C. Hughes, S. J. Lemon, D. Main, C. Snyder, C. Durham, S. Narod, H. T. Lynch, "What you don't know can hurt you: Adverse psychologic effects in members of BRCA1-linked and BRCA2-linked families who decline genetic testing," *Journal of Clinical Oncology* 16 (May 1998): 1650–1654.

Nancy Etcoff, *Survival of the Prettiest: The Science of Beauty* (New York: Anchor Books, 2000).

David Lykken and Auke Tellegen, "Happiness is a stochastic phenomenon," *Psychological Science* 7 (May 1996): 186–189.

Studies of what makes people happy: Ed Diener, Eunkook Suh, Shigehiro Oishi, "Recent findings on subjective well-being," *Indian Journal of Clinical Psychology*, (March 1997): 25–41; D. M. Smith, K. M. Langa, M. U. Kabeto, P. A. Ubel, "Health, wealth, and happiness," *Psychological Science* 16 (September 2005): 663–666.

On recovering from disfiguring accidents: N. E. Van Loey, M. J. Van Son, "Psychopathology and psychological problems in patients with burn scars: Epidemiology and management," *American Journal of Clinical Dermatology* 4 (April 2003): 245–272.

Mastectomy studies: P. Hopwood, A. Lee, A. Shenton, A. Baildam, A. Brain, F. Lalloo, G. Evans, A. Howell, "Clinical follow-up after bilateral risk reducing ('prophylactic') mastectomy: Mental health and body image outcomes," *Psychooncology* 9 (November–December 2000): 462–472; Mal Bebbington Hatcher, Leslie Fallowfield, Roger A'Hern, "The psychological impact of bilateral prophylactic mastectomy: Prospective study using questionnaires and semistructured interviews," *British Medical Journal* 332 (January 13, 2001): 76; Michael Stefanek, Lynn Hartmann, Wendy Nelson, "Risk-reduction mastectomy: Clinical issues and research needs," *Journal of the National Cancer Institute* 93 (September 15, 2001): 1297–1306.

The history and practice of hysterectomies in the United States are described in Christiane Northrup, *Women's Bodies, Women's Wisdom: Creating Physical and Emotional Health and Healing* (New York: Bantam Books, 1998); and The Boston Women's Health Book Collective, *The New Our Bodies, Ourselves* (New York: Simon and Schuster, 1992).

Unnecessary oophorectomies leading to early death: W. H. Parker, M. S. Broder, Z. Liu, D. Shoupe, C. Farquhar, J. S. Berek, "Ovarian conservation at the time of hysterectomy for benign disease," *Obstetrics & Gynecology* 106 (August 2005): 219–226.

Studies of the effects of surgical menopause: A. K. Farrag, E. M. Khedr, H. Abdel-Aleem, T. A. Rageh, "Effect of surgical menopause on cognitive functions," *Dementia and Geriatric Cognitive Disorders,* March 2002, 193–198; R. E. Nappi, E. Sinforiani, M. Mauri, G. Bono, F. Polatti, G. Nappi, "Memory functioning at menopause: Impact of age in ovariectomized women," *Gynecologic and Obstetric Investigation* 47 (January 1999): 29–36; D. Kritz-Silverstein, E. Barrett-Connor, "Hysterectomy, oophorectomy, and cognitive function in older women," *Journal of the American Geriatric Society* 50 (January 2002): 55–61.

Studies of additional markers of ovarian cancer: J. O. Schorge, R. D. Drake, H. Lee, S. J. Skates, R. Rajanbabu, D. S. Miller, J. H. Kim, D. W. Cramer, R. S. Berkowitz, S. C. Mok, "Osteopontin as an adjunct to CA125 in detecting recurrent ovarian cancer," *Clinical Cancer Research* 10 (May 2004): 3474–3478; B. Ye, D. W. Cramer, S. J. Skates, S. P. Gygi, V. Pratomo, L. Fu, N. K. Horick, L. J. Licklider, J. O. Schorge, R. S. Berkowitz, S. C. Mok, "Haptoglobin-alpha subunit as potential serum biomarker in ovarian cancer: Identification and characterization using proteomic profiling and mass spectrometry," *Clinical Cancer Research* 9 (August 2003): 2904–2911; S. C. Mok, J. Chao, S. Skates, K. Wong, G. K. Yiu, M. G. Muto, R. S. Berkowitz, D. W. Cramer, "Prostasin, a potential serum marker for ovarian cancer: Identification through microarray technology," *Journal of the National Cancer Institute* 93 (October 2001): 1458–1464.

Intersex conditions and varying views on the treatment of intersex babies are described on the Web site of the Intersex Society of North America, http://www.isna.org/faq/ conditions (accessed September 13, 2007). Approaches to the treatment of intersex babies are described in Alice Domurat Dreger, "Ambiguous Sex—or Ambivalent Medicine?" *Hastings Center Report* 28 (May/June 1998): 24–35.

The paper that shattered the myth of David Reimer's successful sex reassignment: Milton Diamond, H. Keith Sigmundson, "Sex reassignment at birth: A long term review and clinical implications," *Archives of Pediatric & Adolescent Medicine* 151 (March 1997): 298–304.

For a discussion of once-routine unnecessary surgeries see Thomas B. Freeman, Dorothy E. Vawter, Paul E. Leaverton, James H. Godbold, Robert A. Hauser, Christopher G. Goetz, MD, C. Warren Olanow, MD, "Use of placebo surgery in controlled trials of a cellular-based therapy for Parkinson's disease," *New England Journal of Medicine* 151 (September 23, 1999): 988–992.

Prognosis for mutation carriers who have cancer: M. E. Robson, P. O. Chappuis, J. Satagopan, N. Wong, J. Boyd, J. R. Goffin, C. Hudis, D. Roberge, L. Norton, L. R. Bégin, K. Offit, W. D. Foulkes, "A combined analysis of outcome following breast cancer: Differences in survival based on BRCA1/BRCA2 mutation status and administration of adjuvant treatment," *Breast Cancer Research* 6 (October 2003): R8–R17; Ilana Cass, Rae Lynn Baldwin, Taz Varkey, Roxana Moslehi, Steven A. Narod, Beth Y. Karlan, "Improved survival in women with BRCA-associated ovarian carcinoma," *Cancer* 97 (May 2003): 2187–2195; John R. Goffin, Pierre O. Chappuis, Louis R. Bégin, Nora Wong, Jean-Sébastien Brunet, Nancy Hamel, Ann-Josée Paradis, Jeff Boyd, William D. Foulkes, "Impact of germline BRCA1 mutations and overexpression of p53 on prognosis and response to treatment following breast carcinoma," *Cancer* 97 (February 2003): 527–536; Pål Møller, Åke Borg, D. Gareth Evans, Neva Haites, Marta M. Reis, Hans Vasen, Elaine Anderson, C. Michael

Steel, Jaran Apold, David Goudie, Anthony Howell, Fiona Lalloo, Lovise Mæhle, Helen Gregory, Ketil Heimdal, "Survival in prospectively ascertained familial breast cancer: Analysis of a series stratified by tumour characteristics, BRCA mutations and oophorectomy," *International Journal of Cancer* 101 (October 20, 2002): 555–559; Y. Ben David, A. Chetrit, G. Hirsh-Yechezkel, E. Friedman, B. D. Beck, U. Beller, G. Ben-Baruch, A. Fishman, H. Levavi, F. Lubin, J. Menczer, B. Piura, J. P. Struewing, B. Modan for the National Israeli Study of Ovarian Cancer, "Effect of BRCA mutations on the length of survival in epithelial ovarian tumors," *Journal of Clinical Oncology* 20 (January 2002): 463–466; Lori J. Pierce, Myla Strawderman, Steven A. Narod, Ivo Oliviotto, Andrea Eisen, Laura Dawson, David Gaffney, Lawrence J. Solin, Asa Nixon, Judy Garber, Christine Berg, Claudine Isaacs, Ruth Heimann, Olufunmilayo I. Olopade, Bruce Haffty, Barbara L. Weber, "Effect of radiotherapy after breast-conserving treatment in women with breast cancer and germline BRCA1/2 mutations," *Journal of Clinical Oncology* 18 (October 2000): 3360–3369.

Mutation carriers behavior study: Jeffrey R. Botkin, Ken R. Smith, Bonnie J. Baty, Jean E. Wylie, Debra Dutson, Anna Chan, Heidi A. Hamann, Caryn Lerman, Jamie McDonald, Vickie Venne, John H. Ward, Elaine Lyon, "Genetic testing for a BRCA mutation: Prophylactic surgery and screening behavior in women 2 years post testing," *American Journal of Medical Genetics Part A* 118A (March 2003): 201–209.

On male breast cancer and genetics: J. R. Weiss, K. B. Moysich, H. Swede, "Epidemiology of male breast cancer," *Cancer Epidemiology Biomarkers & Prevention* 14 (January 2005): 20–26; A. Joseph, K. Mokbel, "Male breast cancer," *International Journal of Fertility and Women's Medicine* 49 (September–October 2004): 198–199; K. Syrjäkoski, T. Kuukasjarvi, K. Waltering, K. Haraldsson, A. Auvinen, A. Borg, T. Kainu, O.-P. Kallioniemi, P. A. Koivisto, "BRCA2 mutations in 154 Finnish male breast cancer patients," *Neoplasia* 6 (September–October 2004): 541–545.

On exercise and diet as modifiers of cancer risk: A. McTiernan, "Behavioral risk factors in breast cancer: Can risk be modified?" *Oncologist* 8 (August 2003): 326–334.

On risk perception: David Ropeik, George Gray, *Risk: A Practical Guide for Deciding What's Really Safe and What's Really Dangerous in the World Around You* (Boston: Houghton Mifflin, 2002).

On women's perceptions of breast cancer risk: N. Humpel, S. C. Jones, "'I don't really know, so it's a guess': Women's reasons for breast cancer risk estimation," *Asian Pacific Journal of Cancer Prevention* 5 (October–December 2004): 428–432; S. Davis, S. Stewart, J. Bloom, "Increasing the accuracy of perceived breast cancer risk: Results from a randomized trial with Cancer Information Service callers," *Preventive Medicine* 39 (July 2004): 64–73.

CHAPTER 6: THE FATHER OF HEREDITARY CANCERS

Aldred Scott Warthin and Lynch syndrome: A. G. Thorson, J. A. Knezetic, H. T. Lynch, "A century of progress in hereditary nonpolyposis colorectal cancer (Lynch syndrome)," *Diseases of the Colon and Rectum* 42 (January 1999): 1–9; Henry T. Lynch, Thomas Smyrk, Jane F. Lynch, "Molecular genetics and clinical-pathology features of hereditary nonpolyposis colorectal carcinoma (Lynch syndrome): Historical journey from pedigree anecdote to molecular genetic confirmation," *Oncology* 55 (March–April 1998): 103–108; H. T. Lynch, T. C. Smyrk, P. Watson, S. J. Lanspa, J. F. Lynch, P. M. Lynch, R. J. Cavalieri, C. R. Boland, "Genetics, natural history, tumor spectrum, and pathology of hereditary nonpolyposis colorectal cancer: An updated review," *Gastroenterology* 104 (May 1993): 1535–1549.

Dr. Lynch's search for genetic clues: H. T. Lynch, R. J. Thomas, H. A. Gurgis, J. Lynch, "Clues to cancer risk: Biologic markers," *American Family Physician* 11 (March 1975): 153–158; H. T. Lynch, R. J. Thomas, P. I. Terasaki, A. Ting, H. A. Guirgis, A. R. Kaplan, H. Magee, J. Lynch, C. Kraft, E. Chaperon, "HL-A in cancer family 'N,'" *Cancer* 36 (October 1975): 1315–1320; M. C. King, R. C. Go, H. T. Lynch, R. C. Elston, P. I. Terasaki, N. L. Petrakis, G. C. Rodgers, D. Lattanzio, J. Bailey-Wilson, "Genetic epidemiology of breast cancer and associated cancers in high-risk families. II. Linkage analysis," *Journal of the National Cancer Institute* 71 (September 1983): 463–467.

Dr. Lynch's first suggestion of prophylactic surgery: H. T. Lynch, R. E. Harris, C. H. Organ Jr., H. A. Guirgis, P. M. Lynch, J. F. Lynch, E. J. Nelson, "The surgeon, genetics, and cancer control: The Cancer Family Syndrome," *Annals of Surgery* 185 (April 1977): 435–440.

Dr. Lynch's criticism of colleagues' failure to catch hereditary cancer: H. T. Lynch, S. J. Lanspa, B. M. Boman, T. Smyrk, P. Watson, J. F. Lynch, P. M. Lynch, G. Cristofaro, P. Bufo, A. V. Tauro, et al., "Hereditary nonpolyposis colorectal cancer—Lynch syndromes I and II," *Gastroenterology Clinics of North America* 17 (December 1988): 679–712; H. T. Lynch, E. K. Bronson, P. C. Strayhorn, T. C. Smyrk, J. F. Lynch, E. J. Ploetner, "Genetic diagnosis of Lynch syndrome II in an extended colorectal cancer-prone family," *Cancer* 66 (November 1990): 2233–2238; H. T. Lynch, T. C. Smyrk, S. J. Lanspa, J. X. Jenkins, J. Cavalieri, J. F. Lynch, "Cancer control problems in the Lynch syndromes," *Diseases of the Colon and Rectum* 36 (March 1993): 254–260.

The story of Dr. Lynch's testifying at a malpractice trial: H. T. Lynch, J. Paulson, M. Severin, J. Lynch, P. Lynch, "Failure to diagnose hereditary colorectal cancer and its medicolegal implications: A hereditary nonpolyposis colorectal cancer case," *Diseases of the Colon and Rectum* 42 (January 1999): 31–35.

Dr. Lynch complained that physicians had failed to act on genetic discoveries in H. T. Lynch, T. C. Smyrk, "Hereditary colorectal cancer," *Seminars in Oncology* 26 (October 1999): 478–484.

Papers that report the identification of the Lynch syndrome chromosomes, genes, and mutations: H. T. Lynch, T. C. Smyrk, J. Cavalieri, J. F. Lynch, "Identification of an HNPCC family," *American Journal of Gastroenterology*, April 1994, 605–609; H. T. Lynch, J. F. Lynch, "25 years of HNPCC," *Anticancer Research*, July–August 1994, 1617–1624 (this is also the paper where the Lynches report that they have started to offer their patients the option of a prophylactic subtotal colectomy); H. T. Lynch, J. F. Lynch, "Clinical implications of advances in the molecular genetics of colorectal cancer," *Tumori*, May–June 1995, 19–29; H. T. Lynch, J. F. Lynch, "Genetics of colonic cancer," *Digestion*, August 1998, 481–492.

Lynch papers that describe his tracking of families: H. T. Lynch, P. Watson, M. Kriegler, J. F. Lynch, S. J. Lanspa, J. Marcus, T. Smyrk, R. J. Fitzgibbons Jr., G. Cristofaro, "Differential diagnosis of hereditary nonpolyposis colorectal cancer (Lynch syndrome I and Lynch syndrome II)," *Diseases of the Colon and Rectum* 31 (May 1988): 372–377; H. T. Lynch, J. Ens, J. F. Lynch, P. Watson, "Tumor variation in three extended Lynch syndrome II kindreds," *American Journal of Gastroenterology* 83 (July 1988): 741–747; H. T. Lynch, M. Kriegler, T. A. Christiansen, T. Smyrk, J. F. Lynch, P. Watson, "Laryngeal carcinoma in a Lynch syndrome II kindred," *Cancer* 62 (September 1988): 1007–1013; H. T. Lynch, T. Drouhard, H. F. Vasen, J. Cavalieri, J. Lynch, S. Nord, T. Smyrk, S. Lanspa, P. Murphy, K. L. Whelan, J. Peters, A. de la Chapelle, "Genetic counseling in a Navajo hereditary nonpolyposis colorectal cancer kindred," *Cancer* 77 (January 1996): 30–35; H. T. Lynch, T. Smyrk, "Hereditary nonpolyposis colorectal cancer (Lynch syndrome). An updated review," *Cancer* 78 (September 1996): 1149–1167.

Paper reporting that half the patients would consider surgery: H. T. Lynch, S. Lemon, T. Smyrk, B. Franklin, B. Karr, J. Lynch, S. Slominski-Caster, P. Murphy, C. Connolly, "Genetic counseling in hereditary nonpolyposis colorectal cancer: An extended family with MSH2 mutation," *American Journal of Gastroenterology* 91 (December 1996): 2489–2493.

Paper reporting that the risk will most often be adjusted down following family testing: P. Watson, S. A. Narod, R. Fodde, A. Wagner, J. F. Lynch, S. T. Tinley, C. L. Snyder, S. A. Coronel, B. Riley, Y. Kinarsky, H. T. Lynch, "Carrier risk status changes resulting from mutation testing in hereditary non-polyposis colorectal cancer and hereditary breast-ovarian cancer," *Journal of Medical Genetics* 40 (August 2003): 591–596.

Dr. Lynch introduces the term "cancer destiny": H. T. Lynch, T. Smyrk, J. Lynch, "An update of HNPCC (Lynch syndrome)," *Cancer Genetics and Cytogenetics* 93 (January 1997): 84–99.

Family information services: H. T. Lynch, S. J. Lemon, B. Karr, B. Franklin, J. F. Lynch, P. Watson, S. Tinley, C. Lerman, C. Carter, "Etiology, natural history, management and molecular genetics of hereditary nonpolyposis colorectal cancer (Lynch syndromes): Genetic counseling implications," *Cancer Epidemiology Biomarkers & Prevention* 6 (December 1997): 987–991.

Alicia Chang, Malcolm Ritter, "11 cousins give up stomachs after tests," Associated Press, June 18, 2006.

CHAPTER 7: THE CRUELEST DISEASE

On disease onset and progression: Sandra Close Kirkwood, Jessica L. Su, Michael Conneally, Tatiana Foroud, "Progression of symptoms in the early and middle stages of Huntington disease," *Archives of Neurology* 58 (February 2001): 273; J. S. Paulsen, H. Zhao, J. C. Stout, R. R. Brinkman, M. Guttman, C. A. Ross, P. Como, C. Manning, M. R. Hayden, I. Shoulson, Huntington Study Group, "Clinical markers of early disease in persons near onset of Huntington's disease," *Neurology* 57 (August 2001): 658–662; S. A. Reading, M. A. Yassa, A. Bakker, A. C. Dziorny,

L. M. Gourley, V. Yallapragada, A. Rosenblatt, R. L. Margolis, E. H. Aylward, J. Brandt, S. Mori, P. van Zijl, S. S. Bassett, C. A. Ross, "Regional white matter change in pre-symptomatic Huntington's disease: A diffusion tensor imaging study," *Psychiatry Research* 140 (October 2005): 55–62; *Understanding Huntington Disease: A Resource for Families*, published by the Huntington Society of Canada, 2002; Jane S. Paulsen, *Understanding Behaviour in Huntington Disease: A Practical Guide for Individuals, Families and Professionals Coping with HD*, published by the Huntington Society of Canada, 2002; Adam Rosenblatt, Neal Ranen, Martha Nance, Jane Paulsen, *A Physician's Guide to the Management of Huntington Disease*, Second Edition, published by the Huntington Society of Canada, 2004.

"True dominance" in Huntington's and lack of it in hereditary cancers: Yolanda Narain, Andreas Wyttenbach, Julia Rankin, Robert A. Furlong, David C. Rubinsztein, "A molecular investigation of true dominance in Huntington's disease," *Journal of Medical Genetics* 36 (October 1999): 739–746; N. S. Wexler, A. B. Young, R. E. Tanzi, H. Travers, S. Starosta-Rubinstein, J. B. Penney, S. R. Snodgrass, I. Shoulson, F. Gomez, M. A. Ramos Arroyo, et al., "Homozygotes for Huntington's disease," *Nature* 326 (March 12, 1987): 194–197; T. Ludwig, D. L. Chapman, V. E. Papaioannou, A. Efstratiadis, "Targeted mutations of breast cancer susceptibility gene homologs in mice: Lethal phenotypes of Brca1, Brca2, Brca1/Brca2, Brca1/p53, and Brca2/p53 nullizygous embryos," *Genes & Development* 11 (May 15, 1997): 1226–1241.

The search for the gene is described in Alice Wexler, *Mapping Fate: A Memoir of Family, Risk, and Genetic Research* (Berkeley: University of California Press, 1996).

Jean Baréma, *The Test: Living in the Shadow of Huntington's Disease* (New York: Franklin Square Press, 2005).

Early studies of attitudes toward predictive testing were described in: B. Teltscher, S. Polgar, "Objective knowledge about Huntington's disease and attitudes towards predictive tests of persons at risk," *Journal of Medical Genetics* 18 (February 1981): 31–39; A. Tyler, P. S. Harper, "Attitudes of subjects at risk and their relatives towards genetic counseling in Huntington's chorea," *Journal of Medical Genetics* 20 (June 1983): 179–188; S. Kessler, T. Field, L. Worth, H. Mosbarger, "Attitudes of persons at risk for Huntington disease toward predictive testing," *American Journal of Medical Genetics* 26 (February 1987): 259–270; G. J. Meissen, R. L. Berchek, "Intended use of predictive testing by those at risk for Huntington disease," *American Journal of Medical Genetics* 26 (February 1987): 283–293.

Statistics on the use of predictive testing prior to 1992 and Nancy Wexler's comments on *60 Minutes* are cited in Alice Wexler, *Mapping Fate*.

Transcript of the 2004 CBS News story quoting Nancy Wexler: http://www.cbsnews .com/stories/2004/11/18/eveningnews/main656527.shtml?CMP=ILC-SearchStories (accessed September 13, 2007).

Study of why people choose not to get tested: K. A. Quaid, M. Morris, "Reluctance to undergo predictive testing: The case of Huntington disease," *American Journal of Medical Genetics* 45 (January 1993): 41–45; study showing that those who learned of their status as adults were more likely to get tested: I. M. Van der Steenstraten, A. Tibben, R. A. C. Roos, et al., "Predictive testing for Huntington's disease: Nonparticipants compared with participants in the Dutch program," *American Journal of Human Genetics* 55 (December 1994): 618–625.

Current rates of presymptomatic testing for Huntington's: Peter S. Harpera, Caron Lima, David Craufurdb, on behalf of the UK Huntington's Disease Prediction Consortium, "Ten years of presymptomatic testing for Huntington's disease: The experience of the UK Huntington's Disease Prediction Consortium," *Journal of Medical Genetics* 37 (August 2000): 567–571; M. P. Solis-Perez, J. A. Burguera, F. Palau, L. Livianos, M. Vila, L. Rojo [Results of a program of presymptomatic diagnosis of Huntington's disease: Evaluation of a six-year period], article in Spanish, *Neurologia* 16 (October 2001): 348–352; P. Mandich, G. Jacopini, E. Di Maria, G. Sabbadini, G. Abbruzzese, F. Chimirri, E. Bellone, A. Noveletto, F. Ajmar, M. Frontali, "Predictive testing for Huntington's disease: Ten years' experience in two Italian centres," *Italian Journal of Neurological Sciences* 19 (June 1998): 68–74; S. Creighton, E. W. Almqvist, D. MacGregor, B. Fernandez, H. Hogg, J. Beis, J. P. Welch, C. Riddell, R. Lokkesmoe, M. Khalifa, J. MacKenzie, A. Sajoo, S. Farrell, F. Robert, A. Shugar, A. Summers, W. Meschino, D. Allingham-Hawkins, T. Chiu, A. Hunter, J. Allanson, H. Hare, J. Schween, L. Collins, S. Sanders, C. Greenberg, S. Cardwell, E. Lemire, P. MacLeod, M. R. Hayden, "Predictive, pre-natal and diagnostic genetic testing for Huntington's disease: The experience in Canada from 1987 to 2000," *Clinical Genetics* 63 (June 2003): 462–475.

CHAPTER 8: THE SCIENCE OF MATCHMAKING

On genetic imperatives in Jewish law: Rosner, "Judaism, genetic screening and genetic therapy."

On leprosy and genetics: A. Alcais, M. Mira, J. L. Casanova, E. Schurr, L. Abel, "Genetic dissection of immunity in leprosy," *Current Opinion in Immunology* 17 (February 2005): 44–48.

As of this writing (September 14, 2007), Connecticut, Georgia, Massachusetts, Mississippi, Montana, Oklahoma, and the District of Columbia still required premarital syphilis tests.

Much of the information on Dor Yeshorim and Rabbi Ekstein's quotes come from the following articles, published and unpublished: "From tragedy to triumph: Victory over Tay-Sachs," manuscript of an article prepared for *Jewish Action* by Howard Katzenstein. Supplied by Howard Katzenstein; Alexandra Shimo-Barry, "Modern matchmaker: Premarital tests help Hasidim avert genetic disease," a paper included in a Columbia Journalism School anthology, *Race and Ethnicity 2002*, available online at http://web.jrn.columbia.edu/studentwork/race/2002/gene-shimo.shtml.

Transcript of "The DNA Files: Prenatal testing: Do you really want to know your baby's future?" Sound Vision Productions, 1998; Gina Kolata, "Nightmare or the Dream of a New Era in Genetics?" *New York Times*, December 7, 1993; Christine Rosen, "Eugenics—Sacred and Profane," *New Atlantis*, Summer 2003, 79–89; Mary Jo Layton, "Love v. science? Worldwide program aims to eradicate Jewish birth defect," *Jewish World Review*, February 14, 2006; Josef Ekstein and Howard Katzenstein, "The Dor Yeshorim story: Community-based carrier screening for Tay-Sachs disease," *Advances in Genetics* 44 (2001): 297; Dor Yeshorim promotional booklet.

On bereavement and coping strategies: W. Middleton, B. Raphael, P. Burnett, N. Martinek, "A longitudinal study comparing bereavement phenomena in recently bereaved spouses, adult children and parents," *Australian and New Zealand Journal of Psychiatry* 32 (April 1998): 235–241.

C. Krauth, N. Jalilvand, T. Welte, R. Busse, "Cystic fibrosis: Cost of illness and considerations for the economic evaluation of potential therapies," *Pharmacoeconomics* 21 (2003): 1001–1024; T. Havermans, K. De Boeck, "Cystic fibrosis: A balancing act?" *Journal of Cystic Fibrosis* 6 (July 2006): 161–162.

F. B. Axelrod, "A world without pain or tears," *Clinical Autonomic Research* 16 (April 2006): 90–97; F. B. Axelrod, "Familial dysautonomia: A review of the current pharmacological treatments," *Expert Opinion on Pharmacotherapy* 6 (April 2005): 561–567.

Yeshiva University comments obtained from a student's blog, http://atowncrier .blogspot.com/2006_04_02_atowncrier_archive.html (accessed July 25, 2006); high school student's comments obtained from http://bubblyblogely.blogspot.com/2006/ 04/sorry-for-being-mia.html (accessed July 25, 2006).

C. Lerman, R. T. Croyle, K. P. Tercyak, H. Hamann, "Genetic testing: Psychological aspects and implications," *Journal of Consulting and Clinical Psychology* 70 (June 2002): 784–797.

The original Colin McGinn article was "Can we solve the mind-body problem?" published in *Mind* 98 (July 1989): 349–366. The article I recalled was "Out of body, out of mind," published in *Lingua Franca* in 1994.

There is no reliable information on the percentage of women who choose surgery following a positive BRCA mutation result. A review article, S. Wainberg, J. Husted, "Utilization of screening and preventive surgery among unaffected carriers of a BRCA1 or BRCA2 gene mutation," *Cancer Epidemiology Biomarkers & Prevention* 13 (December 2004): 1989–1995, found a number of studies with the estimates of those electing surgery ranging from 0 to 54 percent. These were studies from the United States and the Netherlands, where both patients and doctors generally favor a somewhat less invasive attitude toward genetic screening and surgical intervention than their Israeli counterparts. It seems reasonable to assume that the Israeli numbers would be at the top of this spectrum.

CHAPTER 10: THE FUTURE THE OLD-FASHIONED WAY

Papers published by Clinic for Special Children staff and others on disorders mentioned in the chapter include:

Kevin Strauss, D. Holmes Morton, "Branched-chain ketoacyl dehydrogenase deficiency: Maple syrup disease," *Current Treatment Options in Neurology* 5 (July 2003): 329–341.

E. G. Puffenberger, E. R. Kauffman, S. Bolk, T. C. Matise, S. S. Washington, M. Angrist, J. Weissenbach, K. L. Garver, M. Mascari, R. Ladda, et al., "Identity-by-descent and association mapping of a recessive gene for Hirschsprung disease on human chromosome 13q22," *Human Molecular Genetics* 3 (August 1994): 1217–1225.

J. Amiel, R. Salomon, T. Attie-Bitach, R. Touraine, J. Steffann, A. Pelet, C. Nihoul-Fekete, M. Vekemans, A. Munnich, S. Lyonnet, "Molecular genetics of Hirschsprung disease: A model of multigenic neurocristopathy," *Journal de la Société de biologie* 194 (2000): 125–128; E. G. Puffenberger, K. Hosoda, S. S. Washington, K. Nakao, D. deWit, M. Yanagisawa, A. Chakravart, "A missense mutation of the endothelin-B receptor gene in multigenic Hirschsprung's disease," *Cell* 79 (December 1994): 1257–1266.

Kevin A. Strauss, Erik G. Puffenberger, Donna L. Robinson, D. Holmes Morton, "Type I glutaric aciduria, part 1: Natural history of 77 patients," *American Journal of Medical Genetics Part C* 121C (August 2003): 38–52; Kevin A. Strauss, D. Holmes Morton, "Type I glutaric aciduria, part 2: A model of acute striatal necrosis," *American Journal of Medical Genetics Part C* 121C (August 2003): 53–70.

J. J. Higgins, D. H. Morton, J. M. Loveless, "Posterior column ataxia with retinitis pigmentosa (AXPC1) maps to chromosome 1q31-q32," *Neurology* 52 (January 1999): 146–150; J. J. Higgins, K. Kluetzman, J. Berciano, O. Combarros, J. M. Loveless, "Posterior column ataxia and retinitis pigmentosa: A distinct clinical and genetic disorder," *Movement Disorders* 15 (May 2000): 575–578.

Kevin Strauss, Erik Puffenberger, Matthew Huentelman, Steven Gottlieb, Seth Dobrin, Jennifer Parod, Dietrich Stephan, Holmes Morton, "Recessive symptomatic focal epilepsy and mutant contactin-associated protein-like 2," *New England Journal of Medicine* 354 (March 30, 2006): 1370–1377.

K. A. Strauss, G. V. Mazariegos, R. Sindhi, R. Squires, D. N. Finegold, G. Vockley, D. L. Robinson, C. Hendrickson, M. Virji, L. Cropcho, E. G. Puffenberger, W. McGhee, L. M. Seward, D. H. Morton, "Elective liver transplantation for the treatment of classical maple syrup urine disease," *American Journal of Transplantation* 6 (March 2006): 557–564; A. Khanna, M. Hart, W. L. Nyhan, T. Hassanein, J. Panyard-Davis, B. A. Barshop, "Domino liver transplantation in maple syrup urine disease," *Liver Transplantation* 12 (May 2006): 876–882.

On the West Virginia T-shirt controversy: Richard Roeper, "A message on a T-shirt? The medium's not so rare," *Chicago Sun-Times,* March 29, 2004.

On Lorenzo's Oil, the following article called early results "disappointing": J. Borel, H. W. Moser, "Dietary management of X-linked adrenoleukodystrophy," *Annual Review of Nutrition* 15 (1995): 379–397. But while the oil did not appear to work as a treatment, used in asymptomatic boys it proved helpful: Hugo W. Moser, MD, Gerald V. Raymond, MD, Shou-En Lu, Ph.D., Larry R. Muenz, Ph.D., Ann B. Moser, BA, Jiahong Xu, MS, Richard O. Jones, Ph.D., Daniel J. Loes, MD, Elias R. Melhem, MD, Prachi Dubey, MD, MPH, Lena Bezman, MD, MPH, N. Hong Brereton, MS, RD, Augusto Odone, "Follow-up of 89 asymptomatic patients with adrenoleukodystrophy treated with Lorenzo's Oil," *Archives of Neurology* 62 (July 2005): 1073–1080.

Articles on Holmes Morton in general-interest publications: Melissa Hendricks, "A doctor who makes barn calls," *Johns Hopkins Magazine,* 1994; Tom Shachtman, "Medical sleuth," *Smithsonian,* February 2006, p. 23; Lisa Belkin, "A doctor for the future," *New York Times Magazine,* November 6, 2005.

Both the conventional and an alternative view of the genetic testing of children are presented in Rosamond Rhodes, "Why test children for adult-onset diseases?" *Mount Sinai Journal of Medicine* 73 (May 2006): 609.

On the impact of age at first period on breast cancer risk: J. Kotsopoulos, J. Lubinski, H. T. Lynch, S. L. Neuhausen, P. Ghadirian, C. Isaacs, B. Weber, C. Kim-Sing, W. D. Foulkes, R. Gershoni-Baruch, P. Ainsworth, E. Friedman, M. Daly, J. E. Garber, B. Karlan, O. I. Olopade, N. Tung, H. M. Saal, A. Eisen, M. Osborne, H. Olsson, D. Gilchrist, P. Sun, S. A. Narod, "Age at menarche and the risk of breast cancer in BRCA1 and BRCA2 mutation carriers," *Cancer Causes and Control* 16 (August 2005): 667–674. Whether and how much exercise and diet influence the age when a girl has her first period is still the subject of much controversy. The available evidence indicates that exercise intensity and dietary restrictions may have to be so extreme to influence the onset of the menstrual cycle that any positive effects would be canceled out. Still, some articles point to a link: C. Castelo-Branco, F. Reina, A. D. Montivero, M. Colodron, J. A. Vanrell, "Influence of high-intensity training and of dietetic and anthropometric factors on menstrual cycle disorders in ballet dancers," *Gynecologic Endocrinology* 22 (January 2006): 31–35; F. Thomas, F. Renaud, E. Benefice, T. de Meeus, J. F. Guegan, "International variability of ages at menarche and menopause: Patterns and main determinants," *Human Biology* 73 (April 2001): 271–290.

Charles P. Hehmeyer, "The case for universal newborn screening," *Exceptional Parent*, August 2001, p. 88.

On the progress and problems of gene therapy development, see: Larry Thompson, "Human gene therapy: Harsh lessons, high hopes," *FDA Consumer* Magazine, September–October 2000; Salima Hacein-Bey-Abina, Marina Cavazzana-Calvo, Alain Fischer, "Gene therapy for severe combined immunodeficiency X1," Presentation at 11th International Congress of Human Genetics; Josephine Johnston and Françoise Baylis, "What happened to gene therapy? A review of recent events," *Clinical Researcher* 4 (January 2004): 11–15; Deanna Cross, James K. Burmester, "Gene therapy for cancer treatment: Past, present and future," *Clinical Medicine & Research* 4 (September 2006): 218–227; Sheryl Gay Stolberg, "The Biotech Death of Jesse Gelsinger," *New York Times*, November 28, 1999, 137.

CHAPTER II: BIOBABBLE

Lyudmila N. Trut, "Early canid domestication: The farm-fox experiment," *American Scientist*, March–April 1999, 160.

Irene Plyusnina, I. Oskina, "Behavioral and adrenocortical responses to open-field test in rats selected for reduced aggressiveness toward humans," *Psychology & Behavior* 61 (March 1997): 381–385.

Publications on catatonic mice and a potential link to depression and schizophrenia research (in Russian): Institut Tsitologii I Genetiki, *Prikladniye Razrabotki: Geneticheskaya model biologicheskoy osnovy predraspolozheniya k shizofrenii;* V. G. Kolpakov, A. V. Kulikov, T. A. Alehina, V. F. Chuguy, O. I. Petrenko, N. N. Barykina, "Katatoniya ili depressiya? Liniya krys GK—geneticheskaya zhivotnaya model psikhopatologii," *Genetika* 40 (2004): 1–7.

The "aggression gene" in humans: H. G. Brunner, M. R. Nelen, P. van Zandvoort, N. G. Abeling, A. H. van Gennip, E. C. Wolters, M. A. Kuiper, H. H. Ropers, B. A. van Oost, "X-linked borderline mental retardation with prominent behavioral disturbance: Phenotype, genetic localization, and evidence for disturbed monoamine metabolism," *American Journal of Human Genetics* 52 (June 1993): 1032–1039; H. G. Brunner, M. Nelen, X. O. Breakefield, H. H. Ropers, B. A. van Oost, "Abnormal behavior associated with a point mutation in the structural gene for monoamine oxidase A," *Science* 262 (October 1993): 578–580; H. G. Brunner, "MAOA deficiency and abnormal behaviour: Perspectives on an association," *Ciba Foundation Symposium 194* (1995): 155–164; discussion: 164–167.

Aggressive mice: A. V. Kulikov, D. V. Osipova, V. S. Naumenko, N. K. Popova, "Association between Tph2 gene polymorphism, brain tryptophan hydroxylase activity and aggressiveness in mouse strains," *Genes, Brain and Behavior* 4 (November 2005): 482–485; Nina K. Popova, Larissa N. Maslova, Ekaterina A. Morosova, Veta V. Bulygina, Isabelle Seif, "MAO A knockout attenuates adrenocortical response to various kinds of stress," *Psychoneuroendocrinology* 31 (September 2006): 179–186; Nina K. Popova, Galina B. Vishnivetskaya, Elena A. Ivanova, Julia A. Skrinskaya, Isabelle Seif, "Altered behavior and alcohol tolerance in transgenic mice lacking MAO A: A comparison with effects of MAO A inhibitor clorgyline," *Pharmacology, Biochemistry and Behavior* 67 (December 2000): 719–727; N. K. Popova, "Geticheskiy Nokaut—Perviye Shagi I Perspektivy dlya Neyrofiziologii povedeniya," *Uspekhi Fiziologicheskikh Nauk* 2 (2000): 3–13; Nina K. Popova, "From genes to aggressive behavior: The role of serotonergic system," *Bioessays* 31 (May 2006): 495.

Ebstein's studies of the dopamine receptor gene: A. N. Kluger, Z. Siegfried, R. P. Ebstein, "A meta-analysis of the association between DRD4 polymorphism and novelty seeking," *Molecular Psychiatry* 7 (August 15, 2002): 712–717; I. Z. Ben Zion, R. Tessler, L. Cohen, E. Lerer, Y. Raz, R. Bachner-Melman, I. Gritsenko, L. Nemanov, A. H. Zohar, R. H. Belmaker, J. Benjamin, R. P. Ebstein, "Polymorphisms in the dopamine D4 receptor gene (DRD4) contribute to individual differences in human sexual behavior: Desire, arousal and sexual function," *Molecular Psychiatry* 11 (August 1, 2006): 782–786; R. Bachner-Melman, I. Gritsenko, L. Nemanov, A. H. Zohar, C. Dina, R. P. Ebstein, "Dopaminergic polymorphisms associated with self-report measures of human altruism: A fresh phenotype for the dopamine D4 receptor," *Molecular Psychiatry* 10 (April 1, 2005): 333–335; B. Dan, C. Hagit, N. Lily, R. P. Ebstein, "An association between fibromyalgia and the dopamine D4 receptor exon III repeat polymorphism and relationship to novelty seeking personality traits," *Molecular Psychiatry* 9 (November 1, 2004): 730–733; R. P. Ebstein, J. Levine, V. Geller, J. Auerbach, I. Gritsenko, R. H. Belmaker, "Dopamine D4 receptor and serotonin transporter promoter in the determination of neonatal temperament," *Molecular Psychiatry* 3 (May 21, 1998): 239–246.

Shoshana Arbelle, Jonathan Benjamin, Moshe Golin, Ilana Kremer, Robert H. Belmaker, Richard P. Ebstein, "Relation of shyness in grade school children to the genotype for the long form of the serotonin transporter promoter region polymorphism," *American Journal of Psychiatry* (April 2003): 671–676.

Ilana Kremer, Rachel Bachner-Melman, Alon Reshef, Leonid Broude, Lubov Nemanov, Inga Gritsenko, Uriel Heresco-Levy, Yoel Elizur, Richard P. Ebstein, "Association of the serotonin transporter gene with smoking behavior," *American Journal of Psychiatry* (May 2005): 924–930.

Rachel Bachner-Melman, Christian Dina, Ada Zohar, Naama Constantini, Elad Lerer, Sarah Hoch, Sarah Sella, Lubov Nemanov, Inga Gritsenko, Pesach Lichtenberg, Roni Granot, Richard P. Ebstein, "AVPR1a and SLC6A4 gene polymorphisms are associated with creative dance performance," *PLoS Genetics* 1 (September 2005): e42.

Claudia Dreifus, "A conversation with Gino Segre," *The New York Times*, August 14, 2007; Henry Fountain, "Ants tend to gravitate to what they do best, researchers show," *New York Times*, August 14, 2007; Ingfei Chen, "The beam of light that flips a switch that turns on the brain," *New York Times*, August 14, 2007; Abigail Zuger, "Sweatology," *New York Times*, August 14, 2007; Carl Zimmer, "Lessons from an insect's life cycle: Extreme sibling rivalry," *New York Times*, August 14, 2007; Gina Kolata, "The myth, the math, the sex," *New York Times*, August 12, 2007; Denise Grady, "Deadly inheritance, desperate trade-off," August 7, 2007; Emily Bazelon, "What autistic girls are made of," *New York Times Magazine*, August 5, 2007; Andrew Pollack, "Initial benefit from genetic engineering likely to be medicine," *New York Times*, July 30, 2007; Julia Reed, "A natural history of love," *New York Times*, July 29, 2007; Anahad O'Connor, "REALLY?; Eating garlic helps repel mosquitoes," *New York Times*, July 24, 2007; David Brooks, "A partnership of minds," July 20, 2007; Nicholas Wade, "Scientists find genetic link for a disorder (Next, respect?)," *New York Times*, July 19, 2007; Peggy Orenstein, "Your gamete, myself," *New York Times Magazine*, July 15, 2007; Shalom Auslander, "Pore me," *New York Times*, August 20, 2007; Cintra Wilson, "A (very poised) dance on the table," *New York Times*, August 9, 2007; Selena Roberts, "Stewards of sport need a lesson from the pit bosses," *New York Times*, July 22, 2007.

CHAPTER 12: WHAT WE FEAR MOST

Yury Verlinsky, Anver Kuliev, *Practical Preimplantation Genetic Diagnosis* (New York: Springer, 2005). Kuliev discusses ethics in chapter 8, beginning on p. 189.

The development of chorionic villus sampling. In the mid-1970s Chinese scientists try it for determining the sex of the fetus: Department of Obstetrics and Gynecology, Tietung Hospital of Anshan Iron and Steel Company, Anshan. "Fetal sex prediction by sex chromatin of chorionic villi cells during early pregnancy," *Chinese Medical Journal* (English) 1 (1975): 117–126. In the early '80s Russian and Hungarian researchers report on their early experiments: Z. Kazy, A. M. Sztigar, V. A. Bacharev [Chorionic biopsy under immediate real-time (ultrasonic) control] *Orv Hetil* 121 (November 1980): 2765–2766 (Article in Hungarian); Z. Kazy, I. S. Rozovsky, V. A. Bakharev, "Chorion biopsy in early pregnancy: A method of early prenatal diagnosis for inherited disorders," *Prenatal Diagnosis* 2 (1982): 39–45. American publications follow, and in 1984 an overview appears with the following abstract (quoted in full): "Of all the methods of prenatal diagnosis the implications of chorionic villus biopsy are the most far-reaching and potentially controversial. DNA analysis of the fetus is now possible at the end of the first trimester and it can only be a matter of time before our knowledge of genetic disease markers makes the perfect race a feasibility." F. E. Loeffler, "Prenatal diagnosis: chorionic villus biopsy," *British Journal of Hospital Medicine* 31 (1984): 418–420. The same year Verlinsky and colleagues report on their first twenty-two successful procedures: A. V. Cadkin, N. A. Ginsberg, E. Pergament, Y. Verlinski, "Chorionic villi sampling: A new technique for detection of genetic abnormalities in the first trimester," *Radiology* 151 (1984): 159–162.

Verlinsky's technique for testing the "discarded" part of the ova: Yury Verlinsky, Norman Ginsberg, Aaron Lifchez, Jorge Valle, Jacob Moise, Charles M. Strom, "Analysis of the first polar body: Preconception genetic diagnosis," *Human Reproduction* 5 (1990): 826–829.

Martha Gellhorn on the Germans: Martha Gellhorn, "Ohne mich: Why I shall never return to Germany," *Granta 42: Krauts!* (1990).

An overview of European laws and attitudes on preimplantation genetics: S. Soini, "Preimplantation genetic diagnosis (PGD) in Europe: diversity of legislation a challenge to the community and its citizens," *Medicine and Law* 26 (June 2007): 309–323.

The deafness gene: Robert J. Morell, Ph.D., Hung Jeff Kim, MD, Linda J. Hood, Ph.D., Leah Goforth, MS, Karen Friderici, Ph.D., Rachel Fisher, Ph.D., Guy Van Camp, Ph.D., Charles I. Berlin, Ph.D., Carole Oddoux, Ph.D., Harry Ostrer, MD, Bronya Keats, Ph.D., Thomas B. Friedman, Ph.D., Theresa San Agustin, MD, Jan Dumon, MD, "Mutations in the connexin 26 gene (GJB2) among Ashkenazi Jews with nonsyndromic recessive deafness," *New England Journal of Medicine* 339 (1998): 1500–1505.

Number of abortions in Russia: I. I. Grebesheva, L. G. Kamsyuk, I. L. Alesina, "Kontratseptsiya glazami zhenshchiny," http://www.owl.ru/win/research/contrac.htm (accessed November 17, 2007).

Sex selection: Y. Verlinsky, S. Rechitsky, M. Freidine, J. Cieslak, C. Strom, A. Lifchez, "Birth of a healthy girl after preimplantation gender determination using a combination of polymerase chain reaction and fluorescent in situ hybridization analysis. Preimplantation Genetics Group," *Fertility and Sterility* 65 (February 1996): 358–360; A. Malpani, A. Malpani, D. Modi, "The use of preimplantation genetic diagnosis in sex selection for family balancing in India," *Reproductive Biomedicine Online* 4 (January–February 2002): 7–9; Judy Siegel-Itzkovich, "Israel allows sex selection of embryos for non-medical reasons," *British Medical Journal* 330 (May 28, 2005).

Chromosomal abnormalities and abortions: B. L. Shaffer, A. B. Caughey, M. E. Norton, "Variation in the decision to terminate pregnancy in the setting of fetal aneuploidy," *Prenatal Diagnosis* 8 (August 2006): 667–671; Arthur Robinson, Bruce Bender, Joyce Borelli, Mary Puck, James Salbenblatt, "Sex chromosomal anomalies: Prospective studies in children," *Journal of Behavioral Genetics* 13 (July 1983): 321–329.

breast cancer. *see also* breast/ovarian
    cancer-gene mutations
  age of onset, 7, 304n
  biopsies and, 76, 84
  BRCA1 and BRCA2 role in, 5–7,
    10–11
  incidence of, 294n
  in men, 7, 104, 289, 297n
  preventive mastectomy and, 11,
    13–14, 76–77, 83, 87–88, 89, 93,
    97–101, 108, 115, 124, 130–131,
    134, 136–137, 191–196, 295n
  risk of, 91, 294n
  risk of recurrence, 106–107
  role of genetic mutations in, 72–73
  screening for, 12, 71–74, 76–77, 78,
    84, 130
breast envelope, 192
breast-feeding, 11, 83, 115
breast/ovarian cancer-gene mutations
  breast and ovarian cancer onset and,
    5–7
  cancer family syndrome and,
    126–128, 130
  among Catholics in the American
    Southwest, 10, 290n
  clues indicating, 15–16
  consanguineous marriage and, 72–73
  decision to have children, 8–9
  ethnic basis of, 9–10, 18, 72–73
  founder mutation and, 9–10, 286
  groups of women sharing, 9–10, 13,
    107–116
  in Holland, 128–129
  identification of, 5–6
  nature of, 6–7
  passing to children, 8–9
  predictive testing for, 81, 86–87,
    88–89, 295n
  preventive mastectomy and, 11,
    13–14, 76–77, 83, 87–88, 89, 93,
    97–101, 108, 115, 124, 130–131,
    134, 136–137, 191–196, 295n
  preventive oophorectomy and, 11, 16,
    75–77, 82–83, 92, 93, 97–106,
    108–113, 115–116, 124, 130–131,
    134, 136–137, 295–296n

psychological approach to, 92–99
  risk of breast/ovarian cancer and,
    10–11, 17, 74, 115, 289n, 290n
  risk of inheriting, 5, 7
  risks of other cancers with, 17, 290n
  screening for, 7–9, 12, 17–18, 71–74,
    87–89, 124–125
breasts
  preventive mastectomy and, 11,
    13–14, 76–77, 83, 87–88, 89, 93,
    97–101, 108, 115, 124, 130–131,
    134, 136–137, 191–196, 295n
  role of, 11–13
Brigham and Women's Hospital
    (Boston), 12
Brunner, Han, 248–249
"bubble boy" disease, 204, 235, 273

CA-125 (cancer antigen 125), 102,
    103
caffeine, 263
CAG repeats, in Huntington's disease,
    141–142, 145, 165
Canavan disease, 20, 21, 179, 187
Cancer (journal), 120
cancer destiny, 123, 295n
cancer excess, 119
cancer family syndrome, 81, 234
  family information services and,
    123–136
  in Green Bay, Wisconsin, 118–136,
    189
  prophylactic surgery and, 119–121,
    130–137. *see also* mastectomy,
    preventive; oophorectomy
Cancer Risk and Prevention Program
    (Dana-Farber Cancer Institute),
    79–83, 101–102
Caplin, Andrew, 86–87
carriers
  defined, 285
  self-knowledge of status, 184–185
  testing, for Tay-Sachs disease,
    184–186
catatonia, 245–247
caudate nucleus, 142–143
cerebral palsy, 211–212, 217

"Mark DeYmaz is one of the most creative thinkers I have had the privilege of meeting. To some extent he is a futurist, and in *The Coming Revolution in Church Economics* he challenges local churches to think innovatively in order to fund and fulfill the Great Commission. Toward that end, I have spoken many times on the subject of generosity; this book, however, takes us a step beyond generosity alone as part of the equation. I highly recommend a thoughtful reading of this engaging challenge."

Ron Blue, cofounder of the National Christian Foundation, founder of Kingdom Advisors, and author of seventeen books on personal finance, including *Master Your Money*

"In *The Coming Revolution in Church Economics*, Mark and Harry give us a thoughtful path to innovation in church finance. But it's more than that—these proven leaders provide a practical plan to pursue greater mission through a better understanding and stewardship of resources. The fact is, churches and ministries can do more to advance the kingdom if they have more resources by which to do so. You and your leadership team will find this guide a helpful resource for effective ministry in the third millennium."

Ed Stetzer, Billy Graham Distinguished Chair, Wheaton College

"With the burning heart of a prophet and the well-worn hands of a seasoned pastor, Mark DeYmaz offers a clarion call to innovate and engage in the task of economic capacity building, without which the local church will not be sustainable, nor will its gospel mission flourish."

Tom Nelson, pastor of Christ Community Church, president of Made to Flourish, and author of *The Economics of Neighborly Love*

"This stimulating book is needed to stretch the financial imagination of every church in America. Immensely practical, it could have been titled *Innovative Ways to Expand Your Church's Mission by Strategically Leveraging Its Assets*. It gives hope to churches uncertain of their long-term sustainability. More important, it offers a fundamental, systemic, and perhaps necessary shift in asset awareness and stewardship to help churches better position themselves for the future."

Warren Bird, PhD, vice president of research,
Evangelical Council for Financial Accountability (ECFA),
and coauthor of thirty-one books, including
How to Break Church Growth Barriers

"I've stood in Mark and Harry's church and seen the reality of what they share in *The Coming Revolution in Church Economics*. Everyone who loves the local church and is concerned with its future funding and sustainability should read this book. It will cause a positive disruption to the way things have been done."

Ron Edmundson, CEO of Leadership Network, Dallas, TX

"The authors break new ground with a much-needed perspective on the financial opportunities and challenges facing the church in the twenty-first century. If the mark of a good book is that you cannot put it down until it is fully read, then this book meets that test! Combining practical wisdom and examples with sound economic thinking and realistic experience in the non-profit and for-profit worlds, this book provides a breath of fresh air for churches that are unsure how to financially sustain their mission. They do not provide an unrealistic 'get rich quick' silver bullet or a sketchy pyramid scheme; rather, they offer relevant, fresh, innovative, and entrepreneurial approaches. My prayer is that church planters and pastors will read this book and that the Holy Spirit will again breathe new vision and vitality into the church to transform society."

W. Jay Moon, director of the office of faith, works,
and economics, Asbury Seminary

"Mark and Harry's book isn't merely about creative ways to fund ministry. Rather, they provide practical pathways to establish more Christ-exalting, socially just, and economically responsible churches—everywhere. Local church pastors and planters need to know how to apply smart economics in today's fast-paced and rapidly changing economy. Those who lead church planting networks and endeavors need to grapple with the ideas and solutions they've developed. This book will show you how to carry mission into the future and, more than that, sustain the effort."

Daniel Yang, director of the Send Institute at the
Billy Graham Center, Wheaton College

"Mark DeYmaz is one of the most innovative, prophetic leaders in the church today. He asks tough questions and does not shrink from challenging conversations or difficult assignments. Rather, he leans into them for the sake of the kingdom . . . and for us. Local church leadership, too, must lean into changing economic realities and not bury its head in the sand by citing empty platitudes of hope that do not require intentional effort on our part. Now is the moment the church must develop additional revenue streams and resources to fund transformative ministry in the community. This book, then . . . Ready. Set. Let's GO!"

Rev. Dr. Stephanie Moore Hand, vitality strategist
for the Western North Carolina Conference
of the United Methodist Church

"Mark and Harry have done something in *The Coming Revolution in Church Economics* that is very rare but extremely important. With sensitivity and through innovation, they not only open the door but provide proven, practical insight to help local church pastors and ministry leaders rethink sustainability and apply forgotten methods of stewardship to maximize kingdom impact."

Santes Beatty, director of multiethnic ministries,
The Wesleyan Church

"Pastors of existing churches and church planters today are looking for additional funding to move the vision of the kingdom forward. In *The Coming Revolution in Church Economics*, Mark and Harry provide proven insight to help overcome stagnant giving and declining budgets by generating additional income to sustain ministry. How can we strategically reposition our churches as agents of transformational change in the community? How can the church develop more than one stream of income and increase its influence in the community? This book answers these questions, and many more."

Noemi Chavez, pastor of Revive Church, Long Beach, CA

"Mark DeYmaz hits the nail on the head! In the future, local churches will not be funded or sustained as they have been in the past. The sooner pastors and planters recognize this and adjust, the sooner their churches will be able to not only survive but thrive in a rapidly changing world."

Dr. Alejandro (Alex) Mandes, leader of the
All People Initiative of the EFCA, Immigrant Hope,
and GATEWAY Theological Institute

"Mark's take on past and present funding models and what will be required in the future to ensure local church sustainability is well worth understanding. I appreciate the book's sound theological reflection, as well as its practical instruction that pastors in every context will find immediately stimulating and applicable."

Bob Roberts, founder and senior pastor
of Northwood Church, Keller, TX

"This isn't theory. What you're about to read in this book are proven principles to bless your community, create multiple streams of income for your church, and better fulfill the Great Commission in today's changing world. Don't miss *The Coming Revolution in Church Economics*!"

Daniel Im, director of NewChurches.com, teaching pastor
of The Fellowship, Nashville, TN, and author
of *No Silver Bullets* and *You Are What You Do: And Six
Other Lies about Work, Life, and Love*

"In *The Coming Revolution in Church Economics*, Mark DeYmaz details a strategy that will enable churches to drive massive socioeconomic transformation in the communities they serve, advance their core mission, and generate income in pursuit of sustainability. This is a must-read for pastors seeking to reverse declining budgets and see their churches financially thrive."

Laura E. Owen, president of UpSkill, LLC, and former Secretary of Commerce, Kansas

"Entrepreneurship is critical to any economy, and it's an essential part of God's plan to restore the world. In *The Coming Revolution in Church Economics*, Mark DeYmaz sets the stage for a looming financial crisis facing the church in America and offers practical wisdom to guide local church leaders into a more sustainable future that benefits the church, the neighborhood, and the narrative of Christian benevolence in our age."

Luke Dooley, president of OCEAN Programs

"The American church stands poised to reclaim its entrepreneurial nature through enterprising endeavors. *The Coming Revolution in Church Economics* is a clarion call to pastors and churches, everywhere, to disrupt the erroneous sacred-secular divide of economics. We look forward to applying the instruction provided through Mark and Harry's prophetic voice."

John Parker, former banker and outreach director of Forest Hill Church, Charlotte, NC

"The local church needs financial resources to accomplish its mission. However, the subject is often difficult to discuss in a society where people perceive the church as only interested in their money. For church planters, *The Coming Revolution in Church Economics* is an instruction manual on how to practice, benefit from, and infuse smart economics into the DNA of your church. For mainline and other existing churches, it is a clarion call by which to recognize we cannot trust any longer in 'business as usual.' I believe that this book will spark a very new conversation and much-needed reflections on the future of local church funding."

Rev. Chip Freed, lead pastor of Garfield Memorial United Methodist Church, Cleveland, OH

"Mark and Harry provide practical insights in a time when a pivot in church economics is not only suggested but required. We don't have to figure this out alone; *The Coming Revolution in Church Economics* points the way."

"I believe Mark is speaking prophetically for today's church. For over fifteen years, I have facilitated sessions for churches going through change. I feel soon churches will have to embrace change at a pace they have never seen before. Mark is sounding the alarm, and I pray pastors pay attention."

"Mark DeYmaz is a cutting-edge practitioner of disruption within the American church. *The Coming Revolution in Church Economics* will help us stop saying we're going to change our cities and actually transform our communities through innovative financial strategies. Mark's new work will help us swim in kingdom hope instead of human fear."

THE COMING
REVOLUTION
IN CHURCH
ECONOMICS

Why Tithes and Offerings
Are No Longer Enough, and What
You Can Do about It

MARK DEYMAZ
WITH HARRY LI

BakerBooks

a division of Baker Publishing Group
Grand Rapids, Michigan

© 2019 by Mark DeYmaz

Published by Baker Books
a division of Baker Publishing Group
PO Box 6287, Grand Rapids, MI 49516-6287
www.bakerbooks.com

Printed in the United States of America

Library of Congress Cataloging-in-Publication Data
Names: DeYmaz, Mark, 1961– author.
Title: The coming revolution in church economics : why tithes and offerings are no
    longer enough, and what you can do about it / Mark DeYmaz, with Harry Li.
Description: Grand Rapids : Baker Books, a division of Baker Publishing Group,
    2019. | Includes bibliographical references
Identifiers: LCCN 2019015821 | ISBN 9780801094798 (pbk. : alk. paper)
Subjects: LCSH: Church finance.
Classification: LCC BV770 .D49 2019 | DDC 254/.8—dc23
LC record available at https://lccn.loc.gov/2019015821

Published in Association with the literary agency of Mark Swee-
ney & Associates.

19   20   21   22   23   24   25   7   6   5   4   3   2   1

In keeping with biblical principles of creation stewardship, Baker Publishing Group advocates the responsible use of our natural resources. As a member of the Green Press Initiative, our company uses recycled paper when possible. The text paper of this book is composed in part of post-consumer waste.

To the elders and the people
of Mosaic Church of Central Arkansas,
who have inspired, encouraged,
and supported disruptive innovation since 2001.
In so doing you have become a beacon of hope for others
daring to walk, work, and worship God together
as one on earth as it is in heaven,
beyond the distinctions of this world
that otherwise divide,
for the sake of the gospel.

CONTENTS

FOREWORD

My grandfather was a cowboy preacher dedicated to serving God and the hardscrabble working poor of the Texas Panhandle. My parents were small family entrepreneurs. Therefore, I grew up thinking that I needed to choose one life or the other . . . so I chose business. Over the next thirty years, I earned a CPA, became a partner in a global accounting firm, and served as the head of a large private investment banking practice. I also spent several years building the country's largest independent distributor of truck parts, sold it to AutoZone, and then retired at an early age.

When I was fifty years old, an Anglican bishop from Rwanda in East Africa challenged me to spend the rest of my life building businesses in his country. He said, "After the genocide, Rwanda had to rebuild every aspect of its society, but in the long run nothing is more important than creating jobs and a vibrant economy." He reasoned, "You're an investment banker and an entrepreneur. You build businesses, and help others build them. Surely God has prepared you to use business to build his kingdom."

That is why I am so excited about this book! Similarly, it will help you recognize opportunities, think like an entrepreneur, and create a vibrant economy in your church—one that will strengthen the financial bottom line and better position it for long-term kingdom and community impact.

In time, I accepted the bishop's invitation and began a sixteen-year journey of building a bridge from the United States to Rwanda that would transform lives at both ends. The organization we founded in 2007, Bridge2Rwanda, has helped encourage entrepreneurship, launch businesses, attract foreign investors, earn international scholarships, and develop a fellowship of globally educated young African leaders in Rwanda as well as in three of its conflict-torn East African neighbors.

As part of this journey, I have also had the opportunity to serve on the advisory council of Rwanda's president, Paul Kagame, and witness Rwanda's extraordinary demonstration of national reconciliation and robust nation building. In 2000, when the post-genocide conflicts had ceased and the nation had been secured, Rwanda's leaders across all sectors—public, private, social, and church—came together and created a vision of the country they wanted to build for themselves and their children. Working together, they have made spectacular progress, so much so that two years ago the United Nations named Rwanda the fastest-developing country in the world. It is a place to see God's hand at work for those with eyes of faith.

The church leaders I have worked with in Rwanda have been notably different from those I have met in traditional American churches. The Rwandans believe they and their churches have a critical role to play in building the country and are whole-heartedly committed to helping fulfill the national vision. Many are fearless social entrepreneurs who plunge into solving problems and meeting the needs of their communities and country, whatever they might be. They zealously search for and befriend widows, single parents, street kids, the sick, the poor, and other marginalized people in their communities. They comfort victims of the genocide and gently reconcile them with its perpetrators. Out of necessity and with scarce resources they build schools, hospitals, churches, homes, farms, and businesses. They mobilize others, innovate, and bring hope to desperate situations.

After working in Rwanda and with its church leaders for several years, my wife, Judi, and I began looking for a church in Little Rock that was not afraid to get out of its comfort zone, pursue fearless social entrepreneurship, and address current realities affecting both the American church and a rapidly changing society. We found these things in Mosaic Church of Central Arkansas. Like the churches in Rwanda, Mosaic does not compartmentalize spiritual, social, and financial concerns. Like church leaders in Rwanda, leaders at Mosaic are pursuing disruptive innovation in ways I believe all churches should and will have to do someday, not only to bring the hope of Jesus to desperate situations but also to remain financially solvent.[1]

With this in mind, I commend to you this important, informed, and groundbreaking work.

Dale Dawson
founder and CEO, Bridge2Rwanda
board of directors, Halftime Institute
Kigali, Rwanda, and Little Rock, Arkansas

INTRODUCTION

On May 17, 2001, Linda and I (Mark) responded in prayer to a very specific call of God on our lives. That day we committed ourselves and our families to a journey of faith, courage, and sacrifice that would lead to the establishment of a multiethnic and economically diverse church in the heart of central Arkansas, a church founded in response to the prayer of Jesus Christ for unity and patterned after the New Testament church at Antioch (Acts 11:19–25; 13:1), a church for all people—a church we called Mosaic.[1] By October 2002, Harry and Melanie Li, along with their three daughters, had joined us as colleagues in pursuit of the dream.

From the beginning, due to the multiethnic and missional nature of our church plant in Little Rock's urban center, tithes and offerings alone have never been enough to pay the bills, fund our vision, or expand our impact in the community. Therefore, we determined early on to get beyond tithes and offerings in order to establish the church and effectively engage the needs of people.

For more than twelve years (2003–15), we rented an abandoned building that formerly housed a Walmart. Today, we own and occupy a nearly 100,000 square-foot former Kmart, as we'll discuss further in chapter 6. At the time of this writing, we make more than $150,000 a year (12–13 percent of our annual budget) by renting parts of our facility to other businesses. The rent checks total $12,000 or so a month, almost enough to cover our $16,000-a-month mortgage payment. With approximately 18,000 square feet of space still available to rent, we are hoping to pick up an additional $4,000 a month in rental income so that soon we can look our people in the eye and say that not one penny of their tithes and offerings services the debt. In addition, $100,000 or so (just over 8 percent of our budget) comes each year in the form of staff support raised outside the church by members of our team and another $100,000 is aggregated through other means.

By shifting works of compassion and justice out from under the church's budget and into the budget of an umbrella nonprofit we launched in 2008, named Vine & Village, Mosaic has also been able to repurpose what would be another $150,000 to $200,000 in related ministry costs. Vine & Village is underwritten primarily through local, state, and federal grants, contributions from other churches, interested donors, and the like. Volunteers from outside the church also serve the city through Vine & Village in ways the people and staff of Mosaic might not otherwise have the capacity to do.

All totaled, approximately 30 percent of Mosaic's $1.2 million annual budget today is funded outside of tithes and offerings.

Conventional wisdom suggests that a church whose budget is not 100 percent supported by tithes and offerings is not sustainable. Recent research from Barna found that *church stability*—defined by church attendance and the percent of the budget comprised of tithes and offerings—is correlated with higher pastor salary, less personal debt, a location with a population that is less dense and sustained funding sources, (i.e., those without an end date). New churches in rural and suburban areas tend to be more stable, and those where the pastor is more amply compensated are more likely to become self-sustaining than those in urban areas or where the pastor subsidizes the ministry with his or her own money. But these indicators are not as common as they should be if we hope to see more new churches become self-sustaining communities of faith.[2]

Barna's right. These indicators are not as common as they should be, particularly where sustained funding is concerned. We believe, however, that a coming revolution in church economics will necessarily redefine the notion of church stability altogether, as an increasing number of congregations find they are not stable or sustainable by this definition. To expect that churches will be so apart otherwise, apart from a revolution in their approach to finance and funding, is to take for granted the status quo.

In addition, churches that rely solely on tithes and offerings unnecessarily limit their capacity. Imagine what your church could do if it generated an additional 30 percent of its revenue through for-profit business enterprise. On the other hand, imagine how much less your church would accomplish if it was forced to cut 30 percent of its budget, resulting in a significant reduction of staff, programming, community engagement, and giving to missionaries overseas. Such a scenario is already playing out in many churches across the United States as tithes and offerings are stagnant and/or in decline. Given the possibility (if not a probability) that local governments may someday assess church real estate to collect property taxes and/or that the federal government could take away tax-exempt status for churches altogether, we believe the time to pivot is now. In part, we have written this book to help you understand the why and how of positioning your church financially in preparation for any future changes to local or national policy.

Before continuing, it may be helpful for you to learn a bit more about Harry and me—that is, who we are and from where we've come. By the way, wherever first person is used throughout the book you should assume it's me (Mark) writing, unless otherwise noted.

Mark's Story

I was born out of wedlock in 1961 and the only child of a single parent, Dorothy, at a time in American history when only 6 percent of children were living without fathers in the home.[3] Because my mother had to work two jobs to support us, I was a "latchkey kid" before the term was popularized.[4] Though she had a full-time job working for the federal government in San Francisco, California, together my mother and I sold and delivered Avon on weeknights throughout a five-block territory in Alameda. In 1972, at eleven years of age and following our move to Phoenix, Arizona, I began sweeping the parking lot of a local convenience store. Two years later I became a dishwasher, and throughout junior and senior high school continued to work in the food industry, first by busing tables and then as a waiter. Simultaneously, I worked in rectories for the priests who lived in them, as part of work scholarship programs at the Catholic schools I attended.

It was from my mother that I learned how to gig and hustle, as well as to remain tenacious and resolved in seeking to overcome financial challenges. She taught me to value frugality, set my sights high, and trust God to provide for the dreams of my heart. Thus, I know what it's like to work hard. I've also learned to work smart, how to stretch dimes to dollars, and to obtain in time what I might not otherwise be expected to achieve or afford.

Through the years, then, I have journeyed from lower middle class to middle class, and following my mother's passing (2016) to upper middle class. But this does not make me a financial expert, though I continue to gain expertise. Rather, I share these things because context is important and, in this case, foundational to my understanding and approach to church economics. For instance, at times in my thirty-six years of full-time ministry, I have had to generate my own personal income, income for others, and the collective income to plant, grow, and sustain a healthy multiethnic and economically diverse church in the urban center of Little Rock, Arkansas. In other words, I have learned how to fund and advance effective ministry in spite of challenging financial circumstances, difficult contexts, and lack of resources. Today, I enjoy helping others overcome these obstacles as well.

Over the past eighteen years, desperation has led to innovation at Mosaic, and in this we are not alone. A still small but growing movement of economic innovation is under way in local churches throughout the United States, one that is indicative of a fundamental, necessary, and systemic shift that is taking place. Embracing this shift can help you lead a church from survival to stability and in time from stability to sustainability, where finances and funding are concerned.

Harry's Story

My father immigrated to this country from China in the late 1940s with $500 in his pocket. His own father died when he was young, and as the only son in a family of five he became the man of the house at an early age. Together, they lived destitute in a small town outside Shanghai. In time, pinning all their hopes and dreams on my father, his family sacrificed and sold nearly everything it had so that he could come to America and receive a Western education.

Years later and for me personally, growing up in the 1960s, the son of first-generation Chinese immigrants living in Tennessee, was quite an experience. Outside our home, I lived a good ol' southern boy life. I learned to fish, walk railroad tracks, play baseball, eat grits, and talk with a southern drawl. But inside our home, my parents maintained a very traditional Asian household. For instance, as children we were required to choose a musical instrument at an early age. Every day we were expected to practice our instruments and to finish our homework after school before being allowed to go outside and play. Likewise, performing poorly in school was never an option.

In terms of household economics, my father was lavishly generous when it came to spending money on things he strongly believed to be essential. Yet he was notoriously practical when it came to spending money on what he considered to be lesser priorities in life. Because of this, I grew up thinking we were poor. For example, we never used air-conditioning in the summer or heat in the winter. Instead, in the humid summers I would take cold showers, cool off in front of a fan, and jump into bed to fall asleep before I started sweating. In the winter it was so cold in our house that at times I could see my breath.

Given his mindset, my dad taught me to live without want, to always shop for the best deal, as well as to fix and maintain things for as long as possible in order to keep from having to buy something new. Many times he would rig up the most absurd Rube Goldberg solutions to maintain an ailing lawn mower or clothes dryer through one more season of service.[5] Only later in life did I learn that we were not poor. On the contrary, my father had invested all of his money in the stock market so that he could pay every penny of our education, for each university and post-graduate degree that I and my siblings aspired to earn . . . and he did so.

Prior to 2002, I was a tenured professor of engineering at the University of Idaho and living comfortably on the position's salary, supporting my wife and three children. So when Mark invited me to join the fledgling staff at Mosaic money was certainly a concern, but I refused to let it be a deal breaker. My first year's salary at the church was just $24,000. Nevertheless, God graciously sustained me and my family in those early years, and the church too, as he does to this day.

Why This Book?

Given our collective life experience and understanding of local church planting, growth, and development, we have written this book to share the why, how, and what of funding local churches beyond tithes and offerings. Ultimately, our purpose is to help your church improve its bottom line in consideration of resources, mission advance, and long-term gospel impact. Since most pastors are not taught economics in Bible college or seminary, we hope to propagate understanding that we believe will be necessary for pastors everywhere to apply going forward, whether they are serving established congregations or seeking to plant churches in urban, suburban or rural settings.

To be clear, we did not write this book to reduce your faith in God; to suggest that tithes, offerings, and generosity are not biblical expectations and therefore primary mechanisms for funding God's work; or to suggest that pastors and churches can get to a point where they are no longer dependent on God for money. Rather, this book is about developing and executing a plan to sustain life-giving ministry in and through the church over a long period of time, directing a greater percentage of tithes and offerings to missional community engagement, and establishing vibrant congregations in areas of a city where they might not otherwise exist. For instance, current understanding has led to an inordinate number of church planters "sensing a call" to start churches in more affluent neighborhoods throughout the United States. We want to disrupt such thinking and open minds to possibilities whereby people in every area of a city can benefit from access to a Christ-centered, exegetically sound, multiethnic, economically diverse, socially just, and economically sustainable church.

Throughout the following pages we will provide and explore seven specific directives to help you strengthen the financial viability of your church. In addition, we have included an afterword from Dr. Christopher Benek, the world's foremost

24

clergy expert on artificial intelligence (AI), whose comments will stretch your thinking even further in consideration of automation, the lives of individuals and, therefore, on church economics in the future.

So turn the page. Let's begin by understanding the need for revolution.

1

Understanding the Need

When any system fails to withstand the inrush of change, it fails to live up to its promise.

Travis G. Mason

Brotherhood Mutual is the second-largest provider of property and casualty insurance in the United States, primarily serving churches, schools, and camps. More than a commercial enterprise, the company views its insurance business as "a platform to accomplish [its] mission: to help America's churches and related ministries build the Kingdom."[1]

In August 2017, Brotherhood Mutual hosted its one hundredth anniversary celebration in Fort Wayne, Indiana, and invited me to participate. Mark A. Robison, its chairman of the board and president, had heard me speak earlier in the year and was specifically interested in my take on economics, sustainability, and the future of the local church, as I first wrote about in my book

Disruption: Repurposing the Church to Redeem the Community.[2] Other speakers at the event included my good friend Albert Tate, founding pastor of Monrovia Fellowship in Los Angeles, California; Brooke Hempell, senior vice president of research at the Barna Group; comedian Michael Jr.; and legendary financial adviser Ron Blue. In 1979, Ron founded a financial planning firm that grew to manage over $2 billion in assets for more than five thousand clients nationwide. He went on to launch Kingdom Advisors, an international network of Christian financial advisers, write seventeen books on personal finance from a biblical perspective, including the bestseller *Master Your Money*, and serve on a number of corporate boards such as Crown Financial Ministries and the National Christian Foundation.[3]

Having never met Ron, I wondered what he thought of my presentations. Following the event, I had the chance to ask him over lunch at the Fort Wayne International Airport. In a very positive and affirming tone he said, "Mark, I've been in financial services for fifty years. In all that time, I have never heard a pastor use the words *church* and *economics* in the same sentence. I'm no prophet, but I'd say you're on to something significant here, perhaps the early stages of a movement!"

Since that day, I've become increasingly convinced that Ron was right. Indeed, there is a growing movement and a need for disruptive innovation . . . a coming revolution in church economics.

Contributing Factors

As to the problem, there are at least four contributing factors driving the need for economic innovation in the local church today.

A Growing Burden on the Middle Class

Any objective consideration of the so-called American middle class faces a challenge when attempting to draw conclusions. The term is used widely to describe anywhere from 25 to 66 percent of households in the United States, depending on qualifiers such as income levels, educational achievements, aspirational values, and more. "One of the narrowest definitions limits it to those who are literally in the middle fifth of the nation's income ladder. A wider characterization includes everyone but the poorest 20 [percent] and the wealthiest 20 [percent]."[4] According to the Pew Research Center, America's middle class includes those making anywhere from two-thirds to two times the national median income ($46,960–$140,900 annually for a household of four).[5]

We should also note that inflation (rising prices for goods and services over time), higher standards of living, the increasing cost of education, the decreasing purchasing power of the United States Dollar (USD) coupled with widely available access to credit, has led Americans in general, and particularly the middle class, to borrow more money than ever before. According to the Federal Reserve Bank of New York (the New York Fed), "Total household debt—a category that includes mortgages, student loans, and car loans along with credit card and other debt—dipped in the wake of the Great Recession, but it has steadily rebounded in the years since. Overall, Americans' debt hit a new high of $13 trillion last year [2017], surpassing the previous record set in 2008 by $280 billion."[6] As one writer recently put it, "Americans have fallen back in love with debt."[7]

Figure 1. The Rise and Fall of Middle-Class Wealth

The share of total US wealth owned by the bottom 90 percent of families, 1917–2012

While debt is up, wealth is down. Mark Gongloff cites research to show that the "middle-class share of the US wealth pie roughly doubled between the Great Depression and the early 1980s. But since then, the middle-class share has shrunk back to its lowest level since 1947"[8] (see figure 1).

On a more personal note, when asked by CNN what being middle class meant to her, Connie White (age thirty-seven) of Turtletown, Tennessee, who owns a cleaning business with her husband, said this:

> Middle class, to me, means making your own way without assistance. I feel like being middle class should mean that you can relax a little, that you are saving for retirement and not still living paycheck to paycheck. I'm afraid that those days are gone, though. I feel I should be middle class. We're able to pay the bills and don't have a lot of debt. We have insurance. I feel lucky in that regard, but I feel we never get ahead. There's barely

anything left at the end of the month. With two people who work hard consistently, you shouldn't have to stress about money.[9]

America's middle class is largely understood to be the demographic group upon whose back churches are built and growth is dependent, through its investment of discretionary time and money. If so, and to the degree that Connie White echoes the current challenges, complexities, and conditions of America's middle class, churches will continue to struggle alongside it in the future.

Considering her words and sentiments we might restate them this way, with the local church in mind:

> I feel like our church should be able to relax a little, have some savings in the bank, and not live offering to offering. I'm afraid those days are gone, though. Sure, we can pay our bills, in spite of our debt, but we never get ahead. There's barely anything left at the end of the month. Needs are pressing, and even more demands on our budget are coming in the future. If everyone would just tithe or give generously, we wouldn't have to stress about money.

A growing burden on the middle class, coupled with a marginal increase in religious giving, as we'll next discuss, suggests that it's unrealistic to expect tithes and offerings alone to fully fund or sustain local churches in the years ahead.

There's the need for Embracing economic diversity - A Pathway to Funing Stability

A Marginal Increase in Religious Giving

As noted in the *Chronicle of Philanthropy*, "Religious giving [as of 2014] remains strong but is losing ground to giving to other causes. While religious organizations gained, generating $114.9 billion or 32 percent of total gifts, more than any other cause, they are slipping from previous years when they commanded 40 to 50 percent of all giving. Giving to religious causes was 56 percent of the total from 1985 to 1989 but just 33 percent from 2010 to 2014. Last year's growth was just a 0.9 percent increase."[10]

According to a more recent Giving USA report, giving to religion increased 2.9 percent in 2017 (0.7 percent when adjusted for inflation).[11] Yet this accounted for the lowest increase in overall charitable giving to the nine major categories of recipient organizations tracked by the report from year to year. Consider that in the same year, giving to arts, culture, and humanities increased by an estimated 8.7 percent (6.5 percent when adjusted for inflation), while giving to environmental and animal organizations increased by 7.2 percent when adjusted for inflation.[12]

As you might imagine, comparatively marginal increases in religious giving impacts local church finances. For example, a Faith Communities Today report found that the median church budget fell from $150,000 in 2009 to $125,000 in 2014.[13] With this in mind, Karl Vaters suggests that there are at least four assumptions pastors can no longer afford to make.[14] One of these involves new believers and their understanding of church funding. He writes, "Many new believers don't think about how the church's bills get paid any more than you spend time thinking about how you get email for free. . . . They think there's some unknown entity paying the main bills, while the money they give is for extras. We can't assume anything anymore. New believers need to be informed of the church's financial realities and their biblical responsibilities."[15]

Yes, they do. However, sustaining innovation based on assumptions rooted in the past will not likely be enough to provide for the financial needs of local churches hoping to grow, multiply, and impact their communities. What is needed is disruptive innovation where church funding is concerned.[16] The entirety of our understanding must change. Indeed, leaders today must learn to think and to act in ways different from the past if they are to see churches well-resourced and sustained in the future.

There are many reasons for only marginal increases in charitable giving to religious institutions. In a 2016 article for *Ministry Today*, Thom S. Rainer suggests at least six, including changing tax laws and declining frequency in attendance.[17] Two other reasons concern differences in generational giving as we'll next explore.

A Shift in Generational Approaches to Giving

According to Rainer, "Those born between 1980 and 2000 (Millennials) are often reticent givers. They want to be certain the church is a good steward of the contributed funds."[18] In addition, he writes, "Giving to the institution is the motive of most of the Builder generation (those born before 1946). Subsequent generations are more likely to give to a cause or a vision. As the Builder generation fades, so does institutionally motivated giving."[19]

Sobell and Co., LLC, is an accounting firm whose report "Generational Differences in Philanthropic Giving" is well worth the time it takes to read. According to its synopsis, the "white paper addresses the observable characteristics of four generations and their impact on charitable giving attitudes. . . . Recognizing the disparity between the generations' decision-making approaches can help a nonprofit determine what types of fundraising initiatives will be most effective for each target audience."[20]

Among other things, authors of the report highlight the following:

> A distinct difference in attitude [toward giving] emerges from a discussion with the different generations. The Silent Generation (born between 1929–1945) and the Baby Boomer Generation (born between 1946–1963) feel they make the biggest difference through the money they contribute. . . . The Silent Generation is more likely to assist religious organizations than any of the subsequent generations. . . . Where 45 percent of Boomers say their financial contribution is key, only 36 percent of Gen Xers (born between 1964–1981) and 25 percent of Gen Ys (Millennials, born after 1981) think that what matters most is a difference made by money. Instead, they believe that volunteering and spreading the word is more impactful.[21]

Given this information, a number of philanthropic consulting firms provide assistance to customers seeking to increase giving by understanding specific generations—how they think, where their passions lie, and what motivates them to give. For example, and specific to Generation X, Abigail L. Coleman shares that since this generation "is thought of as self-reliant, skeptical, savvy, and more traditional, a blended fundraising approach is likely to be most effective."[22] Citing a report titled "The Next Generation of American Giving," she provides strategies to consider when approaching Gen Xers for funding. They include:

1 • Demonstrate impact.

2 • Blend method of appeals.

3 • Solicit more than once per year.

She goes on to say, "The top two ways Gen Xers feel they can make a difference are by donating money and volunteering. Donating goods and spreading the word are other important methods of engagement. Fundraising and advocating are less important, each selected by just 4 [percent] of respondents."[23]

The fact is that "Fundraising is undergoing rapid and irrevocable change."[24] Understanding that emerging generations see, define, and value investment differently than previous generations means "that nonprofit organizations have to be ready for a whole new world of giving [see figure 2]. They will need to be innovative, positioning themselves to attract and engage with this global, impulsive, technology savvy generation . . . [as] fundraising [grows increasingly] reliant on a complicated and comprehensive mix of options."[25]

Throughout this book, we'll refer to the complicated and comprehensive mix of options as "multiple streams of income." Whether yet validated by your experience or not, churches will become more reliant on multiple streams of income in the future and soon need to leverage their assets in order to generate a financial return on investment (ROI).

A Rapidly Changing Population and Demographic

When it comes to understanding and adapting to rapidly changing demographics, the American church remains largely aloof, resistant, and unprepared for what lies ahead. Failure to

Figure 2. Priority Cause by Generation[26]

Cause	Gen Y	Gen X	Baby Boomers	Silent Gen
Local social services	19%	29%	36%	37%
Place of worship	22%	36%	38%	37%
Health charities	20%	24%	19%	23%
Children's charities	29%	28%	22%	20%
Education	17%	16%	14%	20%
Emergency relief	11%	18%	11%	19%
Animal rescue/ protection	16%	21%	18%	11%
Troops/veterans	6%	10%	15%	22%
Environment	4%	10%	9%	9%
First responders	6%	8%	10%	9%
Arts	4%	7%	6%	14%
Advocacy	6%	4%	5%	10%
Election campaigns	2%	2%	2%	5%
Human rights	12%	10%	4%	6%
Victims of crime/abuse	7%	6%	3%	4%

recognize, validate, and adjust to seismic shifts already afoot will lead to even more churches losing ground in the years ahead. Again, the time to pivot is now. Over the next fifteen to twenty-five years, a rapidly changing population will have a significant impact on local churches in the United States. Among other things, church planting, growth, and development must be necessarily altered if local congregations are to remain credible and viable in the years ahead.

For example, according to a U.S. Census Bureau report titled "Demographic Turning Points for the United States: Population Projections for 2020 to 2060," "The rate of population growth is slowing."[27]

- 2.3 million per year—projected growth rate of US population 2010–30
- 1.8 million per year—projected growth rate of US population 2030–40
- 1.5 million per year—projected growth rate of US population 2040–60

In addition, the population is aging.[28]

- By 2030, one out of five Americans will be sixty-five years or older.
- By 2035, older adults are expected to outnumber children for the first time in US history.
- Between 2016 and 2060, the population under age eighteen is projected to grow by only six million people, compared with a growth of forty-six million for the population sixty-five years and over.

Finally, the report demonstrates that the population is changing in terms of racial and ethnic pluralism (see figure 3).[29]

- "By 2020, fewer than one-half of children in the United States are projected to be non-Hispanic White. . . . The non-Hispanic White-alone population is projected to shrink over the coming decades, from 199 million in 2020 to 179 million in 2060—even as the United States population continues to grow."
- "By 2028, the foreign-born share of the US population is projected to be higher than any time since 1850. . . .

Just two years later, by 2030, net international migration is expected to become the primary driver of population growth in the United States—another demographic milestone for the country."

- "'Two or More Races' is projected to be the fastest-growing racial or ethnic group over the next several decades, followed by Asians and Hispanics. . . . The share of children who are 'Two or More Races' is projected to more than double in coming decades, from 5.3 percent today to 11.3 percent in 2060."

Figure 3. Foreign-Born People Living in the US 1850–2010, Projected 2020–2060

By 2028, the foreign-born share of the US population is expected to be higher than at any time since 1850.

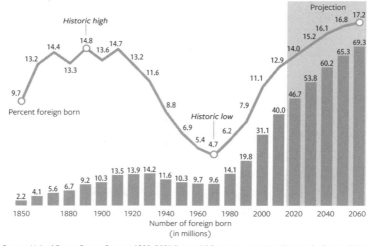

Source: United States Census Bureau, 1850–2000 Decennial Censuses, American Community Survey 2010, 2017 National Population Projections for 2020–2060

In short, the "2030s are projected to be a transformative decade for the U.S. population. The population is expected to grow at a slower pace, age considerably, and become more racially and ethnically diverse. Net international migration is projected to overtake natural increase in 2030 as the primary driver of population growth in the United States, another demographic first."[30]

As with the growing burden on America's middle class, marginal increases in religious giving, and a shift in generational approaches to giving, rapidly changing demographics—given income disparity between people groups—will greatly affect church budgets. Such things will likely lead to a loss in revenue and to a fundamental reshaping of the way congregations are funded.

Lean In

As we've shared, we are not formally trained economists. Rather, what we've learned about church economics has been biblically and experientially informed over the past eighteen years. Desperation led to innovation on our part in seeking to establish a multiethnic and economically diverse church in an underresourced area of our city, an area to which organizations rarely send church planters due to the challenges of funding and sustainability. In fact, due to the financial risks associated with new church starts as well as erroneous definitions and metrics of success, such as the myth of sustainability that we'll discuss in chapter 4, most denominations and church-planting networks remain highly skeptical and/or reluctant to partner with pastors seeking to do what we have done. This leaves people in many parts of a city disenfranchised from the hope and promise of the gospel that churches like ours can deliver—the very hope and promise that those stuck in cycles of injustice and generational poverty need most. Surely we can and must do better in the future, for the sake of the gospel.

In this opening chapter, we've identified and discussed some of the contributing factors that are driving the need for economic innovation in the church today. With that as a goal, to lean into the coming revolution you must first free your mind.

Free Your Mind

The First Directive

There is a reason this chapter is not titled, "Think Outside the Box." Let me tell you why.

Sometime in the late 1960s, my mother came home from work one day and challenged me to solve a geometric puzzle. The puzzle was one that had been introduced to her earlier that day as part of an exercise in professional development. Today known as the *Nine Dot Problem*, it is a classic thinking exercise that gained widespread popularity in the 1970s and 1980s. "Participants are presented with a set of dots arranged in a 3x3 grid and challenged to connect all nine dots without lifting their pencil from the paper using the fewest possible number of straight lines."[1] If you have never seen it before, give it a try.

Figure 4. The Nine Dot Problem

As you might know, the only way to solve the puzzle (we're told) is to think outside the box—that is, to recognize that in attempting to connect the dots we are not required to remain within the perceived box formed by the nine dots to which our minds naturally default. Therefore, the solution requires at least one of the lines to extend beyond this self-imposed restriction.

While there are a number of ways to solve the puzzle, here's how my mother taught me to do so many years ago.

Figure 5. The Nine Dot Problem Solved

The *Nine Dot Problem* is commonly used to encourage unrestricted reasoning in the face of challenges—that is, outside-the-box thinking—in order to overcome obstacles. It turns out, however, that simply telling someone to think outside the box does not necessarily lead them to effective solutions.

The Need for Detailed Strategic Instruction

In his article, "Why Thinking 'Outside the Box' Doesn't Work," Mark McGuinness writes, "The trouble with the usual way of presenting the *Nine Dot Problem* is that it contains an unexamined assumption: i.e., that all we have to do is tell people they can go outside the box and they will find it easy to solve the problem."[2] Yet, as research has shown, this is not true.

Cognitive psychologist Robert W. Weisberg, in his book *Creativity—Beyond the Myth of Genius,*

> describes two experiments in which people were told that the only way to solve the (nine dot) problem was to draw lines outside the square. Contrary to the "outside the box" school of thought, this did not make the problem easy to solve. In fact, only 20–25 percent of subjects were able to solve the problem, even though all of them allowed themselves to draw outside the box. And even the ones who did solve the problem took a long time to do so, and used trial and error, making many different drawings, rather than any special form of "creative thinking." Researchers went on to show that the success rate could be improved by giving subjects prior training in solving simpler line-and-dot problems, and also by giving them "detailed strategy instructions" about how to solve the problem.[3]

Did you catch that?

- When faced with a challenge, simply "thinking outside the box" is not enough to overcome it. Apart from an informed approach to perplexing problems, most people are not able to imagine or implement effective solutions.
- Attempts to solve problems by thinking outside the box typically fail and/or solutions are discovered only after considerable trial and error. In a professional setting, valuable time, effort, and resources can be squandered in this way and lead to a lessening of confidence in leadership.

- Experience in solving similar yet simpler problems, together with detailed strategic instruction on the front
  end, improves the odds of success in problem solving.

In short, to overcome the problem of stagnant or declining
tithes and offerings—a problem your church may be facing at
this moment or perhaps one it will face in the future—it's not
enough to think outside the box. Rather, you must free your
mind.

- Consider the detailed strategic instruction provided in
  this book, rooted in experiential knowledge gained from
  financial problem solving over the course of eighteen
  years in the challenging context of Mosaic.
- Imagine without restriction what a better understanding of economics might mean for your church
  today and what opportunities might be worth pursuing
  in light of it tomorrow.
- Solve similar but simpler financial challenges first before pursuing bolder (riskier) initiatives in order to
  increase the odds of success in learning how to overcome the problem of stagnant or declining tithes and
  offerings.

Overcoming Common Objections

No discussion of what is possible in this regard should begin
apart from or in the absence of sound theological reflection.
Specifically, let's take a moment to address a few of the more
common objections, assumptions, and biblical passages that

well-meaning believers will likely cite to denounce leaders who propose leveraging church assets to generate income beyond tithes and offerings, no matter the motive.

Money Is (Not) the Root of All Evil (1 Tim. 6:10)

Assumption: Pursuing profit through a church is intrinsically and morally repugnant.

In this often-misquoted verse, the apostle Paul actually writes, "For the love of money is a root of all sorts of evil, and some by longing for it have wandered away from the faith and pierced themselves with many griefs" (1 Tim. 6:10). With this in mind, we should pause to note the following truths:

- The love of money is *a* (but not *the* only) root of all sorts of evil.
- *Some* who long for money *may* (but do not always) wander away from the faith.
- *Some* who long for money *can* (but do not always) bring grief upon themselves.

Along this line, Chris Brown writes, "Our attitude toward money is the problem, not the money itself. Money is morally neutral. Throughout the Bible, both the impoverished and the wealthy are considered godly. Job very well could have been the wealthiest person in the world during his time, but the Bible is clear that he was a godly man. . . . In fact, sometimes those who were materially blessed, like Job, were also blessed by God with the power to carry out His work. God clearly doesn't equate money to evil."[4] As Brown suggests, money itself is not evil, nor does God see it that way. It is the *love* of money that can be problematic. Therefore, we are instructed and will need to avoid the *love of* money and/or a *longing for* money in terms of our attitude toward it . . . but not money itself.

To be clear, we are not encouraging pastors or churches to pursue money at the expense of faith, to become double-minded, or to acquire money in order to build their own kingdoms. Far from it! Rather, we are encouraging a more thoughtful, honest, and smarter approach to money where church economics are concerned—that is, a better and wiser stewardship of assets in order to advance the common good, strengthen the local church, and bring the hope of the gospel to people and communities who need it the most.

Beyond its use as an expression of love and devotion, or as given in obedience to God, money is not a necessary evil, it's a necessary tool. As carpenters need hammers, saws, tape measures, and more to do their work, so too a local church needs tools to fulfill its general mission, unique calling, and specific vision . . . and money is certainly one of those tools. In many if not most places throughout the world today churches depend on money to exist. Despite how some prefer to discuss it (somewhat idealistically, as if material and spiritual things are mutually exclusive) or not to discuss it (as if to do so is somehow sinful or self-serving), there is nothing wrong with admitting that a local church needs money, and in many cases more not less of it. This is why we must proactively engage the coming revolution in church economics.

The fact is that material and spiritual things are not mutually exclusive, nor is any frank or pragmatic discussion of money necessarily self-serving.

- Jesus and his disciples needed money (Luke 8:1–3).
- The apostle Paul needed money (2 Cor. 11:7–9; Phil. 4:14–18).
- New Testament churches needed money (Acts 11:27–30; 1 Cor. 16:1–3).
- Similarly, pastors and churches today need money.

Can such thinking become problematic? Can it be leveraged for evil? Can it lead some pastors to wander away from the faith or bring upon themselves much grief? In all fairness, yes it can. As Jesus himself said, "No one can serve two masters; for either he will hate the one and love the other, or he will be devoted to one and despise the other. You cannot serve God and wealth" (Matt. 6:24). Following Luke's account of these words of Jesus (Luke 16:13), we gain even more insight and context. He writes, "Now the Pharisees, who were lovers of money, were listening to all these things and were scoffing at Him. And [Jesus] said to them, 'You are those who justify yourselves in the sight of men, but God knows your hearts; for that which is highly esteemed among men is detestable in the sight of God'" (Luke 16:14–15).

So, yes, we should and must proceed with caution in any attempt to leverage church assets to generate sustainable income. But we should also recognize that such dangers as we've just discussed or that otherwise stem from misunderstanding, mismanagement, or misappropriation of money, are already real and present in the American church. Fear, then, of what some pastors today are doing with money, what some have done with it in the past, or what some might do with it in the future to build their own kingdoms, should not keep the rest of us from leveraging assets to bless the community and help fund mission. Like the men of Issachar, we must understand the times and do what is right in this moment (1 Chron. 12:32). In fact, a greater understanding and application of church economics can help pastors avoid such problems altogether by way of increased expectations and accountability.

Ask and It Will Be Given to You (Matt. 7:7)

Assumption: A church is faith based and should wholly depend on God for its funding.

Suggesting that the local church should pursue additional income beyond tithes and offerings does not in any way imply a lack of faith in God to supply for its needs.

Think about it.

- We go to a doctor when we're sick, take prescription drugs as prescribed, and consent to surgery to improve our health. These actions do not imply a lack of faith on our part, do they? Surely, God does not expect us only to pray for healing as we'll inevitably do in such moments, though prayer is an essential act of faith, a petition of hope, and an expression of surrender to his will.

- We apply for jobs, work hard for our paychecks, and look for ways to make even more money in order to better provide for our families. Do such things necessarily imply we are depending on ourselves, and not on God, for "daily bread?"

- We acquire debt to buy a car or home, cosign school loans to help our children pay for college, and/or invest money in the stock market. Does such borrowing and investing mean we do not trust God?

Therefore, as with individuals, churches should not be saddled by erroneous thinking concerning money and faith. Taking intentional steps for one's livelihood and the well-being of his or her family does not render faith inoperative. On the contrary and by design, God expects us at once to ask, seek, and trust him for all things in life while at the same time expecting us to be deliberate, responsible, and accountable for the resources he entrusts to us. Anything less runs contrary to the teaching of Christ in what is commonly referred to as the parable of the talents (Matt. 25:14–30). Here's the story in our own words.

A man leaving on a journey entrusts to three of his servants money (talents), expecting each of them to steward it well according to their ability—that is, to generate for him ROI according to their experiential knowledge and capacity. To one servant he gives five talents, and to another he gives two talents. These servants wisely trade (invest) these talents, seeking ROI on behalf of their master. To a third servant the master gives only one talent. Unlike the other two servants, the third servant does not act wisely in stewarding the talent. Rather, he digs a hole in the ground and buries it.

Upon the master's return, the first two servants report ROI. Through wise stewardship, these servants doubled what was entrusted to them, to the great delight of their master. Having proven themselves "good and faithful" servants and as a reward for their efforts, they are given responsible authority for many more things in their master's household.

The third servant, however, informs his master that he buried the one talent given to him in fear of a negative reaction should it have been lost through bad investment, stolen, or otherwise squandered.

In the master's rebuke of the third servant, there are at least three key principles that can help us better understand what God expects of those similarly entrusted to steward church assets.

- The master expected the servant to leverage, not bury, the talent.
- The master expected the servant to steward the asset in such a way that it would generate ROI.
- The master took away from the servant what he had previously entrusted to him and gave the asset to another—a good and faithful servant who had proven himself more effective, wise, and responsible.

Jesus concludes by saying, "For to everyone who has [acted wisely by leveraging the assets entrusted to them], more shall be given [to them], and [they] will have [favor and blessing in] abundance; but from [those] who do not have [the wisdom to steward what I have already entrusted to them], even what [they have been given to steward] shall be taken away" (Matt. 25:29, additions ours).

To date, conventional wisdom and teaching in the church about money, stewardship, and investment have been primarily concerned with individuals. They have focused on helping people prioritize spending, pay down debt, and give generously. This is a good thing. In the future, such concern for stewardship and investment should also apply to the collective local church.

The American church typically describes good stewardship as taking care of facilities, properly accounting for income and expenses, and clearly communicating with members and donors. Going forward, our understanding of good stewardship should also include the biblical expectation of knowing what to do with the assets entrusted to us. The collective church, itself, must prove to be a good and faithful servant by leveraging assets to bless the community and generate sustainable income. Indeed, pastors and churches can no longer afford to sit on their assets!

Of course, there will always be people who reject this notion and suggest that all we must do is pray, have faith, and be patient while waiting on God to provide for the church in his way and in his time. To so suggest, however, is to overspiritualize the matter, dismiss complexities, and fail to recognize nuance. Yes prayer, faith, and patience are essential to everything we pursue in, for, and through the church. Yet more is involved, and there is more that can and should be done by leaders tasked with fiscal responsibility.

When you hear people talk this way keep in mind that they have likely never had to carry the burden of meeting fixed expenses where others (outside their families) are concerned, making payroll twice a month, or maintaining a public facility. Tremendous pressure exists on every leader tasked with these demands and with the responsibility of aggregating money to sustain a church. Like the master who praised two of his three servants for being good and faithful, people in the body should bless and support church leaders for their efforts and ingenuity in this regard, and not otherwise criticize or undermine them.

As expressed in the parable of the talents, God expects church leaders to leverage and wisely steward the assets they've been given in order to generate spiritual, social, and financial ROI. In so doing, churches will bear more fruit than tithes and offerings alone might otherwise yield. This doesn't mean that pastors should play the stock market with church finances! It does mean that pastors are free to consider collective stewardship and should do so in order to position the church for long-term impact beyond their tenure at the helm.

My House Will Be Called a House of Prayer (Matt. 21:12-13)

> Assumption: A church should not conduct
> profitable business on its property.

No other incident in the Bible seems to so clearly suggest that a church should not engage in profitable business enterprise—certainly not on its premises—as the one in which merchants and money changers are expelled from the temple by Jesus.[5] In terms of pushback, this account (as recorded in all four Gospels) may be the primary root of any objection you encounter from those reluctant to embrace the thinking or outcomes envisioned in this book.

Upon closer (exegetical) examination, however, Jesus was not reacting to the making of fair or reasonable profit. Rather, he was protesting unfair and greedy profiteering on temple grounds—namely, economic injustice.

Carefully consider this account of the incident as recorded in the Gospel of Mark: "On reaching Jerusalem, Jesus entered the temple courts and began driving out those who were buying and selling there. He overturned the tables of the money changers and the benches of those selling doves, and would not allow anyone to carry merchandise through the temple courts. And as he taught them, he said, 'Is it not written: "My house will be called a house of prayer for all nations" [Isa. 56:7]? But you have made it 'a den of robbers' [Jer. 7:11]" (11:15–17 NIV).

In and through prophetic action, Jesus spoke truth to power by condemning the merchants and money changers, and more directly temple leaders, who were perpetuating economic injustice. How so? By taking financial advantage of others—worshipers who had traveled great distances to be there and especially the poor. Without question, the greedy practices of the merchants and money changers would have been sanctioned by temple leaders and conducted with their blessing for their own financial gain, as well. Yet the inequitable trade placed undue burden on good people seeking to worship God in the temple where they had come to pray, sacrifice, and give alms according to Jewish (religious) law.

In those days, Judea was ruled by the Romans, and Roman coinage was commonly used throughout the region for monetary transactions. At the temple, "Jewish law required that every man should pay a tribute to the service of the sanctuary of 'half a shekel'—a Jewish coin. Therefore, it became a matter of convenience to have a place (in the temple) where Roman coins could be exchanged for Jewish half shekels. The moneychangers provided this convenience."[6] But in so doing, they charged inflated fees and used rates of exchange well beyond what was reasonable. In part, this is why Jesus labeled them "robbers." "Because so many thousands came up for the great feasts in

Jerusalem, exchanging money was a very profitable business—one that resulted in fraud and oppression of the poor."[7]

You may have experienced something similar (though not necessarily fraudulent or oppressive) when traveling overseas, for instance, if you waited until the last minute to convert foreign currency and had to do so on your way home at an airport. In that moment, you're a captive audience. This is why airport-based currency exchange shops typically calculate the cost of converting money at above average rates and will charge higher fees for their services than you might otherwise find at reputable banks inside a country. Without any other options, travelers lose in the transaction. It is no different than being at a professional sports arena on game day and having to pay $5 for a hot dog that might elsewhere cost $1.50.

In addition to the money changers, merchants were also in the temple selling specific animals that would have been required for varying sacrifices, as offerings to atone for sin. Like the money changers, these merchants were taking advantage of worshipers by significantly and unjustifiably marking up the merchandise. Their greed was impeding the kind of worship that God intended in what was to be a house of prayer for all the nations and thus for everyone, regardless of ethnic origin, social standing, or any other cultural distinction.

Certified financial coach Joe Plemon, addressing the question "Why did Jesus drive the money changers from the temple?" writes, "Two things were essential for these foreign worshippers: an animal to sacrifice and Temple coins which local merchants would accept. There was probably nothing wrong with selling animals or operating a legitimate money exchange, but, according to Jesus, these particular vendors were a 'den of robbers' who undoubtedly charged exorbitant rates, thus taking advantage of those who seemingly had no other options."[8] In short, "Jesus occupied the temple because the people in charge had lost the script and were perpetuating injustice."[9]

Insofar as such insight affects the church today, I have previously written: "Based on a misunderstanding of this passage, concerns over how for-profit business enterprise may affect a church's tax-exempt status or to avoid the appearance of impropriety, churches have been reluctant to consider ways in which to [generate profitable income. . . . Yet by doing so], and to the degree that its heart and purposes are not self-serving, a local church can help bring systemic change and redemptive hope to a community in need."[10]

Indeed, when leaders leverage church assets to stimulate economic recovery, repurpose abandoned property, encourage small business, create jobs, and more, they help to reduce local crime, generate tax revenue, advance the gospel, and promote the common good, all in the name of Christ while simultaneously producing fair, reasonable, and sustainable income (profit) to help fund long-term mission.

In addition (or perhaps in response) to such misunderstanding and depending on cultural norms, denominational affiliation or theological sensitivities, many churches consider their property and buildings to be "holy ground." In these churches, specific spaces are often set apart solely for spiritual purposes and deemed off-limits to other activities, certainly to profitable ones that might otherwise be conducted in them.

Of course, there's nothing wrong with churches reserving, designating, or even dedicating specific spaces, such as a sanctuary, for spiritual purposes. The people of Israel did so—first in the wilderness and later in the temple. However, it's important to note that the New Testament does not impose any such expectations on a local church concerning the space or place in which it meets. Therefore, wherever the use of church facilities is restricted in this way we should recognize that limitations have been set by people in the past who once held responsible positions of authority in the congregation. As necessary or desired, people in the present should be able to request or initiate a change in policy.

With this in mind, let's be clear. The sacred nature of the church rests not in a specific space, place, or building but rather in people of Christ-centered faith "called out" (Gr. *ekklesia*) and "gathered" (Gr. *synago*) for his purposes whenever and wherever they meet. For instance, Jesus did not say "If you seek me, you will find me only in a church's sanctuary." He said, "Where two or three have gathered together in My name, I am there in their midst" (Matt. 18:20). In other words, Jesus is with us in whatever place we gather to fulfill sacred purposes—as in coming together to seek and to save the lost, to restore the fallen, to prayerfully petition the Father, and to extend forgiveness in his name (Matt. 18:10–35). Later in the New Testament, Luke includes teaching, fellowship, the breaking of bread, prayer, and worship as sacred purposes for our gathering as well (Acts 2:42–47).

So, yes . . . technically, holy ground is established whenever and wherever the people of God gather for such purposes. Practically, however, people perceive church facilities as sacred ground and these perceptions cannot be ignored in any adaptive use of a building to generate sustainable income. We learned this the hard way at Mosaic.

In the summer of 2018, we began renting the facility we own under the name Rock City Events and Conference Center. Having developed a partnership with a local event planner through an MOU (memo of understanding), we quickly began hosting events and receiving rental income. Of the twelve or so events we hosted by the end of that year, all but two went according to plan. In those two cases more than a little confusion was experienced by some in the church, particularly by our Hispanic leadership team. Without going into detail, suffice it to say that we were told one thing by the people renting from us when, in fact, what actually took place went well beyond the contractual agreement. To make things right, we had to address problematic issues from a business standpoint as well as the personal, cultural, and spiritual concerns of others in our church regarding what amounted to a misuse of the space. In so doing, Harry and I met with our diverse staff and later with lay leaders from the Hispanic community. In that meeting we listened, lamented, apologized, and washed the feet of every person in the room.

Throughout the years, Harry and I "have washed the feet of many others at Mosaic too, both in good times and bad, sometimes in heartfelt humility, and at other times when hurt feelings, raw emotions, and misunderstandings still lingered in the room. By way of practical advice, [we] can tell you that washing the feet of another in the midst of tension is not an easy thing to do. It is an act of submission and humility that reveals more of your own pride than you might otherwise imagine. Yet every single time that [we] have done so in the midst of conflict, it has defused anger and silenced the room. Every single time it has honored God."[11]

In spite of inevitable challenges along the way, we should not dismiss as too difficult the task of leveraging church assets to bless the community and help fund sustainable mission, as we'll discuss more specifically in subsequent chapters.

Render to Caesar What Is Caesar's and to God What Is God's (Matt. 22:21)

Assumption: A church should avoid paying taxes and jeopardizing its nonprofit status.

Jesus made this familiar declaration in response to a question of the Pharisees (Matt. 22:15–22). In short, the question was asked to set a trap by which the Pharisees hoped to discredit Jesus and ultimately justify his arrest: "Is it lawful to give a poll-tax to Caesar, or not?" (v. 17)

Jeannine K. Brown, professor of New Testament at Bethel Seminary, notes, "The tax in view is the census tax, a per person tax of a denarius (Matt. 22:19). The conundrum for Jesus is this: If he answers yes, then he could be perceived as in collusion with Rome, justifying Roman occupation and oppression of the Jews. This would not be a popular answer among the Jewish people. On the other hand, if Jesus answers no, he could be suspected of revolutionary sentiment against Rome."[12]

In reply, Jesus explained that living under and in submission to governing authorities on earth does not necessarily imply a lack of submission to God, as if allegiances are mutually exclusive. "The first clause on its own indicates that the tax should be paid, since the emperor's image and inscription on the coin would cause it to fall under things that are the emperor's. On the other hand, the final clause places a question mark on what belongs to whom! . . . The beauty of Jesus' answer is that he both concedes payment of the census tax while subverting the reach of the emperor."[13]

In Romans 13:1, Paul writes, "Every person is to be in subjection to the governing authorities. For there is no authority except from God, and those which exist are established by God." Of course, this holds true for churches too. In other words, if and when required to do so, pay the taxes.

That said, Western governments have provided tax advantages to the church in one form or another since the time of Constantine "based on the rationale that the church relieved the state of some governmental functions, and therefore deserved a benefit in return."[14] Throughout history, at various times and in various places, these tax advantages have included one or more of the following and/or other similar benefits:

- exempting churches from paying property taxes
- exempting churches from paying federal income tax
- exempting individuals from paying taxes on monies donated to the church

In the United States, it wasn't until 1894 that Congress formally acted to exempt churches from paying federal income tax, a law that was later declared unconstitutional but "reinstated by the Revenue Act of 1913, which defined the modern American income tax system. On January 14, 1924, the US Supreme Court interpreted the reason for the exemption in *Trinidad v. Sagrada Orden*: 'Evidently the exemption is made in recognition of the benefit which the public derives from churches' 'corporate activities.'"[15]

Here's the point. What government grants, it can retract. Therefore, now is the time to ask if the church is prepared for any future changes in policy that might affect its current economic model and proactively adjust. Given population growth, rising costs, and inflation across the board as well as the country's commitment to entitlements such as Social Security and Medicare—which make up approximately two-thirds of the federal budget—responsible authorities at all levels (local, state, and federal) are today looking for ways to increase revenue, not because it would be nice but because it's increasingly necessary to do so. For example:

> By 2034, the Social Security Trustees project that both the OASI and DI trust funds (the Old-Age and Survivors Insurance Trust Fund, and the Disability Insurance Trust Fund—separate accounts in the United States Treasury which provide the automatic spending authority to pay monthly benefits to those eligible) will be fully depleted. Once the trust funds are exhausted, the Social Security Administration will be limited to spending only as much as incoming revenues; that will put 86 million beneficiaries at risk of an across-the-board benefit cut. Retirees could face an immediate 23 percent cut in their scheduled benefits upon depletion; the average retiree would incur an immediate benefit cut of around $6,100 in today's dollars. The DI trust fund is expected to be exhausted in 2032, at which point it would only be able to pay 96 percent of its obligations (a 4 percent reduction). On a combined basis, cuts to the two programs would be 21 percent.[16]

Couple this with the Tax Cuts and Jobs Act (TCJA) of 2017, and "ministers and other church leaders [wonder] about the new law's implications for ministry [in the future]."[17] With this in mind, editors of the *Church Law & Tax Report* recently addressed the question "Will charitable contributions suffer?" In the May/June 2018 issue of the report, senior editor Richard R. Hammar wrote:

The Tax Cuts and Jobs Act retains a deduction for charitable contributions, but the deduction will be available to a smaller number of donors because of a substantial increase in the standard deduction. The significantly increased standard deduction will reduce the number of persons who are able to itemize deductions on Schedule A (Form 1040) from 30 percent to as few as 5 percent of all taxpayers. The result will be a significant decrease in the number of taxpayers who can claim a deduction for contributions they make to churches and other charities. The bottom line: 95 percent of taxpayers have lost their ability to receive a charitable deduction. Many speculate that this will negatively affect giving. Will it?[18]

To address growing financial needs and with an eye to increasing tax revenue, local governments are already rethinking the tax-exempt status of churches in relation to buildings and property they own. In Long Beach, California, for instance, the city all but prevents churches from acquiring and repurposing available or abandoned property, in large part due to the potential loss of tax revenue. Such regulation stifles new church starts and challenges pastors seeking to grow existing congregations in the fifth-largest city in California where, as a result, the largest sanctuary seats less than one thousand people. Likewise, in Pennsylvania, local churches pay property taxes on all but their designated sanctuaries. Ask yourself:

- How much would it cost your church annually if taxes were assessed on the property it owns?
- From where would the money come . . . tithes and offerings?

- How many more people would have to join your church and begin contributing to cover the annual bill?
- Who on staff and/or how much programming would have to be cut to meet the new financial burden?

With such things in mind, we should recognize that there is a fine line between faith and foolishness . . . between wise stewardship and mercenary greed. Sadly, an increasingly secular and cynical society tends to view the church in this way—as greedy. If anything, so as not to be perceived as disengaged, insensitive, or unwilling to contribute, a local church should leverage assets to become an economic player and growth accelerator in its community. In this way, a local church can foster goodwill among nonbelievers concerned more with wages, crime, and education than with spiritual things. By so doing, it can help mitigate any hostilities that might otherwise be directed toward Christians in general and/or the church for one reason or another. Yes, in the twentieth century effective evangelism was rooted in explanation. In what remains of the twenty-first century effective evangelism will be rooted in economics.

For too long Christians, denominations, churches, and church leaders have been perceived as wanting to receive as much benefit as possible while spending as little as possible for it or, as the old saying goes, "To get something for nothing." I have found this true along the way:

- In seeing coworkers at a restaurant avoid waiting on Christians because "Christians pray over their meals but don't tip well."

- In seeing churches pay male pastors a full-time salary while expecting their wives to run the women's ministry, work with children, or otherwise serve the church for free, as part of the deal.
- In seeing churches purchase prime real estate because they could, all the while disregarding negative feedback from the community and/or the disdain of civic leaders who see them negating the potential for tax revenue in the process.

We're talking Christian testimony, the nature of our collective witness, and the credibility of the church. We put it all at risk when, primarily seeking to benefit—from buying land and building new buildings or from increasing attendance and donations—churches don't consider or help the local economy too in ways people (particularly nonbelievers) understand and embrace. The status quo of tax law is not likely to last if pastors and churches continue to think and to act this way. For instance, some pastors have built extravagant mansions by taking advantage of a tax law that allows for a significant portion of income to be deducted by ministers for their primary home-related expenses (commonly called the "housing allowance"). In response, citizens and government agencies are taking notice and decrying what they see in this as an unfair tax advantage for the church and its pastors. Increased scrutiny will likely continue to reveal abuse and greed and lead to additional calls for change. Similarly, churches that challenge local, state, or federal government at every turn in opposition to taxes will surely add to growing frustrations. At some point, something will have to give, and that something may be us . . . ministers losing the housing allowance deduction and local churches losing their tax-deductible status and/or the ability to avoid paying income and property taxes by way of exemption.

Additional Objections

In addition to the passages and assumptions already discussed, here are a few more to consider in brief.

Sell Your Possessions and Give to the Poor (Matt. 19:21)

Assumption: A church should freely give and expect nothing in return.

Much of our ministry is and should continue to be freely given away. But think about it. If we give it all away without attention to sustainability, there will soon be nothing left for us to give. In the future, idealistic and unrealistic views of church economics will lead to even more congregations closing their doors, that is, in addition to those we might otherwise expect to close based on normal attrition.

Fix Your Hope on God Who Supplies All Things (1 Tim. 6:17–18)

Assumption: A church should be rich in good works and store up treasure in heaven.

This is true. However, the Bible does not expect or demand that our good works should be compartmentalized or restricted to those that are spiritual, as without any thought or care being given to those that otherwise advance the common good or promote equitable systems of justice or benevolent economics in the community. "Act justly . . . love mercy . . . and walk humbly with your God," wrote the prophet Micah (6:8 NIV). Jesus himself tells us to let our light so shine that others see our good works and (in so doing) "glorify your Father who is in heaven" (Matt. 5:16).

Shepherd God's Flock Eagerly and Willingly, Not for Money (1 Pet. 5:2)

Assumption: Pastors should focus on people not on profits.

Since pastors have been called in this passage not only to "shepherd" (Gr. *poimaino*) but also to "oversee" (Gr. *episkopeo*) God's flock, they must exercise due diligence and look carefully into things that might help or harm it (the church) down the road. Even so, pastors should not feel or give in to any pressure to directly involve themselves in establishing profitable enterprise on behalf of the church. Rather, those lacking specific experience in business, entrepreneurial gifting, or a passion to work on such things should empower others to run point and/ or to partner with them in the effort—proven and qualified men and women whom God has brought near perhaps for such a time as this.

How Far Can You See?

It was a difficult season. Abram—called by God to leave his country, relatives, and family in obedience to God (Gen. 12:1–4)—found himself in the Negev Desert with his wife and all that belonged to him. His nephew, Lot, was with him as well. According to Genesis 13:6–7, "The land could not sustain them while dwelling together, for their possessions were so great that they were not able to remain together. And there was strife between the herdsmen of Abram's livestock and the herdsmen of Lot's livestock." Given the need to separate, Abram allowed Lot to choose. "'Is not the whole land before you? Please separate from me; if to the left, then I will go to the right; or if to the right, then I will go to the left.' Lot lifted up his eyes and saw all the valley of the Jordan, that it was well watered everywhere. . . . So Lot chose for himself all the valley of the Jordan, and Lot journeyed eastward. Thus they separated from each other. Abram settled in the land of Canaan" (vv. 9–12).

In the moment Abram would have surely been discouraged, as the account suggests. "The LORD said to Abram, after Lot had separated from him, 'Now lift up your eyes and look from the place where you are, northward and southward and eastward and westward; for all the land which you see, I will give it to you and to your descendants forever. . . . Arise, walk about the land through its length and breadth; for I will give it to you'" (vv. 14–15, 17).

Why did God ask Abram to lift up his eyes? Because Abram was looking down. His head was hung and his spirits were dampened. He must have been thinking, *How can we survive on limited resources? If I only had more!*

Sound familiar?

In church after church it's a common refrain: leaders bemoaning a lack of resources and wondering what might be if they only had more. To think in such a way, with eyes turned down and heads hung low, is to give in to a scarcity mindset. Simply stated, "The scarcity mindset is the belief that there will never be enough—whether it's money, food, emotions or something else entirely—and as a result, your actions and thought stem from a place of lack."[19] Grayson Bell suggests that a scarcity mindset keeps people focused on the following:

- maintaining what they have instead of pursuing new opportunities
- staying afloat and being reluctant to take chances
- worrying about bank balances and putting off financial decisions in hopes that issues will go away (typically, they do not, and inaction only makes things worse)[20]

According to John Maxwell, "Leaders who allow a scarcity mindset to work its way into their culture pay a high price. When resources (money, opportunity, recognition) are perceived to be limited, paranoia, fear and politics thrive. In this environment, people become nervous and afraid to make a mistake. As a result, teamwork and innovation suffer."[21]

Contrast this with a mindset of abundance—one first described by Stephen R. Covey in his book *The 7 Habits of Highly Effective People*. An abundance mindset is characterized by a belief that enough resources exist for you to pursue dreams and opportunities, find success, and see others succeed too. It describes an optimistic and proactive approach to possibilities. For Christians, such a mindset stems from our faith in God, the fact that he owns cattle on a thousand hills. In the words of Jesus, "Truly I tell you, if you have faith and do not doubt . . . if you believe, you will receive whatever you ask for in prayer" (Matt. 21:21–22 NIV). Of course, this assumes we ask for things according to his will (1 John 5:14).

As we've said, there is a fine line between faith and foolishness. Nevertheless, to advance the coming revolution in church economics and position our churches for long-term sustainability we must avoid unnecessary fears and the intrinsic limitations of a scarcity mindset.

Ministry leaders embracing a mindset of abundance are able to envision possibilities, inspire optimism, capitalize on opportunities, and unleash entrepreneurial creativity in the church. Concerning the financial future of their congregations, they do not sit idle, hoping good things will happen. Instead, they "get after it" and make good things happen. By leveraging church assets they establish profitable business enterprise, create jobs, reduce local crime, produce tax revenue, resurrect abandoned property, and provide emotional hope to their communities. In so doing, the gospel, the local church, and the kingdom of God are advanced through the application of wise and benevolent church economics.

To be such a revolutionary leader you must first free your mind. Only then will you be able to embrace a mindset of abundance and position your church for long-term sustainability through the disruptive innovation of its economics.

Take a moment to imagine the possibilities.

How far can you see?

IDEAS AT A GLANCE

- When faced with a challenge, such as stagnant or declining donations in a church, thinking outside the box is not enough. Research shows that detailed strategic instruction on the front end will improve the odds for success in overcoming the challenges, especially when first applied to solving simpler but similar problems.

- Well-meaning believers will cite any number of seemingly spiritual reasons or passages from the Bible to denounce a church generating income beyond tithes and offerings, no matter the motive. Sound exegesis is required to get beyond erroneous beliefs regarding Scripture, church, and money. We must be better informed.

QUESTIONS FOR REFLECTION

1. What new insights did you glean from a consideration of specific biblical texts in this chapter?

2. What objection(s) might people raise to keep your church from establishing for-profit business enterprise? How might you counter these objections?

3. What possibilities currently exist or what opportunities might your church initially explore in pursuit of profitable income?

3

Stop Begging for Money

The Second Directive

On December 26, 1980, just a few months after becoming a Christian, I joined a group of young adults for my first experience on a short-term mission trip. Our team traveled to Roatan, part of the beautiful Bay Islands in Honduras, to share the love of God in word, song, and personal testimony with the people living on the island. As part of the effort we also engaged in physical labor, moving and setting blocks in place for what was to become a church building.

In those days I was more interested in play than work. So when a young boy showed up at the job site with a soccer ball, it was all too easy for me to begin kicking it back and forth with him. Within minutes a few other kids joined us as we continued to play in the area. Soon even more children were attracted to the fun, and the space in which we were playing was no longer big enough to accommodate us. We then moved to an adjacent area and resumed kicking the ball.

As you might imagine, increasing numbers of children continued to be drawn to the laughter that now permeated the venue, and once again we outgrew the space. Picking up the ball, I led them out to a dirt street whereby we could kick it even farther between us to the great delight of the children. By this time approximately twenty-five kids were involved, and it didn't take long for that number to double.

Within thirty minutes or so of first engaging the little boy, more than fifty children were now jostling for position, going after the ball and competing with one another in a makeshift game of soccer. This too became problematic. It was then that a man from the island told me about a nearby school with a large courtyard. He suggested it could accommodate the entire group.

Holding the ball high over my head so as to be clearly seen and easily followed, I led the throng of children down the middle of the street on a march to the schoolyard, approximately half a mile away. As we walked they were hanging off my arms, leaping about, and shouting joyfully as they playfully tried to knock the ball from my hands. In the moment I remember thinking, *This must be what Jesus felt like when the crowds pressed in on him!*

By the time we entered the schoolyard nearly one hundred children were with me, and still I only had one ball. In spite of the challenges, I arranged the kids into teams and organized a simple soccer tournament with adult coaches from our group. Following the tournament we shared the gospel.

In Roatan I learned a timeless truth, one that still governs my understanding of responsible leadership to this day: *If you can't do it with a ball, you can't do it with a building.* In other words, church leaders must first learn to use resources in hand before expecting God to entrust them with even more resources to leverage for his purposes.

Nearly ten years ago I heard Rick Warren say, "You can't keep begging people for money."[1] He was right then, and his statement is still true today. Views and practices related to giving are changing, as explained in chapter 1. In addition, church buildings remain largely empty from Sunday to Sunday. These and other assets should be leveraged not only to advance the gospel but to bless the community and generate sustainable income. For these and other reasons, leaders must lean into the coming revolution in church economics.

To be clear, asking others to contribute to the church is not the same as begging for money, which itself is more driven by worry, doubts, and fears. Begging for money might also refer to actions that manipulate potential givers or become the driving force behind what is done under the guise of ministry. Sure, we must make financial needs known, as the New Testament demonstrates. It seems Jesus and the apostles made financial needs known, as we read "some women, who had been cured of evil spirits and diseases . . . were helping to support them out of their means" (Luke 8:1–3). Certainly, the apostle Paul made financial needs known (1 Cor. 16:1–4; 2 Cor. 8:1–7). In so doing church leaders abide in the vine, encourage the faith of others, and bear fruit in time that only God can produce.

At Mosaic, we have always told our people, "If we do what we can for God, God will do for us what we cannot."

With such things in mind, we desire to advance the kingdom of God in bold ways that compel us to let financial needs be known *and* to pursue additional income by leveraging church assets. Along the way, we've tried things to generate revenue that didn't work, other things through which we only broke even, and still other things that continue to help fund the church today. In pursuit of sustainable income and to get away from begging for money, don't be afraid to try and fail from time to time in leveraging resources God has entrusted you to steward. According to *Forbes*, nine out of ten start-ups fail, meaning only one out of ten succeeds.[2] We have a feeling this applies to ideas too. That said, it may only take one idea to become reality and reverse the trajectory of a stagnant church budget or declining tithes and offerings.

When More Is Not Enough

Hard as it is to believe, you might be surprised to learn that there are churches today that have plenty of money. Beyond maintaining what they have through management and accounting, however, some of these churches don't know what else to do with it. For instance, I know of a struggling church with well over $2 million in the bank, the result of endowments. But, as its new pastor recently told me, the church is in decline. Only seventy-five people or so attend worship on Sunday mornings. Those who remain, he said, take great pride in having so much money in the bank. Yet the church is not healthy. Souls are not being saved, the community is not being engaged, and the common good is not being advanced. It's hard to imagine that God is pleased with any such church sitting on its assets while needs abound, opportunity exists, and potential is squandered. The fact is that access to resources may not be the problem for some pastors and churches.

Here is a case in point.

As the student ministries pastor of a suburban megachurch in the late 1990s, I had the privilege of helping to design and build a thirty-six-thousand-square-foot, $3.5 million student center in which to house my ministry. Senior church leadership led a successful fund-raising effort that allowed me to do so, and for which I was grateful. Among other things, the facility had three full basketball courts, two jumbotron screens, a climbing wall, motorized bleachers with one thousand individual pop-up seats, and a stand-alone café. It was also outfitted with the latest in sound, lighting, and technology. In terms of resources once at my disposal, this is what I left behind to start Mosaic in the summer of 2001.

Near the end of that summer, I met another pastor who had recently moved to Little Rock. He too was planting a church, so we met for lunch. Following our time together he asked me to join his mailing list, and I did. Keep in mind that in those days, AOL was all the rage.[3] The company's automated email alert, "You've got mail," even became the title of a Hollywood movie starring Tom Hanks and Meg Ryan. Nevertheless, internet speed was slow, websites were in the earliest stages of development, and getting any amount of substantive content to download seemed to take forever on what today are considered antiquated desktop computers.

One day, I received a newsletter via email from this pastor. In it, he asked supporters for $30,000 to help him launch the church. The money was to pay for what he described as necessary equipment including an LCD projector, a screen, a laptop computer, lighting and sound systems, as well as office space he desired to rent in a wealthier part of town. His inherent assumption was that such things would be required on the front end to successfully plant a church.

In contrast, and in spite of all the technology that I once had at my disposal, I went to a local retailer and purchased a traditional overhead projector for $150, a box of clear transparencies, and a Sharpie marker. After writing song lyrics by hand, I kept the transparencies in a file and used this "technology" to project lyrics during Mosaic's initial worship services. I also bought a microphone, a microphone stand, and cables to patch the microphone and a guitar into an old floor monitor that I repurposed to use as a speaker. As for an office, our initial team of five staff met in restaurants, coffee shops, and homes. We printed copies as needed at what today is a FedEx store. It was six months after launching preview services that we purchased a sound system for $10,000, that is, only after saving most of the first $60,000 we received in tithes and offerings as well as in donations from those attending other churches wanting to help us get established. Around that same time, I found office space for just $400 a month, four hundred square feet that our team of five shared for the next eighteen months. It was a full two years before we transitioned from the $150 overhead projector to a $1,500 computer and purchased a $3,000 projection unit through which to show slides during services. In other words, as a new church start, we prioritized saving over spending, people and programs over facilities and equipment.

Today, only one of these churches still exists. In fact, the other church never got off the ground. But it wasn't for a lack of trying on the pastor's part or even a lack of resources. The church's failure to launch was precipitated by his lack of understanding of church economics. Rather than seeking funds for equipment in hopes of attracting people he should have been saving funds to build a team or cash reserves in anticipation of future needs—and in the meantime, making do with less. Again . . . if you can't do it with a ball, you can't do it with a building.

As we shared in chapter 1, times are changing. Likewise, philosophies and models of church funding must also change. Without disruptive innovation, churches and church plants will continue to struggle financially or soon be forced to close, and valuable resources will be needlessly squandered. To minimize risk going forward, seminaries should provide MA and MDiv students specific instruction in church economics, such as this book defines, as should denominations and church planting networks.

Yes, pastors and planters must learn the why, how, and what of spending, saving, and leveraging church assets in pursuit of sustainability. In addition, they must learn to do more for less. For instance, in spite of limited resources, we distribute approximately 12,000 pounds of quality food every Tuesday afternoon to approximately 55 percent of those in need living in the 72204 ZIP Code—that is, enough to provide them with three to four days of healthy food and groceries each month, all for an average of just $0.03 a pound. In part, we do this by participating with the United States Department of Agriculture (USDA) and building partnerships with other food vendors, such as Panera Bread and Kum & Go, who supply free products to us, including those with a limited shelf life or that are approaching their "sell by" date. By picking up our weekly order in the morning on the same day we distribute it (in the afternoon) we are also able to receive many additional items free of charge from the Arkansas Foodbank, such as perishable fruits and vegetables that need to be moved immediately. In these and many other ways, we have learned through the years to do more for less.

The Local Church: An Economic System

As mentioned in the introduction, I used to work in restaurants ... first as a dishwasher, later as a busboy, and then as a waiter in an upscale steak house. In those days, there was nothing more satisfying than receiving a generous gratuity from a grateful patron, and nothing worse than being "stiffed." When customers pay the bill without leaving a tip, waiters and waitresses are stiffed. In more than one place I worked, waiters and waitresses tried to avoid serving Christians who would often pray before meals but leave very little in the way of a tip. Not good! Other customers would occasionally "dine and dash," meaning that after their meal they would try to slip away unnoticed, without paying anything at all. Some were caught and others got away. Of course, no restaurant could survive if this was the rule and not the exception.

So consider this:

Q. Why are you expected to pay for a meal at a restaurant?
A. In choosing to eat at a restaurant, you become part of an economic system.

Think about it. A restaurant performs for you a service in providing a meal you did not prepare, facilitated by people you do not know, in an environment you do not create or maintain.

In response, you pay for services rendered. The bill you receive at the end of the meal includes not only the cost of the food but also additional costs accrued by the restaurant to acquire and satisfy you as a customer. More than that, it includes a margin of profit without which owners would not be able to pay employees, sustain the enterprise, scale the business, or generate ROI.

This is an economic system.

Similarly, let's consider the local church in terms of economics and more specifically as its own economic system in which there is, likewise, an expectation that those benefiting from services rendered will pay a portion of the total costs incurred in the process.

By choosing to become members or regular attenders of a local church, people benefit from services (ministries) rendered to them or on their behalf, such as:

- music, teaching, and preaching on Sunday mornings
- age-appropriate opportunities for children and students to grow in their Christian faith
- organized activities for relational connection, discipleship, community outreach and service
- baptisms, weddings, hospital visits, and funerals
- home delivery of meals in seasons of new birth, prolonged illness, or death
- local compassion, mercy ministries, and overseas missions
- backend services including accounting and legal compliance

In light of services rendered, church members and regular attenders are expected to contribute financially through the consistent giving of tithes (literally, "a tenth;" historically 10 percent of one's income "as unto the Lord") and offerings (additional financial gifts received by the church from members and from nonmembers desiring to contribute). Money collected is also used to pay for recurring operational costs including the church's monthly rent or mortgage payment, the salaries of administrative and maintenance staff, security, office supplies, advertising, website hosting, and more. In fact, over the past fifteen hundred years, this is how the local church has been financed and sustained.[4] Tithes and offerings have fueled the church's economic system.

As previously mentioned, pastors are not typically trained to think about the local church in this way—that is, as an economic system. If anything, discussing pragmatic economics in the church is largely frowned upon by vocational ministers and lay members alike. Instead, we remind one another that we are a people of faith who should trust God to provide for our needs.

How the church is funded, therefore, is generally presumed . . . more "caught than taught" by pastors only after they have attained positions of responsible fiscal authority. Since we are not taught to think of the church as an economic system we don't usually speak of it in this way, nor are most of us comfortable discussing church finances. We are prone to do so only if and when necessary—for instance, when appealing for increased tithes and offerings due to budget deficits, launching a new building campaign, or seeking approval of the annual budget in a congregational-led church. In all such times we rightly encourage people to give freely, cheerfully, and not out of compulsion (2 Cor. 9:7) while also employing all manner of spiritual rationale to suggest that the church is not about money. In fact, as if to prove the church is not about money, many pastors tell Sunday morning visitors that they are not expected to give anything at all.

Imagine, however, a restaurant telling first-time diners they are not expected to pay anything for their meal. How many restaurants would go out of business if such a policy was in place? If this was true for restaurants across the board, it would be fantastic! We might never pay for a meal out again. Of course, to imagine this scenario is not to suggest a direct comparison to the church or to suggest there is anything wrong with pastors choosing to give visitors a financial pass. Rather, it is to challenge pastors to better understand the dynamics of church economics, to apply economic wisdom, and to speak more plainly with people about the church's need for money.

For example, some pastors speak piously about money, calling it "a necessary evil." Such understanding, however, is more assumed than accurate, more ill-advised than informed. Yes, the Bible says, "The love of money is a root of all sorts of evil" (1 Tim. 6:10), as we considered in chapter 2. But when money (or more broadly, church assets) is recognized and leveraged as a tool to glorify God, advance the common good, and generate sustainable income, it's not only possible but preferable to see it as something good for both the church and the community.

Therefore, we should acknowledge before others what is true in this regard. Among other things the local church is an economic system, and for a number of reasons—first and foremost, to advance the gospel—it must be about money.

Pragmatic Transparency

Similar to a restaurant, a church cannot survive if members and regular attenders who benefit from its services leave each week without paying some portion of the collective bill. Pastors know this to be true, as do mature believers and those responsible for church finances. With this in mind, pastors should be more open and direct by discussing the pragmatics of church economics in addition to making the case for giving as an act of worship and tithing as a biblical standard. People appreciate transparency.

As things currently stand, however, pastors who announce an upcoming sermon on giving, finances, or stewardship are often met with polite nods but not necessarily genuine enthusiasm from the body. If you are a pastor, be honest: Do you really get excited to talk to your church about money? Sure, like Dave Ramsey, you might enjoy discussing finances if it involves helping people learn to better manage personal finances, avoid debt, or grow in faith as part of the discipleship process. That's a good thing. Yet while all such teaching is important, we should also admit that much of it is done with an additional motivation— namely, to increase tithes and offerings, or perhaps ahead of an upcoming building campaign.

Make no mistake. Pastors know how to weave subtle and not so subtle appeals for tithes and offerings into messages—that is, to talk about giving, generosity, finances, and/or stewardship whenever possible with the church's need for money in mind. To be clear, we believe that pastors should teach about these things, as we do at Mosaic. But we have learned there are seasons in a church when doing so results in diminishing returns.

Stated another way, there are times when the collective body experiences "financial fatigue" and when it does, tithes and offerings may stagnate or decline.

LifeWay Research asked Protestant senior pastors how often they talk about tithing.

- Two-thirds had preached on tithing at least once in the past year.
- Nine percent had preached on tithing in the last month.
- Twenty-eight percent had preached on tithing in the past six months.
- Thirty-one percent had preached on tithing between six months and a year earlier.
- Five percent had preached on tithing in the last two years.
- Six percent had preached on tithing three or more years ago.
- One out of five had never made tithing a primary focus of a sermon.[5]

For me (Harry) personally, talking about money or approaching people about money has always been difficult. I am generally uncomfortable, sometimes embarrassed, and I always feel a bit awkward in doing so. Yet it is one of the many costs of leadership. In fact, if you can't will yourself to let financial needs be known to people who are your strongest advocates, you will not likely be able to share them with others. Becoming the senior pastor of Mosaic has forced me to rely on Christ more than ever, and talking to people about money continues to be an avenue for growth and maturity. Of course, there will always be those who suggest we should simply pray and ask God to provide. However, I'm not aware of an effective leader anywhere who is responsible for a local church that is not trusting God by faith *and* making the financial needs of the church known to others. The two approaches—prayer and petition—are not mutually exclusive.

Are there other ways to increase needed revenue beyond merely talking about giving, generosity, finances, or stewardship? The answer is yes, and we'll show you practically how to do so in the following chapters. But before moving on, let's talk a bit more about tithing.

Tithing, Then and Now

Have you ever wondered why from the beginning of human history people have offered sacrifices to what they perceive or believe to be governing deities or supernatural forces in the cosmos? These gifts range from flowers and herbs, to money and jewels, to innocent animals and (tragically) human beings. Throughout time, people have done so for one or more of the following reasons:

- to act in obedience according to their religion
- to appease the gods and solicit goodwill in hopes of earthly prosperity
- to atone for sin
- to demonstrate faith as an act of maturity
- to provide for the needs of others, including the religious community

- to maintain a system of worship in conformity to societal expectations
- to ensure a favorable outcome in the afterlife

Biblically speaking, tithing was expected of God's people in ancient Israel. In fact, the Old Testament provides at least five purposes of the tithe.[6] It was (is) to be given as:

- an act of gratitude and worship
  "After his return from the defeat of Chedorlaomer and the kings who were with him. . . . Abram gave [to Melchizedek] a tenth of everything" (Gen. 14:17–20).
- an act of dependence and obedience
  "All the tithe of the land . . . is the Lord's" (Lev. 27:30).
- an act of provision for the priests and their ministry
  "To the sons of Levi, behold, I have given all the tithe" (Num. 18:21).
- an act of mercy and compassion in support of aliens, orphans, and widows
  "Bring out all the tithe . . . [and these shall] be satisfied" (Deut. 14:28–29).
- an act of faith with a promise of blessing
  "Bring the whole tithe into the storehouse" (Mal. 3:8–10).

In light of Old Testament practices, we might expect tithing to have been instituted in the New Testament, for the church as well. Yet the New Testament does not express it as a requirement for Christians in spite of what some would have us believe. According to the *New Catholic Encyclopedia*, "The payment of tithes was adopted [in time by the church] from the Old [Testament] Law, and early writers speak of it as a divine ordinance and an obligation of conscience. The earliest positive legislation on the subject seems to be contained in the letter of the bishops assembled at Tours in 567 and the canons of the Council of Macon in 585."[7] This same council "threatened excommunication to those who refused to comply [by tithing]."[8]

In the eighth century, Charlemagne made tithing a civil law under threat of imprisonment. Centuries later, the Council of Trent (1545–63) effectively affirmed the tithe, which became standard practice in Protestant churches following the Reformation. In fact, "John Calvin studied the early church's use of tithe and argued that there were four parts for church finance, roughly equal except in periods of dire need: 'one to the clergy, another to the poor, another to the repair of churches and other edifices, a fourth to the poor whether strangers or natives' (this last being given to the bishops to do good works as needed); see his *Institutes of the Christian Religion* IV.4.5–8."[9] For these and other reasons in the past, Christians were taught and expected to tithe to their local churches.

Due in part to history and tradition some pastors today seek to justify tithing from the New Testament and impose the practice on believers. For instance, in an article titled "The Bible Commands Christians to Tithe," William Barclay, writing for the Gospel Coalition, states, "Jesus upholds the tithe in Matthew 23:23 (cf. Luke 11:42). . . . Those who don't believe tithing is for today argue that Jesus is only addressing scribes and Pharisees still under the old covenant. Yet this misses the context. These words are for Jesus' followers. Elsewhere he doesn't shy away from setting aside those parts of the law that no longer apply to his disciples (cf. Mark 7:19). But in teaching his disciples, Jesus upholds the tithe."[10]

However, in considering Barclay's argument, we should note the following:

- Jesus upholds the tithe insofar as tithing was expected in the Old Testament, and thus he expects the Pharisees (and other teachers of the law) not only to teach the principle but also to practice it themselves.

- Jesus is not speaking to his disciples but directly to the scribes and Pharisees here, challenging their hypocrisy, as Luke's account makes clear. This is the context (compare Matt. 23:1, 13).

- Jesus' words are for his "followers" only insofar as Matthew and Luke record the conversation. Yes, today we read his words, but in them there is not a specific command to tithe.

- To suggest that in other places Jesus "doesn't shy away from setting aside those parts of the law that no longer apply to his disciples" is to "appeal to ignorance." Thus, dogma should not be presumed or imposed.[11]

- This is the only passage in the entire New Testament that Barclay cites as directly speaking to the issue in support of his claim.

Indeed, to impose tithing on New Testament believers today requires a measure of theological gymnastics. Worse yet, it can call into question the motives of those who espouse it, no matter how well-meaning.

Still, we must not dismiss tithing entirely, since it can be a worthy and fruitful discipline of mature believers who choose to conform their giving to this biblical (Old Testament) standard. In fact, there are many good reasons to encourage the practice and to tithe ourselves, without misrepresenting it as a New Testament command. Leith Anderson, former president of the National Association of Evangelicals, recently commented, "Many evangelical leaders do not believe there's a biblical requirement to give exactly 10 percent to the local church. Giving generously out of what God has done for us is the main message."[12]

Beyond this, past research has shown that most tithers are highly active in their congregations and read the Bible regularly. According to a 2013 State of the Plate report:

- Ninety-six percent of tithers attend services every weekend (unless sick or traveling).
- Fifty-four percent serve on a church committee.
- Fifty-three percent participate in a small group or Sunday school class.
- Forty-three percent read their Bible frequently (four to six times per week).[13]

In addition to these benefits to the local church, tithers are more likely to be debt free and less likely to owe on a mortgage or to carry credit card bills greater than thirty days.[14] Concerning tithers, Pushpay, a company that provides mobile giving and engagement solutions for churches and nonprofits, puts it this way: "Whether people who are already disciplined are typically also tithers in their tithing or people who tithe become more disciplined with their money, the fact is that tithing and fiscal responsibility go hand in hand. Tithing is more than meeting the church budget: it's about encouraging believers to discover the joy of kingdom living available to those who learn to give faithfully and sacrificially."[15]

TEN NEW TESTAMENT PRINCIPLES TO INSPIRE GIVING

1. **Give God's will priority over money.**

   Consider the rich young ruler (Matt. 19:16–26).

2. **Give sacrificially and not simply from a surplus.**

   Consider the widow giving her mite (Mark 12:41–44).

3. **Give to support those who proclaim the kingdom.**

   Consider the ministering women supporting Jesus and his disciples (Luke 8:1–3).

4. **Give to God all that he asks of us, not just a portion of it.**

   Consider the account of Ananias and Sapphira (Acts 5:1–11).

5. **Give to and through the local church in support of various needs.**

   Consider the church at Antioch taking up an offering for others (Acts 11:29–30).

All such information provides pastors a reason to encourage tithing in their congregations without being legalistic or seeming to beg for money.

How, then, should we encourage people to give? In keeping with their own conscience, to move from minimal and occasional giving to sacrificial and faithful (consistent) giving over time, as figure 6 illustrates.

6. **Give according to your ability and beyond.**
   Consider the churches of Macedonia giving from affliction and poverty (2 Cor. 8:1–5).

7. **Give with a sense of loyalty in support of fellow believers.**
   Consider the churches of Macedonia begging to support the saints (2 Cor. 8:1–5).

8. **Give generously.**
   Consider that "he who sows sparingly will also reap sparingly" (2 Cor. 9:6).

9. **Give purposefully as God directs you individually to do.**
   Consider that "each one [should] do as he has purposed in his heart" (2 Cor. 9:7).

10. **Give cheerfully.**
    Consider that "God loves a cheerful giver" (2 Cor. 9:7).

Figure 6. How People Should Give

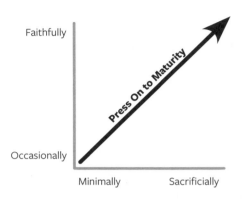

Unless the Lord Builds the House

Like most pastors, we never considered the why, how, and what of church funding, donor development, or such associated challenges on the front end. We simply embraced our calling to full-time pastoral ministry and in time became lead pastors of a local church. What we were not told about is the inevitable worries, doubts, and fears intrinsic to the role that must be overcome where church finances and funding are concerned. Similarly, we did not have any idea that the man or woman in the lead chair automatically becomes the CFO of the church—that is, its chief fund-raising officer.

Among other things, what we have learned along the way is that a church's need for money never goes away, and the pressure for more of it that senior leaders feel and live with is a constant. As soon as one financial challenge is met the next one presents itself, and you find yourself right back in the game considering who to approach, what to ask, and how in the world to get people (again) to give. It can be mentally, emotionally, and relationally exhausting.

Therefore, instead of working only on ways to *attract* more money, we chose many years ago to work on ways to *generate* more money for the church beyond simply asking for and waiting on tithes and offerings. Desperation in our case led to innovation as well as to embracing prayer, patience, and persistence in pursuit of dreams God continues to lay upon our hearts. At the end of the day, "Unless the LORD builds the house, they labor in vain who build it" (Ps. 127:1).

When God Comes Through

On May 31, 2001, I completed nearly eighteen years of full-time service as a student ministries pastor. The next day my wife, Linda, and I set off on a journey to plant a multiethnic church in the urban center of Little Rock. Filled with excitement, we began by talking to anyone and everyone willing to listen. There were meetings morning, noon, and night. In those days, my primary motivation was to chase what John W. Kennedy, writing for *Christianity Today*, would later call a "Big Dream in Little Rock," and not at all to raise money.[16] Yet in the back of my mind, I knew the clock was ticking. I had a mortgage and tuition payments for three children due each month. Having been given a three-month severance package from the church I left, there was some breathing room. Yet after that, I did not have a liquid penny to my name or any guarantee there would be a paycheck come September 15. Joyfully, though, I continued to share the vision and to ignore financial concerns.[17]

Soon, a friend of mine expressed interest in the work. He had once sold a business and established a $3 million foundation through a local church, desiring to make a difference in the inner city. Through the foundation, a team of four to five individuals could perhaps be funded to help get the church started, or so he implied.

Near the end of June, I finally broke away from vision casting to meet with him at a local sandwich shop. I just knew God was going to use him and his foundation to meet initial financial needs, if not for the entire team then surely my own. Having established the foundation, however, my friend was not solely in control of how the money was spent. He could only recommend or present opportunities for projects and funding to the board, of which he was only a part. In this case, it was not interested. To my surprise and disappointment, he told me that the foundation's board had rejected his request on our behalf for financial support.

As I left the restaurant and got into my Honda Odyssey van, the sheer risk and enormity of our decision to strike out as church planters finally hit me. Without the promise of funding, connection to a denomination, or any affiliation with a church-planting network, I was in a word . . . well, I can't write that word here! But I sure shared it with God that day, as I did many other choice words right then and there in the van, in the parking lot of a sandwich shop, as I cried out to him in a fit of frustration, anger, and fear.

"What in the world have you gotten me into?! I'm out here doing what you said to do, following your lead, starting over, leaving everything and everyone I know behind, and now what? I'm like a goose with a long neck, and it's about to get cut off! I've got kids.

I've got a family. What about them? How in the world are we going to make it?!"

Make no mistake: this was not a contemplative moment. It was a full-on emotional meltdown! To this day, I thank the Lord for how he handled it.

- He listened.
- He did not immediately speak.
- He waited patiently.

Then in my spirit, I heard him say, "Are you finished . . . ready to get back to work?"

What choice did I have? What would you have done?

Pulling myself together, I next drove to Philander Smith College, a historically black college and university (HBCU) in Little Rock. I went there hoping the school would let me use its chapel on Sunday nights to meet for worship with those showing interest in the vision of Mosaic, as a prelude to our official launch. The school's chaplain, however, said no to my request. It was another disappointment.

On my way out of the administration building, I heard someone call my name from the top of a staircase in the lobby. Looking up, I recognized William Powell. He was a professor at the school and for a short time had worked as a part-time choir director at the church I had just left. He came downstairs and greeted me, asking why I was on campus. After I shared the vision of Mosaic, and without any expectation of contribution or prompting on my part, William said, "Mark, I think God sent you here today as an answer to my prayer. Recently, my wife and I received some unexpected money in the mail. It's not a lot, but we feel led to give some of it away. In fact, just this morning we were praying about who we might bless and here you are! Would you come up to my office? I'd like to give you something."

That very day, and having previously been only acquaintances, William Powell put a check in my hand. Thanking him, I left his office, walked outside, and unfolded the check . . . one made out to Mosaic Church in the amount of $50. As you might imagine, it was for me a holy moment and a significant one too. William Powell's $50 check was the first money ever given to Mosaic Church. More than that, his unsolicited gift came on the heels of my freak-out in the van. Therefore, I received it as a sign from God—one letting me know that everything would be all right. In that moment, I began again to cry . . . but this time, with tears of joy. Once more that day, God spoke to my spirit: "You see? You do what you can for me, and I'll do for you what you cannot."

In part, my approach to church economics remains rooted in that experience so many years ago. As a collective church, we do what we can for God—which includes the giving of tithes and offerings *and* the leveraging of our assets—all while trusting him to do for us what we cannot do for ourselves. It's a both/and, not an either/or approach to church economics.

What we don't do is sit idle and beg for money.

Conclusion

For much of its history, the church has been funded by tithes and offerings. But tithes and offerings alone will not likely be enough to fully fund most churches in the future. The fact is that many churches today are not receiving enough money through the collection plate to support the entirety of their projected budgets. Yet churches continue to approve budget deficits from year to year and to survive from one offering to the next, while leaders hope and pray they will have enough next week to make payroll. Furthermore, budgets based solely on tithes and offerings limit what might otherwise be accomplished through the leveraging of church assets.

At some point, due to stagnant giving or declining offerings, church leaders must do more. They must face reality. And when they do, there will be only three options.

1. Pursue worn-out church-growth strategies of the past, assuming that increased size will yield increased giving. As discussed in chapter 1, however, this assumption is no longer a given.
2. Make budget cuts, lay off staff, curtail ministry, sell property, or close the church.
3. Engage the coming revolution in church economics.

Yes, it is time to do more than simply hope and pray for resources. In the twenty-first century pastors must act with intentionality not only to promote worship, evangelism, and discipleship, but also to generate profitable income. Indeed, as good and faithful servants we can no longer sit on our assets while begging for money. We must learn to create multiple streams of income.

IDEAS AT A GLANCE

- Worry, doubts, and fears can shape our understanding of church finances, determine priorities, inform practices, lead to erroneous teaching of biblical expectations concerning giving, or manipulate potential givers by playing on their emotions. Views and practices related to giving are changing and should be embraced by local churches.

- Many churches today are not receiving enough in tithes and offerings to fund budgets or advance bold initiatives. Encouraging church members to be more faithful and sacrificial in their giving is important, yet it is not enough. Churches can and should do more than hope and pray to strengthen their financial position and to ensure long-term sustainability.

QUESTIONS FOR REFLECTION

1. What resources or assets does your church have in hand that might be leveraged to advance mission?

2. What is your understanding of tithing? Is it informed by sound exegesis or more by history, tradition, personal beliefs, or assumptions? How and in what ways do you instruct your church to give?

3. Why and how might your church become more intentional so as to encourage giving *and* pursue sustainability through profitable business enterprise?

4

Create Multiple Streams of Income

The Third Directive

From Nice to Necessary

Following World War II, one paycheck was all that was required for people to live a generally middle-class life in the United States. At that time, a stereotypical family (though not every household) included a husband with a full-time job and a stay-at-home wife who cared for the children. When the kids were old enough for school, she volunteered discretionary time to the PTA (Parent Teacher Association) or other charitable organizations in the community, not otherwise needing to earn a paycheck.

Beginning in the 1960s, however, things began to change. An increasing number of children were being born out of wedlock and living in single-parent homes (as was my experience).

The feminist movement encouraged women to pursue occupations in the marketplace, and a global energy crisis during President Jimmy Carter's time in office (1977–81) significantly raised the price of oil and gas. For these and other reasons, the number of two-parent households—those in which only the father was employed and with children under the age of eighteen—dropped by approximately 25 percent between 1960 and 1980. Simultaneously, there was a percentage increase in dual income households (see figure 7).[1] Thirty years later, nearly 60 percent of households could be described as dual income, while only 31 percent of fathers and 6 percent of mothers were the sole wage earners in their homes. By 2017,

Figure 7. The Rise in Dual-Income Households

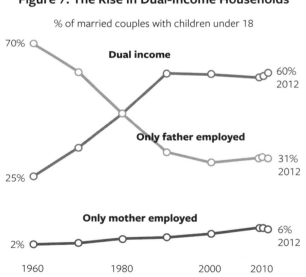

% of married couples with children under 18

Source: Pew Research Center analysis of the Decennial Census and American Community Surveys (ACS) Integrated Public Use Microdata Sample (IPUMS) files the percentage of two-parent households with children under eighteen had risen to 61.9 percent and not surprisingly, 48.3 percent of married couples without children were living in dual income households.[2]

Today, dual income households are normative. In a majority of situations, two or more paychecks are necessary to maintain present standards of living and pay the bills. Thus, one might assume that financial challenges in the home are readily overcome. Yet keep in mind the words of Connie White (referenced in chapter 1), who describing her own dual income household said, "We're able to pay the bills and don't have a lot of debt, [but] there's barely anything left at the end of the month."[3] In other words, to get ahead financially and beyond living from paycheck to paycheck, people must now develop multiple streams of income. In fact, the author of a 2010 *U.S. News & World Report* article begins by saying, "Building multiple streams of income is no longer a luxury, it has become a necessity."[4] He goes on to provide examples "of how alternative income sources can be used to lessen the risk of living on a single income."[5]

In addition, Thomas C. Corley, writing for *Business Insider*, notes:

In my five-year study of the daily habits of the rich and poor, I learned that most self-made millionaires generated their income from many baskets:

- 65 [percent] had three streams of income
- 45 [percent] had four streams of income
- 29 [percent] had five or more streams of income

Having multiple income streams makes a lot of sense. When one stream is negatively affected by systematic economic downturns, of which you have no control, the other streams can come to the rescue and help you survive the downturn, without seeing your lifestyle dramatically affected.[6]

In the introduction, I mentioned my journey from lower to upper middle class over the past fifty-eight years. What I didn't tell you is that my wife and I have developed five income streams to provide for the needs and desires of our family these days, where two adult children, two children still in college, and three grandchildren are concerned. Beyond this, we are looking ahead to what will surely be for us a slowing down in the years to come, dealing with end-of-life issues we will someday face, and leaving something of ourselves and our assets behind in terms of legacy. In all such considerations, we are not alone.

Here's the point. Living within the spectrum of middle class today requires most households to have two or more income streams. As households have gone in this regard so too is going the local church. In other words, tithes and offerings alone—as if one paycheck in the collective church household—are no longer enough to fully fund, grow, or sustain bold vision and effective ministry in an increasing number of congregations. In order to do so, local churches—like most households and self-made millionaires—will need to create multiple streams of income.

Still not convinced?

Consider the following excerpt:

There is increasingly strong debate in the United States over whether or not churches should be exempt from paying federal income and property taxes, allowed to continue receiving tax-deductible donations, and providing a housing allowance to ministers. Opponents provide at least twelve reasons the federal government should abolish these privileges.[7] Currently, the Internal Revenue Service can revoke the tax-exempt status for churches "directly or indirectly participating in, or intervening in, any political campaign on behalf of (or in opposition to) any candidate for elective public office."[8] In addition, many feel the US Supreme Court's landmark 2015 decision guaranteeing same-sex couples the fundamental right to marry leaves the church at risk as well, believing that churches could someday have their tax-exempt status revoked for "practices [that are] contrary to a compelling public policy."[9] If this were to happen—and it could—think of the cumulative effect. For instance, according to Erwin Chemerinsky, "if the parsonage exemption on religious ministers' housing costs were revoked, American clergy members would cumulatively lose an estimated $2.3 billion over five years."[10] Thus, the American church today is woefully unprepared for wholesale changes in this regard, potential changes that should not be summarily dismissed as if nearly impossible.[11]

Given the rise of dual income households and in consideration of possible changes to federal, state, or local tax laws in the years ahead, leadership has a responsibility to consider what can and should be done by the church to improve its financial situation, create multiple streams of income, and get beyond sole dependence on tithes and offerings.

By way of illustration and throughout the rest of this chapter, we'll discuss the need to create multiple streams of income through the lens of church planting. Why? Because conventional wisdom concerning what it means for a church to be sustainable begins here. Furthermore, denominational and network leaders today advising church planters and pastors typically draw from their understanding of nonprofit finance and funding, which itself is largely informed by conventional wisdom or their own past experience. Yet sustaining (incremental) innovation is no longer enough to address the financial challenges existing churches or emerging church plants are today facing. Rather, it takes disruptive innovation to establish gospel-centered, socially just, multiethnic, and economically diverse churches where they are needed most, in underserved and underresourced communities.

The Myth of Sustainability

In an article titled "The Math Doesn't Work: Why the Future of Church Planting Is Bi-Vocational," pastors Jimmy Scroggins and Steve Wright write, "Well-connected, networked church planters can often raise several hundred thousand dollars in initial support, typically pledged over three years. At the end of three years, supporting churches normally expect the church plant to be self-supporting or close to it."[12]

In a subsequent response to the article, pastors J. D. Greear and Mike McDaniel suggest, "The problem with this model . . . is that it is expensive. Many church plants start with a yearly budget of $200,000 or more, which means that before they've even planted the church, they need to grow the church to 200+ just to become self-sustaining. For many planters, especially in difficult contexts, this is simply unrealistic. Unfortunately, most don't recognize the mistake until year three when their funding begins to run out."[13]

In other words, conventional wisdom typically encourages church planters to do the following:

- Develop a three-year budget as part of a plan to plant.
- Raise actual cash or pledged financial gifts prior to launch in order to cover expenses for three years.
- Begin immediately to collect tithes and offerings from regular attenders and members.
- Grow to 200+ by the end of year three and before committed funding runs out.
- Hope and pray that after three years tithes and offerings will be enough to pay for all related church expenses.

Should things go as planned, church plants are deemed to be "self-supporting" or "self-sustaining" by denominational and network overseers looking to report successes, who then quickly shift attention and resources to fresh efforts. Yet can a church plant rightly be considered sustainable after only three years, even if tithes and offerings are covering all related expenses? In short, the answer is no. The best that can be said of such a church is that its tithes and offerings are meeting budgetary needs in the moment. But what about tomorrow, next week, in six months, one year later or beyond?

- What if the church loses its lease, needs to move, or has its rent increased?
- What if the church grows and needs to hire additional staff?

- What if tithes and offerings remain stagnant and the cost of providing health insurance for full-time employees rises?
- What if a big donor moves away or others leave the church and attendance decreases?
- What if after five years expenses are higher than predicted, income is lower than expected and the church, though it "may be running around 150–200 . . . still cannot pay [the planter] enough to support [his or her] growing family?"[14]

Concerning this last question, Scroggins and Wright write, "At this point the planter declares himself a 'catalytic' leader who specializes in 'starting new things' and moves on to a different ministry job that can support his middle-class lifestyle. The church is then taken over by someone else. . . . Despite the success of the new church, the pastor had to leave. Why? Because from day one the math didn't work."[15]

Beyond these reflections statistics too confirm that in large part the math doesn't work. For example, in an exclusive LifeWay Research and NewChurches.com 2015 report entitled, "The State of Church Planting in the US," authors Ed Stetzer, Micah Fries, and Daniel Im asked approximately 1,200 pastors who had started churches between 2008 and 2012 this question: "During which year (if at all) did the new church become financially self-sufficient (no financial support received from outside sources)?"[16]

After surveying the data the authors wrote, "Financial self-sufficiency happens in varying ways, but from a statistical perspective, if your new church plant isn't financially self-sufficient by its fourth year of existence, it is increasingly unlikely that it will become self-sufficient."[17] In addition, they found:

> Nearly half (44 percent) of all churches started in 2012 or earlier became financially self-sufficient within three years. That percentage dwindles past that critical three-year threshold: and some 31 [percent] of the churches started in 2012 never reached financial self-sufficiency. Of this group of recent church plants that remain financially dependent on other organizations or churches, nearly half began in 2010 or earlier (five or more years ago). Obviously, that is not a sustainable percentage over the long-run. Many of those churches contained in the 31 [percent] may not succeed and may eventually close.[18]

As a hypothetical, let's consider how much money will have been spent and lost if the entire 31 percent of these attempted church plants never reach financial self-sufficiency and close their doors after five years. In so doing, let's also consider money to be "lost" when initial investment fails to produce ROI . . . in this case, a self-sustaining church. Sure, we might spiritualize, rationalize, or ultimately excuse any failed attempt to plant a church knowing that "God causes all things to work together for good to those who love God, to those who are called according to His purpose" (Rom. 8:28). Nevertheless, let's not do so at the moment. Rather, let's consider failure and money the way those in business do. In fact, since most of the money invested in attempted church planting comes from funds donated "as unto the Lord," we have both a spiritual and a fiduciary responsibility to think about failure in this way. In so doing, we should also ask whether or not we are being wise stewards of these relatively large capital investments before pumping even more into the model, especially since we know that at least 31 percent of attempted church plants will never reach financial self-sufficiency and likely fail within three to five years.

To calculate the amount of money lost in just one failed attempt to plant a church, let's use the $200,000 number that Greear and McDaniel suggest is reflective of annual need and budget. Keep in mind, as the authors allude, this is a lower estimate of what is typically needed to fund a new church plant. Let's also assume that our hypothetical church plant never breaks two hundred in attendance, since "the Barna Group pegs the average Protestant church size in America at 89 adults. . . . According to Carl George and Warren Bird, fully 85 [percent] of all Protestant churches in North America never break the 200-attendance mark."[19]

With such things in mind, this is how years one, three, and five might look.

FY1 (fiscal year 1)

» Budgeted: $200,000

» Secured: $175,000

○ Received from sources outside the church: $143,800

○ Received from attending members of the church: $31,200

○ Weekly average donation per person: $10

» Average Sunday morning attendance: 60

FY3 (fiscal year 3)

» Budgeted: $225,000

» Secured: $175,000
  o Received from sources outside the church: $87,360
  o Received from attending members of the church: $93,600
  o Weekly average donation per person: $14
» Average Sunday morning attendance: 120

FY5 (fiscal year 5)
  » Budgeted: $250,000
  » Secured: $200,000
    o Received from sources outside the church: $59,600
    o Received from attending members of the church: $140,400
    o Weekly average donation per person: $18
  » Average Sunday morning attendance: 150

At the end of this five-year scenario, assuming similar but modest gains were made in years two and four, here are the results:

- Sought and budgeted: $1.1 million
- Secured and spent: $925,000
- Average annual budget: $220,000
- Average annual income: $180,000
- Average annual deficit: $35,000

Again, this hypothetical is based on a projected need of $200,000–$250,000 annually over the course of a church plant's first five years. It might also be considered an overly optimistic view of growth in giving. In fact, things could be worse. I asked three-time effective church planter Matt McGue to look over these numbers to see if they were realistic based on his experience. In response, Matt wrote, "Looks good to me, a pretty realistic projection. The thing is, new converts and members don't typically give well for at least a year or two. So I think your projected per person giving increase is too high. I think the actual increase in most cases would be only $1 a week, $52 a year, per new member. Just my opinion."[20]

So where in our scenario did all the money go . . . nearly $1 million in cash?

- to salaries and benefits: $647,500 (70 percent)
- to rent payments: $180,000 (19 percent)
- to programming and other expenses: $97,500 (11 percent)

In this hypothetical yet realistic scenario, 89 percent of the total revenue received by the church over a five-year period would have been spent on payroll and rent. This is consistent with what business owners will tell you: their two biggest expenses are employees (people) and rent or mortgage (facilities). Assessing things the way those in business do, and assuming our hypothetical church plant was forced to close its doors at the end of five years, the significant amount of money spent and ultimately lost would be considered a bad investment having failed to produce ROI—in this case, a financially self-sufficient, sustainable church. Keep in mind, this analysis considers only one such attempt to plant.

In the 2015 NewChurches.com report some 1,200 church planters were surveyed. Since 31 percent of 1,200 is 372, let's extrapolate what we've learned and apply it to this number. Using a five-year scenario, the economic loss to the American Church considering its investment in 372 failed church plants would look something like this:

- Secured: $344.1 million
- Spent (payroll and rent): $307.83 million
  - » Payroll: $240.87 million
  - » Rent: $66.96 million

We should pause to ask, "Is this the best we can hope for in terms of money, time, and effort?"

Simply Unrealistic

Notice the words that pastors we've quoted use to describe this model of funding church plants, one that depends on a large capital investment from denominations, networks, other churches, family members and friends—that is, those who are willing to invest (risk) discretionary income in support of church-planting efforts despite a 31 percent chance of failure within three to five years. Scroggins says it's a model that depends on church planters being "well-connected [and] networked." Greear and McDaniel say that it is "expensive," and therefore, "especially in difficult contexts . . . simply unrealistic." These sentiments reflect "code" (albeit unintentional) and suggest that church planters must know the "right" (more often than not, white) people, have good relationships with the "right" people, and plant churches in the "right" part of town if they hope to succeed.

The problem is, for all the talk of taking a city for Christ, church planting and multiplication, and of developing outwardly focused or missional communities of faith, the entire model often forces well-meaning men and women to abandon ideals in favor of pragmatics. Unaware of alternative approaches, lacking connections, and without any other encouragement or the resources to plant in "difficult contexts," most church planters will choose a path of least resistance . . . and another well resourced, affluent, predominantly homogenous church is born.

Furthermore, this model creates a need to "feed the beast."

- It's what often drives churches in their push for numbers.
- It's the determinant factor in how tithes and offerings are spent.
- It's helped establish size as the primary metric of success.
- It's behind decisions to invest heavily in lighting, sound, production, staging and more, to attract an ever-growing crowd of people and retain them.
- It's given rise to the industry of generosity, though being generous is otherwise a good thing and biblical.
- It's why annual budgets are often discussed and determined by assigning a projected dollar amount to every new person or family expected to join the church in a coming year.

- It's reduced households to "giving units," philosophical convictions to pragmatic choices, and what elevates ends over means in order to grow.

Harsh words? Perhaps. Yet are we so far off the mark?

As an alternative, Scroggins and Wright suggest a bi-vocational approach to overcome such challenges. They write, "We must find pastors who are willing and able to support their families without taking a salary from the church—maybe ever."[21] Likewise, Greear and McDaniel suggest that bi-vocational strategies are needed to reach our cities. Many others in recent years have come to similar conclusions. Increasingly popular is the term "co-vocational," which suggests employment outside the church is intentional and not merely functional or a means to an end, as the term bi-vocational can (but does not always) imply. Co-vocational pastors embrace a calling to serve God in the church *and* in the marketplace.

Yet Greear and McDaniel also believe that high-capacity pastors and planters should not be discouraged from pursuing full-time work and funding through the church. For such leaders, "the most appropriate strategy is to enter the game full-time. If we put up a strategy that is bi-vocational only, many of them will conclude that church planting is not for them. We need to put before these people a compelling vision for urban centers and why church planting is a best use of all of their gifts."[22] According to Greear and McDaniel, bi-vocational ministry has challenges of its own. "For too long, we've relied on one strategy. However, the answer is not to replace one 'one-size-fits all' strategy for another, but to expand it. We need multiple models of sending to fulfill the Great Commission."[23]

Yes, we do. That's why we're talking about the need for local churches to create multiple streams of income so as to help fund full-time leadership, strengthen the bottom line, advance the common good, and help ensure sustainable gospel-centered congregations in every area of the city. Conventional wisdom, however, suggests there are only two paths for church planters to pursue, as these articles discuss: raise full-time financial support through well-established connections or seek bi-vocational employment. A third option is the co-vocational approach or perhaps a hybrid of it, whereby pastors supplement their base salary via the "gig" or "System D" economy.[24]

But what if there is a more comprehensive approach, another way to fund pastors and planters called to serve full time, for church revenue to be increased, and for communities to benefit at the same time? In fact, there is . . . and that way is to create multiple streams of income by establishing both nonprofit and for-profit enterprise as part of a disruptive approach to church economics.[25] This is a necessary and vital next step for church plants to take, as it is for existing churches struggling to pay the bills. In more affluent churches, such an approach can provide even more discretionary income that can in turn be invested to support good and needed work in the "difficult contexts" Greear and McDaniel imagine.

Church planting is difficult, and urban church planting is especially challenging. The ground is hard packed, the resources are limited, the problems are messy and complex. In virtually every way, our own work in Little Rock has been marked by the organic, arduous, and labor-intensive process the words of Jesus describe: "The kingdom of God is like a man who casts seed upon the soil; and he goes to bed at night and gets up by day, and the seed sprouts and grows—how, he himself does not know. The soil produces crops by itself; first the blade, then the head, then the mature grain in the head. But when the crop permits, he immediately puts in the sickle, because the harvest has come" (Mark 4:26–29).

In this passage, Jesus speaks of patience and expects persistence from laborers working in his vineyard. These are the same character qualities uniquely required of pastors who labor in difficult contexts, such as we do in Little Rock. Galatians 6:7–8 implies at least two laws of sowing and reaping: we reap what we sow (the first law) in a different season than when we sow (the second law). In our case, it took five years of consistent sowing into our beloved 72204 ZIP Code before neighbors began to express genuine appreciation to us for bringing diverse people together and for the impact we were having in the community. Similarly, patience and persistence are required of churches and church leaders seeking to create multiple streams of income.

A Disruptive Approach

After winning the 2012 Super Bowl as head coach of the National Football League's (NFL) New York Giants, Tom Coughlin provided insight on what it took for his team to return from an ambiguous 7–7 start to finish the season 9–7 and then go on to win the rest of their games to become the league's champion. Coughlin said, "Offense, defense and special teams doing their job. Each group having different objectives and motives but playing in harmony for each other, for the good of everyone."[26]

Let's break this down.

A football team is actually composed of three separate teams, or groups, as Coughlin calls them: offense, defense, and special teams. While the entire team has a head coach, each separate group also has a coach of its own. For the entire team to win a big game, such as an NFL Super Bowl or, on the collegiate level, a National Championship, each separate group must perform with excellence and minimize mistakes. For example, on any given day a team's offense may play well, but the game will likely be lost if its defense can't stop the run. Similarly, its offense and defense may play well for nearly sixty minutes, but a game will be lost in the final seconds of play if a late kickoff is returned for a touchdown or a field goal is missed by a kicker. In such cases, the special teams group fails and the entire football team loses the game.

Similarly, a well balanced approach is needed today in the local church to advance a credible, compelling, and sustained witness of the gospel in an increasingly diverse, painfully polarized, and cynical society . . . a disruptive model for church growth and development in the twenty-first century.[27]

As with a football team, think of your local church as one team composed of three separate teams, "each group having different objectives and motives but playing in harmony for each other, for the good of everyone."[28] The three teams can be designated spiritual, social, and financial.

- The spiritual team exists to evangelize, baptize, disciple, and multiply the Christian faith.
  - » It gathers people for worship, connects them relationally, and deploys them to work with children,

with young adults, overseas as missionaries, and in many more ways, according to their varying passions and giftings.

» The spiritual team (the church itself) is organized as a nonprofit and therefore able to receive tax-deductible donations.

- The social team exists to advance compassion and justice in the community.

» It distributes food and clothing to those in need, mentors at-risk children, provides low-cost legal services to immigrants, assists families who have adopted or foster children, works with teen mothers of preschoolers, and teaches English as a second language (ESL), and the like.

» The social team is organized by the church as a separate nonprofit (think two sisters living in the same house). It is therefore able to receive local, state, and federal grants not otherwise available to churches; engage and solicit resources from other churches or Christian organizations not typically inclined to give money directly to a local church; and tax-deductible donations from individuals.[29]

- The financial team exists to encourage and develop small business and generate sustainable income.

» It leverages church assets, including property or facilities, money, and people, to revitalize abandoned space, create jobs, establish rent roles, reduce the likelihood of crime, generate taxes for the city, and advance the common good.

» The financial team is organized as a for-profit entity (or entities) and therefore not able to receive tax-deductible donations.

We can picture this structure as a three-legged stool (see figure 8).[30]

Figure 8. The Local Church

Spiritual   Financial

Social

Assuming your church is structured this way, as we are at Mosaic, consider the various ways that each team might secure or generate revenue by creating multiple streams of income.

- The Spiritual Team
  Organized as a nonprofit; specifically, a church
  Primary method of income: tithes and offerings
  Led by a senior pastor

  » Collecting ongoing tithes and offerings from members

  » Boosting tithes and offerings by:
    o Teaching members the principles of budget building, debt reduction, and building a legacy, as, for instance, Dave Ramsey encourages through his ministry[31]

    o Encouraging members to be faithful, sacrificial, and generous in their giving

» Allowing vocational workers to raise their salaries (as do missionaries), supplemental financial support (beyond what the church can otherwise pay them), or to participate in the gig economy, and in so doing attracting staff it might not otherwise be able to afford[32]

» Promoting planned giving in the congregation through wills and living trusts

» Launching a capital campaign or in other seasons emphasizing special financial needs and challenging the church to give in addition to regular tithes and offerings

- The Social Team

  Organized as a nonprofit such as a CDC (community development corporation); legally distinct from the church
  Primary method of income: grants and donations
  Led by an executive director

  » Soliciting local, state, and federal grants not otherwise available to local churches

  » Enlisting other churches to support its programs by giving money and/or sending people to help lead them, something one church will not typically do for another church

  » Attracting charitable gifts from companies, organizations, and foundations that cannot give to a church

  » Receiving in-kind gifts and services from local businesses that support and benefit from its direct community engagement

» Partnering with the Corporation for National & Community Service to attract and employ AmeriCorps VISTA workers[33]

- The Financial Team
  Organized as for-profit business enterprise that generates ROI for owners and that benefits the church; legally owned by or separate from the church
  Primary method of income: for-profit ROI
  Led by a chief executive officer (CEO)

  » Renting space in an existing church facility to local businesses or nonprofit agencies

  » Monetizing existing services already being provided to others by the church

  » Starting a church-owned business such as a Limited Liability Corporation (LLC)

  » Partnering with investors to create a holding company whereby church facilities can be leveraged to house multiple businesses

  » Developing a strip mall or shopping center so as to embed the church in the community, rent space to local businesses, and from the income earned cover the monthly mortgage

In subsequent chapters we'll explore such possibilities in greater detail. More specifically, we'll focus on the financial leg of the stool by discussing church assets and how to leverage them; the concept of benevolent ownership; what it means to monetize existing services; and how to start new businesses—essentially, the how and what of disruptive innovation in church economics.

Mending Holes in the Net

In November 2010, at Mosaix's 1st National Multiethnic Church Conference, in San Diego, California, David T. Olson presented findings from the American Church Research Project. At the time, it was one of the largest research studies ever undertaken on the American church.[34] The comprehensive, multiyear study of 200,000 churches revealed some disturbing trends. For instance, the research showed that while the general population of the United States grew by 56 million people over a twenty-year period between 1990 and 2009, church attendance in those same years increased by just 446,540.

Equally disturbing was research concerning the growth of evangelical churches as a function of median household income (see figure 9). Between 2000 and 2009, churches of affluence (with a median household income of more than $60,000) grew

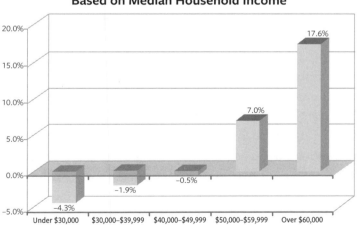

Figure 9. 2000–2009 Evangelical Growth Rate Based on Median Household Income

© 2010 by David T. Olson. www.theamericanchurch.org

at a very healthy 17.6 percent. As median household income dropped, however, so did the growth rate of churches. The study found that evangelical churches with a median household income of $50,000 to $59,000 experienced 7 percent growth, while churches on the far end of the graph with a median household income of less than $30,000 shrank by -4.3 percent. This information hit home because the median household income in Little Rock's 72204 ZIP Code, where we had established Mosaic, was about $30,000 at the time.

Having shared the information, Olson then asked the four hundred pastors and ministry leaders in attendance a significant question: "What do you think will be the long-term implications for the church in America if these trends continue?" An awkward pause followed, one that seemed to last forever. Looking back, I (Harry) consider it a defining moment in my own ministry journey, a revelation of sorts that makes my heart burn with deep passion even today as I relive it. Let me explain.

In that moment, as I reflected on Olson's question, I recalled the words of Christ in Matthew 13:47. In this passage, Jesus compares the kingdom of heaven to a fisherman's dragnet. With the American church in mind, I realized one horrific characteristic about our dragnet: it contains significant holes in the most poverty-stricken areas of our cities. Therefore, a significant number of people are dissociated from the church and from the hope of the gospel based on socioeconomics, due to traditional approaches to church planting, development, and funding. So right then and there, I committed my heart to do my best and to ensure that any holes in the net near Mosaic would be repaired . . . that is, that in our area of town the net would be mended, strengthened, and as effective as possible.

Of course, finances remain a challenge for any church regardless of where it's located. However, for churches in the most underresourced areas of a city as well as in rural communities, lack of funding remains a constant threat to survival and growth. I, for one, have spent many agonizing hours of prayer seeking God's provision for our church. True to his character, he has continually given us what we need and blessed Mosaic in unexpected ways through the years. Rarely, however, has God supplied for us in such a way that we did not need to depend on him in one way or another to come through financially. As with most other things in the Christian life, he expects us to be involved with him in the pursuit of finances by praying fervently, taking intentional steps, working smart, dreaming big, and being wise with money.

To shore up the largest holes in the net, those that are in the most poverty-stricken areas of our cities, we must not be afraid to examine the motivations of church planters and/or of church planting leaders in denominations and networks seeking to establish congregations in a disproportionate number of communities that are generally affluent or upwardly mobile. Increasingly, we must challenge pastors and collective local churches to become repairers of the breach in difficult contexts (Isa. 58:12). Of course, it's much easier to plant churches in trendy or growing neighborhoods, those that provide a better potential for attracting white-collar workers and professionals to the church. But honestly, how many people from the lower to lower-middle classes might actually become part of such a church? Would the homeless and/or those caught in cycles of addiction, poverty, or dysfunction ever truly be welcomed as valued members, seen as people to love and not problems to address? It's not likely.

Figure 10. Distribution of Mosaic Adult Attenders, Little Rock Campus

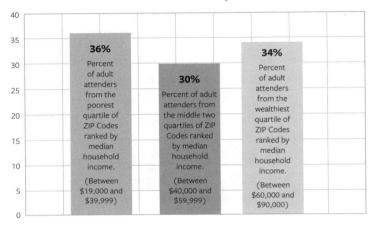

Seeking to plant in communities where the potential for numeric and financial growth exists is only natural. But God calls us to live in the supernatural. In other words, there will always be planters who gravitate toward affluent areas. What is needed are those who would dare to plant churches in areas of great need, in difficult contexts. Without them, great numbers of fish (so to speak) will continue to swim right through large holes in the net, while Jesus weeps.

Six years later, I marveled at what God had done at Mosaic. By way of analysis, I ranked thirty-one of the thirty-four ZIP Codes in central Arkansas by median household income and cross-listed them with the residences of our core members. Here's what I discovered:

- Just over one-third (36 percent) of our people live in the poorest quartile of ZIP Codes.

Figure 11. Mosaic Adult Attenders' ZIP Codes
Ranked by Median Household Income

Rank (Low to High)	ZIP Code	Median House-hold Income	Percent of Attenders	Running Sum	Reverse Running Sum
1	72114	$19,088	1.0%	1.0%	100.0%
2	72202	$29,289	4.1%	5.1%	99.0%
3	72209	$30,014	8.2%	13.3%	94.9%
4	72204	$30,984	17.4%	30.7%	86.7%
5	72117	$34,042	0.5%	31.2%	69.3%
6	72206	$36,221	4.6%	35.8%	68.8%
7	72072	$38,029	0.3%	36.1%	64.2%
8	72118	$40,107	2.1%	38.2%	63.9%
9	72106	$41,607	1.5%	39.7%	61.8%
10	72076	$42,141	0.5%	40.2%	60.3%
11	72143	$46,897	0.3%	40.5%	59.8%
12	72103	$47,850	1.3%	41.8%	59.5%
13	72201	$48,542	1.3%	43.1%	58.2%
14	72205	$49,446	8.2%	51.3%	56.9%
15	72086	$50,632	0.5%	51.8%	48.7%
16	72210	$51,449	5.9%	57.7%	48.2%
17	72065	$51,703	0.5%	58.2%	42.3%
·18	72011	$54,015	0.5%	58.7%	41.8%
19	72002	$54,468	0.8%	59.5%	41.3%
20	72142	$56,477	0.5%	60.0%	40.5%
21	72023	$57,408	0.8%	60.8%	40.0%
22	72120	$57,720	2.8%	63.6%	39.2%
23	72227	$59,839	2.1%	65.7%	36.4%
24	72022	$60,620	3.6%	69.3%	34.3%
25	72116	$61,833	1.5%	70.8%	30.7%
26	72207	$63,848	3.6%	74.4%	29.2%
27	72211	$66,559	8.5%	82.9%	25.6%
28	72019	$67,112	1.3%	84.2%	17.1%
29	72113	$71,012	3.8%	88.0%	15.8%
30	72223	$76,704	7.7%	95.7%	12.0%
31	72212	$88,533	4.3%	100.0%	4.3%

Note: approximately 3 percent of Mosaic attenders did not list a ZIP Code and were not included in this analysis.

- About one-third (34 percent) of our people live in the wealthiest quartile of ZIP Codes.
- The remainder (30 percent) live in the middle two quartiles of ZIP Codes.

In other words, we have been able to mend holes in the net associated with our geographical area of the city.

These results have not come apart from intentionality regarding church economics. Among other things, we have sought and nurtured outside investment from people who don't attend Mosaic but who believe strongly in our mission and vision. We have stimulated economic development in our community, so much so that people from the suburbs are no longer afraid to venture into an area of town they once avoided. We have also established a sister nonprofit, Vine & Village, through which we have launched numerous programs focused on the people of 72204. Through Vine & Village we have developed entrepreneurial strategies for community engagement, applied for and received local, state, and federal grants, partnered with more affluent churches willing to send their people and money to help, and ultimately progressed beyond rhetoric to results for the glory of God. In addition, we leverage our facility and property to generate income, and most of our staff members (at one time or another) have supplemented their salaries by raising financial support as urban missionaries.

In short, we have developed multiple streams of income in order to effectively fund our work, not so much because we can do so but because doing so is essential, given our context. Each of these revenue streams plays an important part in supporting the work of Mosaic beyond what our congregation's tithes and offerings might otherwise be able to support. As we said, tithes and offerings today account for just 70 percent of our total annual budget.

Here's another interesting dynamic about the demographics of our church. Approximately 260 core families populate thirty-one of the thirty-four ZIP Codes in central Arkansas. Mark has previously explained why a healthy multiethnic church has the potential for significantly greater influence in a community beyond what its size seems to suggest, even greater influence than has a homogeneous church ten times its size, and our analysis proves this is true.[35] Due to the ethnic and economic diversity of Mosaic, our people scatter on Sundays to do life, to work, and to fellowship with others in virtually every part of the city: from homeless camps to rural areas of the county, from the barrios and urban neighborhoods to gated communities. In fact, there are very few places in central Arkansas in which Mosaic's influence is not recognized or influential.

Chris Rice once wrote, "God is using race to heal the church."[36] I too now believe that in the twenty-first century those who dare to establish such churches can and will bring renewed faith, peace, hope, and love to the most impoverished, underserved, and underresourced areas of a city. Here's the question: Is your own church willing to help mend holes in the net? If so, create multiple streams of income not only to ensure its own stability but to help establish sustainable churches or campuses of its own in difficult contexts.

Needed: Disruptive Innovators

In an opinion piece for CNN, Mel Robbins, an expert on human behavior and motivation, wrote about people and organizations known in the business sector as disrupters. Disrupters "break the mold, change our thinking about the mold and then hand us the new rules for how things work. . . . They don't fix what's broken because they don't innovate from inside the system."[37] Rather, they operate outside conventional wisdom, turning systems upside down to effect systemic change. Stated another way, disrupters can both see and sense what lies ahead, around the corner, long before others have even arrived at the intersection of present and future. As such, disrupters define, refine, and ultimately create new realities by changing the way we see things, think about things, and get things done. Once disrupters gain momentum, they do more than simply describe the future, they create and establish it. They frame the questions, shape the narrative, and influence the conversation. In so doing, they challenge what is and inspire what is yet to come.[38]

That's what we're doing in this book: challenging what is in order to inspire and inform what is yet to come. By sharing our understanding and a model for disruptive economic innovation, we are providing you with detailed strategic instructions. In the process, we are encouraging your church and churches everywhere to create multiple streams of income so that they will no longer be limited by stagnant or declining tithes and offerings, or entirely dependent on them. Indeed, we believe it will be necessary to do so in the future as both global and local economies continue to morph and subsequently affect jobs, wages, and giving. Only by applying such understanding will pastors be able to lead churches from survival to stability, and from there to sustainability in the years ahead.

IDEAS AT A GLANCE

- Times have changed. The demands on individuals and families have increased in terms of the work required and the money needed to live a middle-class life. Resources, both human and financial, are spread thin, and this affects the church. Tithes and offerings alone, like single paychecks in the past, are no longer enough to provide for a church's needs and wants.

- It is natural to plant churches where growth potential exists. But doing so keeps people in the most poverty-stricken areas of our cities disenfranchised from the help and hope they so desperately need. Pastors, planters, denominations, and networks must reexamine motives and methods.

QUESTIONS FOR REFLECTION

1. How have the financial demands on families, even your own, changed throughout your time in ministry?

2. What revenue streams currently exist in your church through which it receives income beyond tithes and offerings?

3. How would you describe the economics of your church at the moment? Are you merely surviving, relatively stable, or financially sustainable? Why is this so? Where might the future lead?

5

Leverage Church Assets

The Fourth Directive

The Arizona Diamondbacks (D-backs) became an official Major League Baseball team in 1998. As part of the National League West, the D-backs compete regularly against the Los Angeles Dodgers, San Francisco Giants, San Diego Padres, and Colorado Rockies. In the summer of 2018, in recognition of the team's twentieth anniversary, *Phoenix* magazine published a special edition, "edited and designed by *Phoenix* staff, to celebrate the achievements of the Valley's Major League Baseball franchise, and to recognize our longstanding partnership with the team."[1] What caught my attention in the issue was not so much memorable individual and team moments from D-backs history, including a World Series win in 2001, but the remarkable impact the team, together with the Arizona Diamondbacks Foundation, has had in its community. In a "Letter from the Team President," D-backs president and CEO Derrick M. Hall shared that more than $55 million had been raised and donated through the years to benefit initiatives related to children, health care, and education. In addition, forty youth baseball and soccer fields had been built throughout Arizona complete with amenities, more than 150,000 youth baseball and softball players had been out-fitted with team-branded jerseys and caps since 2014, and more than 1,200 nonprofit programs in the state had been awarded grants totaling $5.5 million. Hall concluded his letter by writing, "While I am proud with all this organization has accomplished as we celebrate our 20th anniversary season, I cannot wait to see what the next 20 years bring on and off the field."[2]

How are the D-backs able to accomplish so much off the field in terms of benevolent impact? By leveraging assets to bless the community. In so doing they build brand awareness, recognition, and loyalty. In turn, people buy tickets to see the team play, which keeps it in business. Without ticket sales, the team could not exist. Without investment in the community, people would not be as motivated to go to a game. In the end, it's a win-win approach, one whereby the community benefits and the organization is sustained.

As we continue to explore the coming revolution in church economics, keep this image in mind: an organization blessing its community *and* in the process helping to fund its mission. The operative philosophy can be stated this way—leverage church assets to bless the community and generate sustainable income. The question is, How can you and your church do so?

Sacred *and* Secular

Unfortunately, many pastors tend to view their work as sacred and the work of those in business as secular. In the eternal scheme of things, some wrongly consider (whether consciously or not) that the role of those in business is to make money and the role of those in ministry is to advance the kingdom of God. In other words, an erroneous distinction exists in the minds of many between work that is sacred and work that is secular.

Such thinking was addressed in a *New York Times* article titled "Their Mission: Spreading the Word through Business." In discussing "business as mission" (BAM), one entrepreneurial faith leader said, "There has been this idea that it's not as spiritual to be a businessperson. . . . The truth is totally the opposite because this is genuinely how most people have to live their lives."[3] The article continues, "Christian-run companies are multiplying in just about every corner of the globe, reshaping overseas mission work. These businesses form a movement known variously as business as mission, kingdom business and great commission companies, after the biblical charge to 'make disciples of all the nations.' . . . Business as mission grew from a 1980's mission movement to reach people in the 'resistant belt' across North Africa, the Middle East and Asia where Muslim, Buddhist or anti-theistic governments made it hard or impossible for religious workers to get visas."[4]

Steven L. Rundle, an associate professor of economics at Biola University and the author of *Great Commission Companies: The Emerging Role of Business in Missions*, also notes, "Evangelical groups are recognizing that mission-minded businesspeople can do things that traditional missionaries cannot. 'The future generation of missionary will be the rank-and-file businessman,' he said. The wheel, he added, has come full circle: many of the first emissaries of the gospel were tradesmen, not priests."[5]

By definition, "Business as mission is demonstrating what the Kingdom of God is like in the context of business—and as we do so, engaging with the world's more pressing social, economic, environmental and spiritual issues."[6] Jo Plummer, writing for *Business as Mission*, states:

> [The] contemporary business as mission movement represents a growing intentionality in the global Church to fully integrate business goals with the call to the whole church to take the whole gospel to the whole world. It is an answer to the prayer, "May Your Kingdom come on earth as it is in heaven," as people and communities are positively transformed through for-profit business activities. . . . BAM is the intentional integration of business and mission. . . . [It is] not "ministry" tacked onto business for convenience or business tacked onto ministry. Instead the mission is worked out in and through the business, through its activities, through the products and services and through relationships.[7]

BAM can be described as profitable and sustainable business that is:

- intentional about advancing kingdom of God purposes and impacting people and nations for this reason
- focused on holistic transformation and the multiple bottom lines of economic, social, environmental, and spiritual outcomes
- concerned about the world's poorest and least evangelized peoples[8]

For too long, secular business and sacred mission have been compartmentalized in the minds of many pastors, if not the American church. In the future, these must be better understood and rightly integrated, as believers around the world recognize. BAM philosophy applied by a local church can exponentially enhance its influence and impact in a community.[9]

Church Assets

Of course, the assets of a church are many. In this chapter, we'll discuss the three we believe to be most prominent: people, facilities, and money. As you'll see, each broad category carries nuanced understanding and possibilities all its own.

- people: members, staff, and interested others with whom a church has a relationship
- facilities: buildings, property and/or land owned by a church or that it can acquire to develop
- money: funds that can be collected, pledged, aggregated or borrowed

People

Business-minded people within a church or who are otherwise sympathetic to it can and should become valued partners in any effort to ensure sustainability. As with facilities, such people are often underutilized by pastors who haven't thought deeply about how best to deploy them within the church or who value them primarily for their money. In churches seeking to get beyond total dependence on tithes and offerings, this will need to change.

From Transactional to Personal

Whether pastors recognize or will admit it, financially successful people in a church are often viewed as a potential and necessary go-to target for money, especially when larger sums are required. In many ways, this is understandable. Pastors are like the captain of a ship and responsible for its course. They can't afford to be bashful but must be bold in this regard if they are to keep the church afloat and moving forward. So when giving is down, needs arise, or a campaign goal is set, pastors quite naturally gravitate toward those in the body who have the ability to make significant contributions or a substantial onetime gift. There is nothing inherently wrong with doing so.

However, pastors who feel the need to spend more time with, cater to, or placate financially successful people risk establishing transactional relationships that can become problematic. Similarly, we may offend or alienate such people by calling on them only when money is needed or in times of financial crisis. We are called to shepherd the flock of God, not to squeeze or play it for money.

The New Testament has something to say about this all-too-common human struggle of treating those with money differently than others. As the New Testament writer James notes:

> My brothers and sisters, believers in our glorious Lord Jesus Christ must not show favoritism. Suppose a man comes into your meeting wearing a gold ring and fine clothes, and a poor man in filthy old clothes also comes in. If you show special attention to the man wearing fine clothes and say, "Here's a good seat for you," but say to the poor man, "You stand there" or "Sit on the floor by my feet," have you not discriminated among

yourselves and become judges with evil thoughts? (James 2:1–4 NIV)

With this in mind, take a moment to reflect honestly on the following questions:

- What are your motives for building relationships with financially successful people in your church?
- Are the relationships you have or that you are forming with them more personal or transactional?
- What might you do to avoid favoring or using financially successful people in ways that are unhealthy?

To avoid such problems, we strive to ensure that leadership throughout Mosaic reflects not only ethnic but also economic diversity. At any given time, for instance, our elder board includes blue-collar workers, white-collar professionals, government employees, and business owners who are all equally empowered to serve as one without distinctions. Typically, this does not happen apart from intentionality.

From Employees to Entrepreneurs

In addition to asking for money, pastors often approach financially successful people and entrepreneurs to be leaders in the church. Because these people are usually friendly, outgoing, and relate well to others, pastors may ask them to be greeters or ushers, teach a class for adults or children, serve coffee in the café, or count the offering on Sunday mornings. Since they want to serve God, love the church, and are loyal team players, these people will gladly agree to do so. In this way, however, pastors all too often mismanage entrepreneurs by positioning them as "employees" (so to speak) in the church.

On the other hand, pastors may recognize the capacity of entrepreneurs and invite them to shoulder even more responsibility by serving on church boards or committees, becoming a small group leader, supervising a member of the staff, or overseeing a ministry team. In this way, however, pastors all too often mismanage entrepreneurs by positioning them as "managers" in the church.

To be clear, we are not diminishing the importance of the above mentioned work or suggesting that pastors should not enlist people to serve responsibly in such roles. What we are saying is that when entrepreneurs are rightly empowered to help the church meet economic challenges and overcome financial obstacles in partnership with pastors, in recognition of their life experience and in the strength of their gifting, they come alive! Yes, when pastors position entrepreneurs as "entrepreneurs" in the church, it will thrive.

Bobb Biehl "is a proven 'behind the scenes' leader . . . mentoring and bringing strength up under the platform of 'up front leaders.'"[10] Years ago, I had the opportunity to meet Bobb and benefit from his counsel. Among other things, his *Team Profile* is an excellent tool for identifying role preferences and best positioning people for service in a church, company, or organization, as part of a team.[11] The self-scoring inventory will "help your team define what they really want to do. Doing what you love is the key to fulfillment and is key to a successful organization."[12] Larry Osborne, pastor of North Coast Church in Vista, California, describes the *Team Profile* as an invaluable tool. "We use it with every potential hire and team member.

It's a profoundly simple tool that helps us quickly identify what roles and assignments will most likely energize or emotionally drain a person."[13]

Biehl's *Team Profile* suggests that there are five essential roles on a team and stresses the importance of rightly positioning individuals according to their passions and preferences. The profile identifies and describes role preferences in this way:

- Design

  A designer prefers to work with abstract theories and ideas.

- Design-develop

  A design-developer prefers to start with a blank sheet of paper, work on original ideas, and make original ideas work the first time.

- Develop

  A developer prefers to work on someone else's great idea by tweaking an existing model and establishing clear goals.

- Develop-stable

  A developer-stabilizer prefers to be task-oriented, will be faithful to the task, and is committed to improving results.

- Stable

  A stabilizer prefers to be people-oriented, will be loyal to the team, and is committed to maintaining control (i.e., order, process, and systems).[14]

For the purposes of our discussion, we suggest that pastors begin by looking to empower and leverage design-developer types to help the church create multiple streams of income. As the *Team Profile* suggests, design-developers not only have ideas but also will pursue them relentlessly until they come to fruition. When empowered by pastors as partners in leading the church to become a benevolent owner (chapter 6), monetize existing services (chapter 7), and start new businesses (chapter 8), design-developers can and will be able to create "something from nothing." In so doing, they will recruit and empower all the rest (designers, developers, developer-stabilizers, and stabilizers) to get the job done.

From Fixed to Flexible

Currently, our payroll accounts for approximately 55 percent of Mosaic's budget, and we have worked very hard throughout the years to raise salaries to market rates. Still, most if not all of us could make more money by serving on staff in other churches, and we've all been offered better-paying jobs. It's our calling that keeps us here. To offset what the church cannot otherwise provide in terms of larger paychecks, we offer much to our staff in the way of intangibles. In other words, while Mosaic does not have the financial resources to pay its staff what they might earn at more affluent churches, we do have flexibility, and we know how to use it.

For example, we encourage and allow time for our staff to generate income outside the church. Some consult for organizations, and others have started businesses or own rental property to supplement the income they earn at Mosaic. Of course, each person has a job to do at the church and is expected to do it with excellence, whether it takes fifteen or fifty hours in any given week. The quantity of time they invest is not as important to us as the quality of their work from week to week. As for time off, the requesting and recording of vacation days is also handled in a very informal way. In general, if and when team members need a break, they can take one as long as they have made arrangements to fulfill responsibilities. Similarly, we don't demand or track fixed office hours. We ask only that staff attend regularly scheduled meetings and shoulder additional burdens from time to time as part of the team. Since most of our spouses have jobs too, we often work from home or attend meetings online to save time and money. In these ways and more, our staff is given flexibility to manage themselves, which they not only appreciate but also recognize as valuable, since time is money.

From Stars to Servants

While the old adage "one person, one job" is ideal, in churches with limited resources this is not often possible. Therefore, it's smart financially to hire servants (not stars) who are willing to do whatever is necessary, whenever it's necessary, to support the team and church without regard to power, position, or privilege. For instance, as a former engineer, I (Harry) consistently find myself atop a scissor lift pulling cable, hanging security cameras, and analyzing weather data to determine the probability of flooding (our building sits in a flood plain *and* a flood way), in addition to being a teaching pastor and shepherd of the flock. Alex Diaz is also a teaching pastor at Mosaic and our designated staff director. He regularly pitches in with graphic design and volunteer management. Maria Cheung, who until recently served as our children's pastor, has a sharp eye for detail and is a trained accountant who has served on our finance team. Paul Kroger, the executive director of Vine & Village, also manages tenant relations on behalf of the church. Even the most recent addition to our staff, Lawrence Hicks, who oversees logistics and operations, as he did in the Air Force (colonel, retired), has become a recognized teaching pastor here as well.

Facilities

It's no secret: throughout the United States church buildings, parking lots, and real estate sit idle from Sunday to Sunday. The increased costs to maintain these spaces, coupled with shrinking attendance and a subsequent decline in giving, have left many church leaders tired of fighting the budget and therefore opting to close the doors or merge with other congregations. In some situations, facilities and land revert to a denomination which may sell them for similar reasons, reassign them to other churches, or make them available to new church planters. Alternatively, leaders under financial pressure can be all too quick to sell underutilized property or undeveloped tracks of land to relieve financial strain, alleviate debt, or fund expensive repairs to existing facilities. Yet this is often shortsighted. In a few years, the money from these sales will be quickly spent . . . then what?

Thom Rainer writes, "Between 6,000 and 10,000 churches in the U.S. are dying each year. That means around 100–200 churches will close this week. The pace will accelerate unless our congregations make some dramatic changes."[15] Yes, more churches today must find a way to stay afloat. But beyond mere survival, the creative use of buildings, property, and/or land can help them thrive. Such things are assets a church can and should leverage to bless the community and generate sustainable income by creating rent roles, for instance, or establishing other profitable business enterprise.

For example:

- What if seldom-used classrooms or entire wings in a building were converted to house a nonprofit center; a co-working space; an innovation or nutrition hub; a health and fitness club; medical, legal, or counseling offices; or a commercial coffee shop from which the church collected rent?

- What if, instead of being sold, church-owned land was leased to a commercial developer who established an office complex or a mall whereby the church would be paid a percentage of the rents collected for the next thirty years?[16]

- What if land was acquired in partnership with investors and developed for the community *and* used by the church?

- What if an empty or underutilized parsonage was converted to a rental house and the pastor's salary increased by a portion of the profits, so that he or she might be able, in turn, to buy a home and build equity?

- What if, instead of determining to buy land and build new buildings, the church bought an abandoned property and repurposed it at a fraction of new construction costs in order to save money and bless the community?

- What if "dark property" was rented by the church or church plant for less than what similar square footage in a better part of town would cost, and then the church leveraged the building to attract small businesses, create jobs, help reduce crime, generate tax revenue, and improve quality of life in an underresourced community?[17]

- What if facilities were developed as a commercial "point of destination" whereby the church helps attract people to the area, creates a sense of pride in the neighborhood, and changes the way others perceive a part of the city?

As I have written elsewhere:

It's long past time for churches to stop building facilities on islands of land entirely to themselves and start planning for mixed-use development from which the entire community can benefit. In so doing, literally, the church can be built into the community, become incarnate with the community, and by leveraging resources and assets generate income that can potentially pay for the project. . . . Whether by building or purchasing a small strip center, then, with just a few stores, or something on a much grander scale with loft apartments, such thinking today makes spiritual, social, and financial sense. It is the type of good work that brings glory to God (Matt. 5:16), advances a credible witness, proclaims resurrection, and demonstrates the power of redemption in tangible ways, especially whenever abandoned property is brought back to life in an underresourced community.[18]

Money

There is another old adage that says, "It takes money to make money." While there is some debate as to whether or not this is always true, what is certain is that a collective group of people has the ability to aggregate funds well beyond what most individuals can raise on their own. For example, an individual who sets off to start a business might have $25,000 to invest in it. But what if the total amount of capital needed to open is $100,000? To get the additional $75,000, he or she would likely approach family and friends, find potential partners, or apply for a commercial business loan. If investors and/or partners are added, the individual's share of the profit is diluted.

However, when two hundred, five hundred, or one thousand members of a local church, for instance, determine that a financial need exists, there is a worthy dream to chase or a compelling reason to pool resources, then $75,000 is much easier to raise and more quickly acquired. In addition, much of the money can be donated to the church as tax deductible and spent in ways that can be leveraged someday under current Internal Revenue Service (IRS) regulations. Therefore, the ability to aggregate financial resources is also an asset of the church.

In addition, members of the church or the church's staff are well connected to a host of others outside the church. These relationships too are an asset that can be leveraged to attract donations or investment in the church. For example, Alex Diaz joined our staff team in the fall of 2015. As part of negotiations, he asked Mosaic to guarantee him a full salary and benefit package. In return, Alex determined to leverage his connections outside the church to raise money for Mosaic. In this way, Alex could transition from raising his own personal support, as he had been doing for many years, to being fully funded by a local church. In agreeing to this plan, both the church and Alex benefited, as did our student ministries pastor at the time, Cai Lane. Because of this arrangement, Mosaic was able to use a portion of the money raised by Alex to hire Cai full-time. Three years later, Cai was fully funded through the church's budget, and Alex's primary donor, though having already fulfilled his original commitment, still sends a significant amount of money to Mosaic each year in support of our mission and vision.

The fact that a church is typically established as a nonprofit means that possibilities like this can be pursued (in addition to tithes and offerings) and money aggregated from donors outside the church who not only believe in the work but can benefit from a tax deduction. Thus, a church's nonprofit status should also be considered an asset.

But be careful. As Carter and Courtney Reum, writing for *Inc.*, remind us, "The more money the better, right? Not so fast. More money might sound and feel better, but it may mean you'll be forced to grow your company prematurely. Do you remember Groupon? They went and expanded across the world, setting up sales teams everywhere. Now, people hardly use them. That's because they grew prematurely. Take the least amount of money required to help find your earliest customers and tune your product."[19]

In other words, as discussed earlier, seek first to solve simpler financial challenges before pursuing bolder, riskier initiatives in order to increase the odds of effectiveness in leveraging church assets.

Avoid Compartmentalization

Like the erroneous compartmentalization of secular and sacred work, the American church today also compartmentalizes justice and Jesus (the gospel). Particularly in evangelical circles, the pursuit of justice is seen as peripheral to the gospel and not intrinsic to the gospel as it should be understood. Micah 6:8 is just one biblical passage that directly connects the two: "He has told you, O [people, presumably those who love God and their neighbors], what is good; and what does the LORD require of you but to do justice, to love kindness, and to walk humbly with your God?"

In the same way, it is likely that some will see church economics, such as we're prescribing, as human means and effort. These will consider our encouragement to leverage church assets in pursuit of sustainable income as advice that stands in direct opposition to what God otherwise expects—namely, that we (the church) should walk by faith and not by sight (2 Cor. 5:7); seek first the kingdom of God, knowing that all the rest will be added (Matt. 6:33); ask of him, knowing it will be given, and knock, knowing doors will open (Matt. 7:7).

Overly spiritual or idealistic thinking, however, cannot establish the long-term sustainability of a local church any more than simply asking God to save individuals will make it happen (in most cases) apart from personal, collective, and intentional evangelism. This is not to take anything away from God. Rather, this is how God himself has determined for the gospel to advance. In other words, as we do what we can for him, he does for us, for others, and for his church what we cannot do ourselves. The apostle Paul, for instance, instructs believers in Ephesus to *do something*—namely, to "walk in a manner worthy of the calling with which you have been called, with all humility and gentleness, with patience, showing tolerance for one another in love, being diligent to preserve the unity of the Spirit in the bond of peace" (Eph. 4:1–3)—just after he acknowledges the One "who is able to do far more abundantly beyond all that we ask or think, according to the power that works within us" (3:20). Thus, God does for us *and* expects us to do for him as well. Yes, for eternal purposes known only to him, God has chosen to make us partners in the efficacious accomplishment of his divine will.

Similarly, churches that are in a good place financially at the moment should not presume they will be so in the future apart from intentionality. Since we intentionally develop plans for evangelism, discipleship, and worship from week to week, we should not shy away from intentionality when it comes to leveraging church assets, blessing the community, and generating sustainable income. As we have seen, good and faithful stewardship involves both management and the investment of assets. To assume or to argue that God intends for a local church to struggle financially while sitting on its assets as leaders wait, hope, and pray for provisions—as if to do so somehow makes them more spiritual—is misguided at best and certainly a poor excuse to maintain the status quo.

Might our recommendations be taken out of context, exploited by self-serving, greedy, or narcissistic leaders, and/or lead some to build their own kingdoms under the guise of building the kingdom of God? Unfortunately, yes . . . it's possible. Nevertheless, we believe that the majority of pastors are well meaning and genuinely want to bless their communities, advance the common good, lead people to Christ, see their congregations mature, and someday leave their churches well positioned for long-term kingdom impact. The fact is that self-serving leaders already exist in the American church and are exploiting people through nefarious business practices. That didn't begin with us, nor will it end with us . . . and, of course, exploitation of the faithful is not at all what we have in mind. Again, the question is, How can a church leverage its assets to bless the community and generate sustainable income? Indeed, if leaders don't address this question today, they may not have a church tomorrow.

Advance the Common Good

Similarly, we are not suggesting that pastors should leverage church assets in ways that are dismissive of any dependence on God, are otherwise mercenary and self-serving, or contribute to problems associated with gentrification. Rather, churches should do so to participate in the local economy, help stimulate recovery, and promote revitalization efforts particularly in neighborhoods that are underserved and underresourced. Indeed, "churches will need to do so in the years ahead not only to survive but also to thrive in uncertain national and economic times, and in the process to advance a credible gospel witness and influence."[20]

In terms of next steps, here are three that your church might take to assess opportunities:

1. Contact local business and community leaders.
   » Establish relationships with those who have a history of involvement with the community and a vested interest in its development. Approach them humbly to learn about current economic realities and to share your heart. Ask questions concerning the following:
      o Industry. What currently drives the local economy? What businesses are needed for it to not only survive but also thrive in the future?

o Infrastructure. What support is lacking or keeping the community from becoming a destination for people and businesses?

o Investment. Who might be willing to invest, create commerce, or in some other way partner with the church in redeeming the community? What incentive is there to do so?

2. Consult an attorney or tax specialist.

» Generally speaking, senior pastors will not have previously purchased commercial real estate, started a nonprofit organization, opened a for-profit business, or studied tax law. For these and other reasons, it is important to consult an attorney or tax specialist regarding local, state, and federal laws before jumping into these waters (Luke 14:28–30). Here are just a few of the questions to ask:

o Can a church or 501(c)(3) actively engage in for-profit business? If so, are there different laws affecting the two types of organizations?

o What is unrelated business income (UBI)? When is a church or nonprofit subject to it?

o If a church can show that income made was used to support its overall mission and expenses, will it have to pay taxes on that income? If so, how much, to whom, and why?

3. Consider hiring or designating someone for profitable business development.

» Traditionally, a church will employ a business administrator or bookkeeper to provide financial oversight and accounting, and it is important to do so. You might go beyond this, however, and hire someone with even greater business acumen, that is, someone who can not only account for revenue but generate it by:

o Attracting small business, pursuing outside investments, and taking advantage of federal or private grants.

o Assisting with contract negotiations, tenant management, and the development of community .

o Accounting for UBI, billing, and legal services.[21]

"Play with Who You Got"

From my sophomore through my senior year in college, I played baseball for Coach Al Worthington at Liberty University (1981–83).[22] Having spent fourteen years as a pitcher in Major League Baseball, Coach Worthington was a man of great wisdom, insight, and life experience, who cared more about seeing his players become godly men than winning baseball games—though win we did, competing in three NAIA (National Association of Intercollegiate Athletics) National Championships during my time on the team.

One day not long after we had rolled out of Liberty for an away game, some of us noticed that our starting pitcher, the one who was scheduled to throw that day, was missing. As word and panic spread, it reached Coach Worthington who was seated near the front of the bus. Completely unfazed, he then stood up to address the team. With arms outstretched, and holding on to the luggage racks above the seats on each side of the aisle, he calmly said, "Fellas, I learned a long time ago . . . you play with who you got."

Similarly, local pastors and churches should play with who and what they've got—that is, they should leverage available assets to bless the community and generate sustainable income. The following three questions and examples can help you more specifically consider this directive.

Are You Thinking Correctly?

Years ago, Mosaic agreed to help a small church of fewer than twenty people in another state. More than fifty years old, the church and its facility had long ago seen better days. Meeting each Sunday in the basement, the people were renting their sanctuary and classrooms one floor above them to a mega-church in the area. The $2,000 a month the larger group was paying in rent was considered a blessing by leaders of the small church. The money enabled them to cover the bills each month, or so it seemed.

While on the surface $2,000 a month sounded good, we calculated that it was actually costing the small church $2,500 a month in associated expenses to house the larger church's campus. In other words, the small church was actually losing $500 a month in the deal. After involving a commercial Realtor to determine the true market value of the property, we learned it was worth $7,500–$8,000 a month. With this in mind, we met with the megachurch's campus leaders, showed them the data, and were able to negotiate a new contract charging a more commercially competitive rate of $6,500 a month in rent. As a result, the megachurch still got a deal by renting the space below its market value, and we helped the small church turn a loss of $500 a month into a net gain of $4,000 a month, which greatly strengthened their financial situation.

We suspect a good number of churches operate with a similar mentality—a ministry mentality—when it comes to renting their existing space to others. How so? By merely guessing at fees to charge and/or by considering any potential loss of revenue as "ministry" rather than as "business." In other words, they fail to consider what they have in terms of an asset, what its fair market value is, and what the actual associated costs both to rent and maintain it are.

Are You Thinking Strategically?

In the mid-1980s, I worked as a youth pastor in Portland, Oregon, for a once thriving church that by then had become stagnant. In addition to its own property, the church owned four houses on a street adjacent to what is now Multnomah University. One house was used for student ministries, and the other three houses were used as parsonages or places where missionaries would stay when home on furlough. In addition to these four houses, the church also owned a large Christian campground that had been deeded to it in 1965.

Throughout the entire three years I was employed by the church, the church faced the same financial challenges that many churches face today—those intrinsic to having an aging congregation, associated with a decline in attendance and giving, and owning property that is expensive to maintain. Still, the pastor had a radio show that the church funded from week to week.

The pastor's solution to turn things around? "Sell everything," he said, "but don't touch my radio show!"

One night, at a decisive meeting of the elder board, the other pastors on staff and I lobbied in opposition to his plan by offering alternative ideas for the use of the houses that we felt would be more beneficial to the church. In the end, the only suggestion embraced by the board was one that I proposed—namely, to keep the student ministries house and make one of the properties an intern house where four students from Multnomah could live free of charge if they agreed to work part-time for me. This would be a win for college students (who would not have to pay for housing while in school) and a win for the church (which would benefit from Christ-centered collegiate young people investing in junior and senior high school students).

By selling the other two houses, and later by turning the camp over to its board to be run as an independent entity, the church surrendered valuable assets and squandered opportunities in return for short-term financial gains. For instance, the houses could have been rented to married couples with families attending Multnomah, so as to generate income for decades to come. Instead, within a couple of years the assets were gone, the pastor was fired, and the radio show was scuttled.

Despite a lack of money to hire additional youth staff, and by thinking strategically to create the intern house, our youth ministry grew from twenty-five students in 1985 to roughly three hundred students in 1988. Better still, the intern house continued to function in this way for more than twenty-five years, allowing scores of Christian college students to benefit by living in the home free of charge, gaining valuable ministry experience, and by serving in a local church, which also benefited in ways too numerous to count.

Are You Thinking Creatively?

In 2018, I spent a weekend with a church in the southeastern United States. The church's senior pastor, together with the elders, had contracted Mosaix to help further their collective effort to build a healthy multiethnic church.[23] In my first session with leadership, the conversation quickly turned to economics. Among other things, the congregation was growing older, giving was stagnant, and the property was increasingly expensive to maintain.

Walking through the building, I learned that various groups in the community were allowed to use certain areas on a recurring basis from week to week without paying much, if anything, in return to the church. Such "ministry" was costing the church money in terms of utilities, maintenance, paper products, and more, not to mention the loss of potential monthly revenue. Even by charging these groups a nominal amount for rent each month would help cover such costs and strengthen the budget.

In addition, the church had an entire wing of classrooms that was no longer in use, one with a door at the end of the hall that opened directly to a sidewalk leading to a parking lot. I suggested that the wing could be easily converted into a counseling center, nonprofit offices, or even a dental clinic to generate rental income. It was when I walked outside, however, that a more significant (in terms of scale) opportunity became apparent.

The church sat on a well-traveled road with a high traffic count and good commerce. Part of its property was undeveloped acreage that was once used as a baseball/softball field. Adjacent to this land was a new strip mall, and among its shops was a well-established restaurant that had been in the community for more than twenty-five years. I asked the pastor if he had ever thought of approaching the developer and offering him a long-term lease of the land in order to add more shops and restaurants to the strip mall. If not this developer, I reasoned, surely someone else would be interested. As I explained, it would be a win for the church, a win for the developer, and a win for the community.

To bring dreams to reality, prayer, patience, and persistence must be combined with creative thinking and a willingness to take informed next steps. By leveraging church assets, we can bless the community and, in the process, generate sustainable income for our churches. In part, this will require us to become benevolent owners.

IDEAS AT A GLANCE

- In the United States, Christians tend to compartmentalize business and mission as if the two are mutually exclusive. Around the world, this is not the case. Christians understand the need to engage and pursue holistic transformation in communities through not only spiritual but also economic means.

- The coming revolution in church economics will be defined by addressing a single question that churches are already beginning to ask: "How can we leverage our assets to bless the community and generate sustainable income?" To rightly answer the question, churches will need to think correctly, strategically, and creatively.

QUESTIONS FOR REFLECTION

1. Is it wise or ill-advised to compartmentalize business and mission? Explain your thinking.

2. Who are the design-developers in your church, and are they rightly deployed? What ideas might they have to bless the community *and* generate sustainable income?

3. How might your facilities be leveraged to do so? Are certain rooms or areas of your building underutilized? In what ways might they be repurposed?

6

Become a Benevolent Owner

The Fifth Directive

In Luke 14:26–33, Jesus addresses the high price of discipleship by describing what is required of fully devoted followers surrendered to his will. In so doing, he poses a rhetorical question: "For which one of you, when he wants to build a tower, does not first sit down and calculate the cost to see if he has enough to complete it? Otherwise, when he has laid a foundation and is not able to finish, all who observe it begin to ridicule him, saying, 'This man began to build and was not able to finish'" (vv. 28–30).

We lived with such ongoing concerns over the course of ten years at Mosaic (2006–16) in pursuit of a building we now call our own. More specifically, we sat down in 2006, calculated the cost, and realized that unless we could someday rent parts of the facility we would not have enough money to complete the purchase, make the mortgage payment, or renovate it for our purposes.

The story of our journey from renting space in a former Walmart to becoming "benevolent owners" of a former Kmart is instructive for pastors seeking to leverage church assets to bless their communities and generate sustainable income.[1]

THE WAYS AND MEANS OF BENEVOLENT OWNERSHIP

Why do we pay so much for a cup of coffee at a local coffee shop? In most cases, it is because the coffee shop proprietor (a small business owner) rents space from someone who owns the building as an investment, someone who seeks top dollar from the tenants. The high rent is passed on to customers (the community), who must then pay higher prices for their coffee. The building owner wins, the small business owner may or may not win (depending on whether they can get customers to pay higher prices to cover the overhead), and the consumers lose by having to pay the higher prices. We can illustrate the situation as follows:

+ Building owner

? Small business owner

– Consumers (the community)

Yet if a benevolent owner has space to rent (such as a church like Mosaic), demanding top dollar from tenants is not essential to its vision or purpose. Rather than asking small business owners

Watching and Learning (2003–6)
Benevolent Owners Leverage Assets

In the summer of 2003, just two years after its launch, Mosaic subleased a 78,000-square-foot former Walmart for just 10 cents a square foot—a mere $650 a month, down from the $18,000 a month we were originally told would be required. For more than three years, we were established in that place

to pay top dollar for rent, we can lease space to them for much less than they would otherwise expect to pay. In return, we can ask them to pass on the savings to their customers, who might now be able to buy coffee at a cheaper price, creating a win for us (generating income), a win for the small business owner (lowering overhead), and a win for the community (paying lower prices). We can illustrate the alternative situation as follows:

+ Building owner

+ Small business owner

+ Consumers (the community)[2]

For nine years, we had been paying rent to the property owners of the old Walmart. The owner also was collecting rent from three stores that had been attracted to the site by our efforts. Learning from this experience, we asked the following questions:

1. What if we owned the building?
2. How could we leverage it to help our work?
3. What if we could do so in a way that blessed the community?

and under those terms. In January 2006, however, the facility reverted to the owner, and it was made clear to us that we would someday need to relocate. In fact, our contract was then converted into a month-to-month lease. At any time, we could be given a thirty-day notice and be forced to move.[3]

Unfortunately, purchasing the facility was not an option. The owner intended to convert the large space into smaller shops, and that is what he did over the next few years. First a Dollar Tree moved into the building, and later two clothing shops. Each time a new store opened we adjusted by moving our worship space, office cubicles, and designated classroom areas. It was from watching this happen that we first recognized the value of a building beyond its use as a church and how its owner profits from dividing space, attracting local business, and charging rent. We began to wonder, *What if we did the same thing?*

Because the old Walmart was not for sale there remained only one other possible location for the church in the 72204 ZIP Code, an area of unusually high crime, poverty, and homelessness to which we felt called. For some time and just a few blocks east of our location, a former Kmart had been for sale. It was the largest and most expensive property available in the area at the time. Its five-hundred-space parking lot had been problematic for years, filled as it was with potholes and being used as a late-night gathering place for criminal activity. In spite of such things, we had been trying for more than a year to buy it. When the building was purchased by someone else, however, in April 2006, and upon receiving the news, I was filled with overwhelming discouragement. The new owners had been previously renting it and selling used office furniture from what by then had become essentially a warehouse.

Nevertheless through remarkable, truly miraculous, circumstances that I have previously documented in *Building a Healthy Multi-ethnic Church* and *Disruption: Repurposing the Church to Redeem the Community*, we signed a contract in December 2006 to purchase the property for $1.7 million by December 31, 2012. To execute the deal, we put $100,000 down and promised to give the owners another $100,000 by December 31, 2009, if necessary, in order to extend the contract through December 2012 should we not be able to complete the purchase before that time.[4]

Dreaming and Pursuing (2007–12)
Benevolent Owners Bless the Community

Counting the cost, we realized that the only way we could someday own and occupy the building was to rent space to local businesses, beyond what the church would otherwise need, as we had learned in practice from the owner of the former Walmart. With the building under contract, I tried over the next several years to attract small business owners and secure leases even prior to our formal purchase of the facility. If the church could bring more money to the table by establishing rent roles on the front end, it was my hope that a bank might loan us money to complete the purchase sooner rather than later.

In theory it was a good idea, and I came close a couple of times to landing tenants. But all such efforts failed to produce results, so we continued to pray and wait. In that season, we also received offers from companies wanting to buy us out of our contract so that they could acquire the building for themselves. We had already determined, however, that any business we brought to the facility had to be good for the community no matter how else it might benefit the church. These offers were not of that nature.

Such decisions help define what it means for a church to be a benevolent owner. Indeed, the process of actually becoming one requires a great deal of time and effort on the part of pastors and lay leaders alike. Some of the early possibilities we chased revealed our inexperience. Yet the Lord honored our hearts and protected us in supernatural ways, sometimes shifting our direction through a single conversation.

In those years and despite noble intentions we often felt like we were climbing a mountain whose summit we might never reach. Yet we knew in our hearts that God had called us to this community and we continued to believe he would some-day make a way for the church to be permanently established within it. Since 2006, we had saved every dime we could scrape together knowing that Mosaic would need it to buy the former Kmart; yet it was still not enough. So, in December 2009 we exercised our option and gave the owners an additional $100,000 to extend the contract through December 2012. In addition, we also paid architects along the way to design and draft plans for redevelopment of the space, apart from which we would not have been able to raise more money or obtain a loan.

Between 2010 and 2012, four banks turned us down. The numbers just didn't work. Nevertheless, we persisted until finally we had saved enough money to secure financing. Since we had already put $200,000 down, and with another $400,000 in hand, we borrowed the additional $1.1 million from Arvest Bank to close the deal. With only three weeks to spare, we completed the $1.7 million purchase on December 12, 2012, to the glory of God.

Owning and Redeeming (2013–15)
Benevolent Owners Generate Income

So we finally owned the nearly 100,000-square-foot facility, but we couldn't afford to renovate or move into it. Continuing my search to find local businesses interested in renting space from us, I contacted Eric Buckner, the owner of a Little Rock–based health club called 10 Fitness with twelve stores throughout Arkansas. Over tacos at a local Mexican restaurant, I shared with him our vision, and he expressed interest. "His concept was simple: ten dollars a month, no contract, for club membership and access to high-quality equipment, classes, and instructors. The enterprise exists to promote three values: 'be yourself, stay healthy, and live well.' Eric's business model works only if he's able to secure facilities at an affordable price, which allows him to control his overhead costs."[5] In short, if I could land the deal, it would be good for Mosaic, good for 10 Fitness, and good for the 72204 community.

After many years and many other attempts to attract tenants, 10 Fitness became our first. They came in big too renting 48,000 square feet for just $2 a square foot—that is, $8,000 a month, which was enough to cover our $7,400 monthly mortgage payment. In total, the contract guaranteed the church approximately $1 million over ten years. In addition, Eric gave the church another $150,000 for improvements to the parking lot, which the city of Little Rock required us to make. The contract also included two five-year options allowing 10 Fitness to remain in the space through 2033. If they exercise these options, 10 Fitness will have paid Mosaic more than $2 million dollars over a twenty-year period.

Given the relatively low price per square foot, Eric agreed to pay for his own tenant improvements. He spent $1.7 million on structural renovation, including the installation of electricity, HVAC, plumbing, restrooms, walls, flooring, and more. In turn, this increased the value of our building/property as follows:

- $1.5 million (December 2012): value at the time of Mosaic's purchase of the building
- $3.1 million (February 2015): value as appraised during the build-out of 10 Fitness
- $3.5 million (May 2015): value at the time 10 Fitness opened to the public

By July 2015, the dramatic increase in the value of our building positioned the church to secure a $1.7 million construction loan in order to renovate approximately 32,000 square feet of it for our own use. By then, the building/property was appraised for $3.9 million. This time, First Security Bank provided the necessary funding, wrapping our original $1.1 million note into a single loan and setting the total amount of our mortgage at $2.8 million.

On January 3, 2016, Mosaic held its first service in the newly renovated facility. At that time, our monthly mortgage payment rose from $7,400 a month to just under $16,000 a month, which included the additional costs for renovations. Half of that amount ($8,000) was covered by the rent from 10 Fitness, as it is to this day. With a goal of someday being able to completely pay the mortgage apart from tithes and offerings, we next rented our loading dock to a small appliance store for

$1,000 a month; office space to a nonprofit for $400 a month; and our parking lot for use by a carnival which generates another $16,000 annually. Later, we began renting our facility to the community, duel branding it as Rock City Events & Conference Center.

Thus, in just over three years, Mosaic had purchased a building, recruited tenants, secured a long-term lease, created $1.1 million in equity, and seen the value of the property increase by more than $2 million! In addition, six thousand people from the community joined "our" 10 Fitness, some fifteen to twenty new jobs were created, and "in a roughly three-mile radius of the church, Part I incidents dropped 19 percent, from a total of 1,361 in 2012 to 1,105 in 2016."[6]

Considering what we did as renters in bringing back to life an abandoned building that once housed a Walmart, and through the purchase and renovation of a former Kmart, Mosaic has played a significant role in blessing the 72204 community by repurposing nearly 180,000 square feet of abandoned or underutilized retail space. The combined amount of square footage equates to 54 percent of the approximately 335,000 square feet of commercial space, described in a 2007 report solicited by the University of Arkansas at Little Rock as showing signs of "significant disinvestment." In regard to the renovations of the former Kmart, our work has been described as a "structural re-use [that] keeps an important community anchor within the neighborhood, and promotes a sustainable approach to neighborhood redevelopment."[7]

At the time of this writing a successful, locally owned coffee shop in downtown Little Rock, Nexus, is retrofitting our café space with plans soon to open what will be the first such store south of I-630 in Little Rock's emerging University District. We are also helping an entrepreneur—Kim Harris, the wife of Joshua Harris who is currently the director of Vine & Village's Fresh2You Mobile Market—start her own wellness business in a part of the building not otherwise being used. Soon, both Nexus and the wellness studio will bring additional revenue to the church and move us closer to our goal of having the entire monthly mortgage paid through unrelated business taxable income (UBTI). With that goal in mind, a member of our church (a carpenter) recently began leasing space from us; and helping us get there too we expect will be Erasmo Garcia.

Erasmo is a young entrepreneur, a professional boxer, and an MMA enthusiast who started a private security firm. In so doing, he began to employ many of his fellow MMA colleagues, most of whom are current or former law enforcement officials and military types. Due to his winsome personality, dependability and quality of service, his business has grown exponentially. In fact, he started a second company to serve a large garment manufacturer in central Arkansas by screening job applicants and potential employees.

By faithfully providing security for our congregation on Sunday mornings over the past two years, Erasmo has also built credibility and trust at Mosaic. Recently, he asked if he could interview a prospective security guard by using a room at the church. This led to a discussion about housing his corporate offices at Mosaic, which we are willing to do. By adding him to the mix, the church would not only benefit from another rent-paying tenant but from the onsite security his company will provide during business hours.

Generating Rental Income (2016–present)
Benevolent Owners Remain Innovative

To be clear, "The church, like any other charity, is legally entitled to rent to a for-profit organization."[8] However, churches seeking to rent portions of their facility to for-profit entities should "consider the possible tax implications in state and local jurisdictions."[9] In addition, they should speak with their own county's tax assessor to determine what, if any, real estate (property) taxes they might incur in so doing. According to the IRS and where churches are concerned, "Rental income from real property received by exempt organizations is normally excluded from unrelated business taxable income" except in certain situations in which the exempt organization is required to pay UBTI, such as when "the property is debt-financed income or leased to a controlled entity."[10] For instance, if a church borrows money to buy or renovate a building and then rents some or all of that property, it will likely be required to pay UBTI. None of this, however, prevents a church from creating rent roles. The main thing is to ensure that it is in legal compliance.[11]

Will We Lose Tax-Exempt Status?

At this point you may be wondering, "If we rent our facilities to generate net profit, will we lose our tax-exempt status, or have to pay real estate taxes?" Indeed, Robert W. Wood asks, "Will churches get taxed here [in the US]? In the case of property taxes, it seems possible. Income taxes are another matter and seem beyond the reach of the federal government. For now."[12]

The fact is that "rental to a for-profit could cause the church to lose its real estate tax exemption, at least on the portion of the property leased to [a for-profit business]. But that can be handled by making [that business, as part of its rental agreement], pay the taxes."[13] This is exactly what we do at Mosaic. Our real estate tax exemption does not apply to the 48,000 square feet we rent to 10 Fitness. Thus, we are assessed approximately $25,000 in real estate taxes annually by Pulaski County (Arkansas); a bill we, in turn, pass on to the company. On top of their rent payment each month, 10 Fitness adds a one-twelfth portion of the expected annual taxes and we hold that money until the payment is due, in October each year, at which time we pay it.

The following excerpt from my book *Disruption: Repurposing the Church to Redeem the Community* bears repeating:

> There is an increasingly strong debate in the United States over whether or not churches should be exempt from paying federal income and property taxes, allowed to continue receiving tax-deductible donations, and providing a housing allowance to ministers. Opponents provide at least twelve reasons the federal government should abolish these privileges.[14] Currently, the Internal Revenue Service (IRS) can revoke the tax-exempt status for churches "directly or indirectly participating in, or intervening in, any political campaign on behalf of (or in opposition to) any candidate for elective public office."[15] In addition, many feel the US Supreme Court's landmark 2015 decision guaranteeing same-sex couples the fundamental right to marry leaves the church at risk as well, believing that churches could someday have their tax-exempt status revoked for "practices [that are] contrary to a compelling public policy."[16] If this were to happen—and it could—think of the cumulative effect. For

instance, according to Erwin Chemerinsky, "if the 'parsonage exemption' on religious ministers' housing costs were revoked, American clergy members would cumulatively lose an estimated $2.3 billion over five years."[17] Thus, the American church today is woefully unprepared for wholesale changes in this regard, potential changes that should not be summarily dismissed as if nearly impossible. . . . [Indeed] churches today must begin to think differently when it comes to money, not only in terms of where it will come from, but how to generate it in the future.

With such things and the future in mind, consider the following questions:

- If your state or local county began assessing congregations annually to collect taxes on buildings or property they own, what impact would this have on your own church's budget?
- Is your church prepared to absorb this additional annual cost?
- Is paying property taxes from tithes and offerings something your church would feel good about doing?

These questions address just one example of what churches might face in the future and why now is the time to become a benevolent owner.

Leveraging Power, Position, and Privilege

The lyrics to a familiar Sunday school song, written by Ann Omley in 1948, remind us that a wise man builds his house upon the rock (see Matt. 7:24–27; Luke 6:46–49).[18] In other words, proactive consideration of what may come and adjusting now, while there is still time and opportunity to do so in order to mitigate risk, is a wise approach that churches would do well to apply.

I (Harry) love what Mark has learned through his relationship with legendary evangelical civil rights leader Dr. John M. Perkins: the aim is not so much to give someone a fish, or even to teach others how to fish. Ultimately, we want to help people own the pond. Benevolent owners are those who help both individuals and entire communities get beyond barriers, and overcome obstacles that might otherwise make it difficult for them to navigate.

In Philippians 2:1–11 the apostle Paul provides an example of someone willing to empty himself in order to help others "own the pond." This someone is none other than Jesus Christ, who, being in very nature God, did not regard equality with God (his power, position, and privilege) as "something to be used to his own advantage" (Phil. 2:6 NIV). Rather, in emptying himself, Christ set aside power, position, and privilege so that others could equally share with him "every spiritual blessing in the heavenly places" (Eph. 1:3) and become "fellow heirs" (Rom. 8:17; see also Eph. 1:5). Though Paul called such humility an "attitude" (Phil. 2:5), it is clear he expects it of us in action as well. More specifically, he expects leaders in the church to leverage power, position, and privilege to advance the common good even at their own expense, as did Christ himself.

When I was in elementary school, we played a game called "King of the Hill." Finding a small hill or raised berm, we would all scramble our way to the top, where only one of us could be "king." To attain or to maintain sole position at the top of the hill, others would have to be fought off or pushed down. It was a competition and only one of us could win. Fortunately, Jesus Christ did not come to be king of the hill but to be king of the world. His aim is not to keep us down but to lift us up.

Think about it. Every demographic group in the United States today is fighting either to attain or to maintain power, position, and privilege. Yet the disruptive example of Christ demonstrates that the way to gain such things is to let them go. It's counterintuitive. In so living we do not lose anything. Rather, we gain friends, influence the community, and win high praise from God (Phil. 2:9) as others see our good works and glorify our Father in heaven (Matt. 5:16).

The reality is that none of us can change how we were born. To humbly and obediently follow in the footsteps of Christ, though, we can and must choose to leverage whatever measure of power, position, and privilege this life has afforded us and/or that we have earned, in order to advance the common good, not merely our own. Furthermore, what is true for us individually is also true for us collectively. Thus, a local church should strive to become a benevolent owner and a strong champion and advocate for the community whose actions benefit others outside its walls, especially the marginalized and disenfranchised.

At the core of this discussion lies the reality of systemic injustice and inequity, or what some call disparity. Those who have never experienced the negative effects of disparity in dealing with legal, educational, or economic systems in the United States can find comprehending the deep-rooted complexities difficult. Systemic inequities largely affected by race, class, and cultural distinctions, particularly impact the underserved, underresourced areas of a city where people are living at or below the poverty line. In such places exploitation is rampant, neglect is commonplace, expectations are low, crime is high, and addictions are seen as a way of coping and surviving—that is, a way of life. These complex issues have historical roots that are at once collective and personal. The problems are not easily solved. Like a tree with branches running in many different directions, so is systemic inequity. Its bad fruit is too often and easily dismissed by those seemingly unaffected by it. But make no mistake. We are all affected by injustice, and the church should position itself to make a difference.

The roots by which such bad fruit is produced run deep in the soil of America, and the strong trunk of the tree is not easily moved. Nevertheless, benevolent owners run to, not from, these complexities and provide hope to all, even at their own expense.

Loving Our Neighbors

Finally, churches seeking to become benevolent owners must practice love for their neighbors and the neighborhood. In other words, whatever decisions are made by the church and deemed good for it should also be good for the community. Likewise, when the community wins the church wins. The prophet Jeremiah said it this way: "Seek the peace and prosperity of the city to which I have carried you into exile. Pray to the LORD for it, because if it prospers, you too will prosper" (Jer. 29:7 NIV). Therefore, instead of becoming an island unto itself, a church should embed itself in the community.

Resurrect Abandoned Space

Similar to conventional wisdom concerning the funding of church planters, successful church plants are widely expected to someday move out of rented facilities, acquire land, and build buildings. In fact, the acquisition of property is seen as a sign of growth and maturity. However, this is problematic on several fronts. First, it takes a lot of money to buy land and build new buildings. In addition, a church cannot assume that doing so will be greeted favorably by the community, one concerned with the potential for increased traffic, the loss of green space or whose officials bemoan the church's involvement in taking off the market good commercial land that could have otherwise generated tax revenue. For example, I know of a church that once bought five premier homes across the street under arguably surreptitious circumstances, only to tear them down and build a parking lot. The neighborhood was furious and threw a fit all the way to city hall. Due to its economic power and privilege, however, the church was able to secure the votes it needed from local government. But at what cost to the community and to its reputation?

Several years later that church sold its property, including the new parking lot, only to buy prime acreage in the suburbs and build an entirely new facility. Why? As leaders shared at the time, they wanted a church campus that was less sprawling, one that was easier for members to navigate. But we should ask, "Is there not a more just and equitable way to leverage nearly $60 million in tithes and offerings, in donations past, present, and future, to ensure the community benefits as well and not merely to make a campus more comfortable for church members?"

In asking such a question, we are not intending to claim any moral high ground. We are, however, suggesting that others consider a more disruptive approach. Mosaic chose long ago not to buy land or to build new buildings, though to do so is not necessarily wrong. Rather, we chose to purchase and repurpose abandoned property in the city at a fraction of the cost. In fact, "the latter garners far more interest and appreciation among nonbelievers and can then be leveraged to redeem a community."[19] Architect and author of *10 Tsunamis Impacting Ministries*, Daniel Cook, asks, "What would happen to our ministries if we cut out the mortgages, reduce the overhead expenses of our buildings by two-thirds or more, and use our buildings seven days a week? What would happen if we make a real estate investment which allows us to interact with lots of

EMBED YOUR CHURCH IN THE COMMUNITY

- If you are planting a church, locate it in an underresourced area of the city where properties are readily available for rent or purchase at relatively low prices.

- If you are pastoring an existing church, consider the purchase of such property in order to develop it as a benevolent owner.

- If you are pastoring a church with multiple sites, look for abandoned space in which to open a new campus.[20]

people, employ persons from within our ministries, and gain sustainable income in the process?"[21]

This is what Mosaic has sought to do in bringing a former Kmart back to life. In recognition of our efforts, the Arkansas Martin Luther King Jr. Commission awarded the church its Economic Development Award in 2016. At the time, DuShun L. Scarbrough, the commission's executive director, wrote:

> As a Little Rock native, I watched the community in and around what is today Little Rock's University District decline over time, from a once-thriving business district with malls and restaurants to one with empty and neglected buildings. That's why it has been inspiring and motivating to see [Mosaic's] vision and work in recent years turn vacant buildings into viable, profitable business ventures, and help bring this community back to life. It gives me hope for what can become of distressed areas throughout the city and our state. It reminds me of what one person and one church can do to improve communities such as Dr. King envisioned. . . . This is the very kind of positive community development that I believe Dr. King envisioned. He believed that churches should not be bounded with four walls, and that churches must reach out to communities and impact the lives of people living and working within them, all people, especially those with the greatest needs.[22]

Include the Marginalized

In 1987, twenty-three-year-old college students Paul Anderson and Clint Bidleman were sponsored amateur freestyle skaters attending what is now Multnomah University in Portland, Oregon, and informally skateboarding with kids in the neighborhood. "One night while studying, Paul met some skaters on campus who had wandered away from a gym night at nearby Central Bible Church. Returning the skaters to the gym, Paul was encouraged by Central Bible youth intern John Stone to come weekly and minister to skaters who didn't fit in. Determined to follow the path Jesus had for them, Paul and Clint then joined forces with youth pastor Mark DeYmaz and founded Skatechurch in the Central Bible parking lot."[23]

Because we were the first church in the United States to launch a ministry to skateboarders, our staff quickly learned that this demographic group was viewed by most adults as more trouble than they were worth. From malls and shopkeepers posting signs that read, "NO SKATEBOARDING," to church members complaining about the occasional broken window or marks left on the building from young people performing a "wallslide," we had to stand up for and engage these students in order to win them for Christ. In those days I often found myself quoting wisdom literature: "Where there are no oxen, the manger is empty, but from the strength of an ox come abundant harvests" (Prov. 14:4 NIV).

Similarly, people who are homeless, are presumably undocumented, or have limited means and lower incomes are often left out of discussions by local businesses, civic leaders, and churches focused on executing plans for property development. Such development is often derided as gentrification to the degree that the marginalized people living in such areas are undervalued and overlooked in the process or displaced once dreams become reality.

By definition, "Gentrification is a general term for the arrival of wealthier people in an existing urban district, a related increase in rents and property values, and changes in the district's character and culture. The term is often used negatively, suggesting the displacement of poor communities by rich outsiders. But the effects of gentrification are complex and contradictory, and its real impact varies."[24]

With complexity in mind, Benjamin Grant, an urban designer and city planner, notes, "Many aspects of the gentrification process are desirable. Who wouldn't want to see reduced crime, new investment in buildings and infrastructure, and increased economic activity in their neighborhoods? Unfortunately, the benefits of these changes are often enjoyed disproportionately by the new arrivals, while the established residents find themselves economically and socially marginalized."[25] Therefore, in all aspects of development benevolent owners account for their marginalized neighbors and as the apostle Paul expects, "do not merely look out for [their] own personal interests, but also for the interests of others" (Phil. 2:4).

Carl Schultz is a good example of a benevolent owner and neighbor. In 2009, he and his sister, Cindy, purchased a low-income apartment complex directly across the street from what is today our property, the former Kmart. At the time, the apartments were known as a safe haven for drug use and prostitution as well as a place where violent crimes were frequently perpetrated, including murder on more than one occasion. Rather than gentrify the complex, however, Carl and his team brought excellence to the facility. They remodeled every apartment and the exterior of the buildings, including the landscaping. Because Carl did not raise the rent many people living there were able to remain while only the more criminally inclined were uprooted. As if an exclamation point to his intentionality, the apartment complex was renamed after one of its long-term tenants, a woman called Big Country—thus, the Big Country Chateau Apartments, LLC.

Collectively, local churches become benevolent owners and neighbors when, for example, they use their facilities to provide day space and free beverages to the homeless, as does the Crossing Church in Costa Mesa, California; to house an immigration counseling and legal services center, as does Vineyard Columbus (Ohio); or to distribute fresh produce and encourage healthy eating among the urban poor and elderly living in food deserts throughout Little Rock, as we do at Mosaic.

Go the Extra Mile

Finally, we should note that benevolent owners and neighbors must at times humble themselves, set aside afforded rights, and bend over backward to ensure good relationships are maintained with others in the community.

Recently, we had the opportunity to practice this aspect of benevolent ownership when one man representing a local Chinese buffet took Mosaic to task on social media. The man's indignation was fueled by the fact that a regional carnival we host twice a year on our property, and one from which we generate income for the church, was interfering with restaurant operations and costing it money. Publicly, he published what follows in multiple posts on social media (that I have aggregated and lightly edited):

Where do I start? My family by marriage are Asian and of the Buddhist Faith. They own a small business selling food buffet style. They employ Asians and Hispanics. Most of their wait staff are single moms and highly dependent on their wages and tips to care for their children and pay their bills. The restaurant has a parking lot for its customers that is posted private property. Since March 2017, Harry Li and the Mosaic Church have hosted Fun Time Shows from Doniphan, MO. Somehow or another Mosaic Church managed to get a zoning permit from the city. . . . THEREIN LIES THE RUB: The Mosaic Church has the carnival using my family's private property parking lot without permission/consent or a "percentage of their gate receipts as compensation." . . . Here is my solution: in my role as security, I will have every car towed that is illegally parked on our property. . . . Each time a tow truck shows up I will call LRPD at 911 and have them show up to prevent a disturbance. I will also have LRPD generate an incident report each time. I will then work with the City Director to have this carnival declared a public nuisance and have the zoning permit pulled at the City Board of Directors Meeting. As for the Mosaic Church, I am ashamed of your lack of tolerance for other religions and faiths like the Buddhist religion and Roman Catholics as evidenced by you continuing to take financial advantage of my family and their employees. You are bad business neighbors and you are certainly not CHRIST-like. As for myself and my family we promise to never step foot in your church because I believe it to be another one of Satan's workshops. In the meantime, I will continue to pray for rain.[26]

As soon as I was alerted to his posts, I immediately contacted the man through social media and arranged a time to talk. Harry also reached out to him. When we finally spoke the following day I simply listened. In so doing, I learned more of the complexities of the situation from both his vantage point and that of the restaurant. He estimated the carnival was costing the business somewhere between $1,000 and $1,500 each time it came to town due to patrons not being able to find a parking spot in front of the restaurant and leaving, people choosing to eat at the carnival, and employees needing to devote extra time to site cleanup and security. I learned later that there was more to the story, but his ire alone warranted a gracious plan to resolve the matter to his satisfaction.

I spoke with Harry and we agreed to pay for additional security at the restaurant until the carnival completed its current run on our property. In addition, I arranged to meet with the man later that day, at which time I offered my sincere apology on behalf of the church and further considered his complaints. Finally, I suggested that we meet again in person before the next time the carnival came to town in order to address the positioning of rides on our property, ways to ensure that the restaurant's parking spaces remained available for the exclusive use of its patrons, how to strengthen the presence of security, and the possibility of financial remuneration to cover any financial loss the business could show was directly related to the event.

Thankfully, the man was more than satisfied. In fact, he asked if I wanted him to delete his previous social media posts that disparaged the church. I told him doing so was unnecessary. Instead, I invited him to consider posting additional comments that reflected his satisfaction with our response, and the next day he did so. He wrote, "The church leadership (Mark and Harry) quickly reached out to me in an effort to resolve this. . . . I as well as my family are very pleased with the resolution. As for myself, I apologize for resorting to the nuclear option. Good luck."[27] Later that day he added, "We are very appreciative of the resolution you and your church leadership put in place to address this recurring economic and safety nightmare we have been experiencing when the carnival comes to town. Thanks again."[28] As a token of our appreciation, we later treated our entire staff to lunch at the restaurant as another way to build relationships and further goodwill with our neighbors.

In the context of encouraging his disciples to avoid revenge, Jesus instructed, "You have heard that it was said, 'Eye for eye, and tooth for tooth.' But I tell you, do not resist an evil person. If anyone slaps you on the right cheek, turn to them the other cheek also. And if anyone wants to sue you and take your shirt, hand over your coat as well. If anyone forces you to go one mile, go with them two miles" (Matt. 5:38–41 NIV). Indeed, "A gentle answer turns away wrath, but a harsh word stirs up anger" (Prov. 15:1). While we certainly do not always get things right, as pastors and churches we should strive for nothing less than to be benevolent owners and neighbors.

Economics as Evangelism

When a prophet named Agabus went down from Jerusalem to the church at Antioch, he "stood up and through the Spirit predicted that a severe famine would spread over the entire Roman world" (Acts 11:28 NIV). Essentially, Agabus warned the people of this multiethnic and economically diverse church to prepare themselves for what was to come. Luke, the writer of Acts, chooses not to tell us how they did so . . . only to share the church's response: "The disciples, as each one was able, decided to provide help for the brothers and sisters living in Judea. This they did, sending their gift to the elders by Barnabas and Saul" (Acts 11:29–30 NIV).

In large part this is why churches must become benevolent owners: to "bless the community, lead people to Christ, encourage the greater body, and fulfill the Great Commission."[29] To suggest that we do such things while generating profitable income is not a mutually exclusive proposition, nor are these competing demands that cannot otherwise be integrated for the reasons we've suggested. In fact, we predict that wise application of benevolent church economics will become one of the most effective means by which local congregations can engage and evangelize the world in the twenty-first century.

IDEAS AT A GLANCE

- There was a time when church facilities were at the center of community life—a gathering spot for town hall meetings, social events, and relational connections. Today, many if not most church facilities have been designed and built more as an island to themselves than with the community in mind. To change perceptions local churches should become benevolent owners.

- By renting existing space to businesses that well serve and are needed in the community, local churches can leverage their facilities to advance the common good and generate sustainable income. In so doing, they become benevolent owners and neighbors by improving the local economy, helping to reduce crime, and restoring hope among the marginalized.

QUESTIONS FOR REFLECTION

1. If neighbors were asked how your church's facility blesses or affects them, what would they say?

2. Why do you think churches buy land and build new buildings when they could purchase and remodel existing space in communities for a fraction of the cost?

3. How might you help people recognize that smart, benevolent economics will become one of the most effective means by which to engage and evangelize the world in the twenty-first century?

Monetize
Existing Services

The Sixth Directive

Over the course of a thirty-six-year ministry career, I have progressed (thankfully) in my understanding of best practices when it comes to handling money in the church. For example, I no longer

- collect money for T-shirt sales in one pocket and camp registrations in another, as I sometimes did when I was a youth pastor
- count the Sunday offering, record it in a ledger, and deposit it in the bank on Monday morning, as I did in the early days of Mosaic

- continue to think that donations alone will cover the cost of some things churches typically provide, now that I better understand church economics

On this last point, perhaps you know what I am talking about.

The Cost of Doing ~~Business~~ Ministry

Say that a church provides free coffee to people on Sunday mornings and next to the coffeepot is a jar with a sign that reads, "Donations Appreciated." While unspoken, here's what those words really mean:

> Do you know how much it costs the church in time and money to have hot coffee ready and waiting for you each week? Come on . . . don't make us charge you for this service; help us out by making a donation. That way we can keep providing free coffee and avoid appearing cheap (by cutting this line item from the budget) or concerned (as if we lack faith in God to provide).

Youth groups do something similar when providing free slices of pizza on Wednesday nights. In either case, assume the cost to provide free coffee is $100 a week, but just $45 a week is collected through the donation jar. This amounts to a loss of $55 each week. Over the course of a year nearly $3,000 would be squandered in this way, that is, money given through tithes and offerings "as unto the Lord." Now consider how many times, in how many ways, and in how many other areas of the church something similar plays out. The aggregate cost to a church can be significant. Yet pastors don't seem to mind. Instead, we'll suggest to the finance committee that needs justify the expense or explain away losses as simply "the cost of doing ministry."[1]

In business, when a company spends money in such a way, for example as churches do in providing free coffee from week to week, the giveaway is considered part of its "customer acquisition cost . . . the cost incurred by an organization while convincing a customer to buy a service or a product. In other words, it is the cost of resources for the business in order to acquire a new customer."[2] With this in mind, church leaders might see a value in providing free coffee, hoping through hospitality to attract new members and retain those who are already involved.

While there is nothing wrong with such thinking, we should recognize that there is a difference. In business, "Acquiring new customers involves persuading consumers to purchase a company's products and/or services."[3] Stated another way, businesses have a plan to convert giveaways into revenue and thus recoup expenses when people who receive something for free go on to purchase other items. On the other hand, churches do not typically sell products to offset the cost of free giveaways, such as coffee on Sunday mornings. Therefore, ministry leaders responsible for church finances can only hope and pray that those drinking coffee will donate money through the jar, join the church, and/or remain in it so as to consistently contribute tithes and offerings.

Thus, some will argue:

- Providing free coffee helps the church acquire new members and retain existing ones.
- The cost of providing free coffee will be covered through current and future giving.

- To offset any loss of income that comes from people leaving the church we should provide free coffee to attract new members.

Nevertheless, successful companies cannot afford such losses nor will they accept them. On the contrary, they track and measure profits from sales against any money spent to acquire new customers in the short term and to retain them in the long term. In other words, "The last thing a company wants to do is spend more on acquiring customers than the customers spend."[4]

In the context of money spent on giveaways not fully recovered through donations, what pastors label "ministry" is more accurately called a financial loss. Yet pastors do not typically think about money in the same way as do those in business. Rarely, for instance, will they have a strategic plan to recoup any differences between money received and the actual cost (per person), for instance, of taking members on a retreat, prepurchasing books for a small group, or providing T-shirts at a special event. Instead, they consider whatever money is not recovered as something that comes with the territory. In fact, pastors are so accustomed to spending and losing money in this way that they'll call it an "investment," presumably in people, to make themselves and others feel better. Where business is concerned, however, financial losses are considered unacceptable. In the future, ministry leaders should consider them to be so as well.

Wanted: Determined Nonconformists

Michael Feuer, CEO of Max-Ventures, a private equity and consulting firm, and cofounder of OfficeMax, writes, "Docile acquiescence of a problem that chronically plagues performance too often is accepted and referred to as 'just the cost of doing business.' These types of problems that mute or stifle performance can usually be partially or completely solved. It may be time-consuming and costly, but an investment to eliminate an issue or even mitigate a heretofore perennial profit drainer can provide a sustained and significant payback."[5]

Did you catch that? Feuer calls "docile acquiescence" a problem too often accepted in business. That's what we're talking about too as it pertains to church leaders tasked with responsible authority for overseeing budgets. Indeed, churches can no longer afford and pastors can no longer accept docile acquiescence when it comes to money needlessly squandered. Sure, give away free coffee; but develop a plan to recoup the costs.

Using antonyms, the opposite of docile acquiescence is "determined nonconformity." Indeed, the coming revolution is already today being shaped through determined nonconformity and, more specifically, by determined nonconformists pursuing paths of disruptive innovation in the area of church economics.

Determined nonconformists are "disrupters."[6] According to Mel Robbins, an expert on human behavior and motivation, disrupters are those "whose entire 'brand' is . . . to turn the way we do things on its head. . . . [Disrupters] break the mold, change our thinking about the mold and then hand us the new rules for how things work."[7] By way of example Robbins cites disruptive leaders and companies who have completely changed the game in their respective industries: from Amazon, that's changed the way we shop; to Uber, that's changed the way we ride; to Apple, that's changed the way we use cell phones; to

Facebook, that's changed the way we connect relationally. "Disrupters," Robbins writes, "don't fix what's broken because they don't innovate from inside the system."[8] Instead, "they operate outside conventional wisdom, turning systems upside down to effect systemic change."[9]

Elsewhere, I've described disrupters in this way:

> Opting to work outside, to leave, or otherwise driven out of established companies by the hubris, insecurity, or incompetence of existing leaders, [disrupters] upend the system and set seismic shift in motion. With an entirely different set of motivations, values, and metrics, they say what needs to be said and do what needs to be done before institutional authorities understand that such things need to be said or done. Thus, disruptive leaders do not co-opt the platform of others; rather, they create their own platforms and compel others to follow their lead, eventually including even those who in the moment do not understand or share their bias. In this way, they speak the future into existence by simplifying complexity, navigating nuance, and articulating the why, what, and way forward for others.[10]

Needed: Disruptive Innovators

Those determined to stop financial losses, avoid docile acquiescence, and pursue disruptive innovation in the church must learn how to monetize existing services. Put simply, there are certain things a church is already doing and for which it is already paying that can otherwise generate income. To effectively monetize an existing service leaders must be willing to discuss possibilities, gain understanding, act with intentionality, and leverage church assets. To illustrate, let's again consider the serving of free coffee.

In December 2016, as we were concluding our first year in the former Kmart, I walked past the small café adjacent to the building's large entry space we call "Town Hall," where people gather before and after services each Sunday. It was a Monday morning, and the smell of coffee permeated the air, as someone from our staff had brewed a fresh pot. Though I don't drink coffee, I wondered in the moment, *What is it costing us to provide free coffee on Sunday mornings?* Later I learned that we were spending approximately $200 a month, some $2,400 a year, to do so.

As we have already discussed, such costs may be justified where hospitality is concerned, and $200 a month will not seem like a lot of money to some churches. However, according to the Barna Group, nearly half of American churchgoers (46 percent) "attend a church of 100 or fewer members [and] more than one-third (37 percent) attend a midsize church of over 100, but not larger than 499."[11] To put this in perspective, I once served in a small rural church of 150 people whose entire youth budget was $50 a month, or $600 a year. In other words, $200 a month is, in fact, a significant amount of money for many churches in the United States and certainly not an amount that should be spent lightly from month to month.

Therefore, determined to stop financial losses, avoid docile acquiescence, and pursue disruptive innovation, I went to Sam's Club. Leveraging church assets (in this case, money), I purchased among other things an espresso machine, a microwave, two waffle irons, and a food warmer. In addition, I bought ingredients to make and serve waffles, twelve-packs of

Jimmy Dean Sausage Biscuits for $10 each ($0.83 per biscuit), and tinfoil squares in which to wrap them. In other words, I was on a mission to monetize an existing service: namely, the giving away of free coffee from week to week.

By selling a few items out of the café on Sunday mornings, I reasoned, we could recoup the $200 a month it was costing us to give away free coffee and perhaps generate additional income in the process. For instance, by selling a warm Jimmy Dean Sausage Biscuit for $2.00, we would net $1.05 a biscuit (our total cost per biscuit being $0.95 after factoring in additional costs for tinfoil squares, napkins, plastic plates, utensils, and more). By selling 2,400 of them in a year (an average of 46 biscuits per week), we could cover the entire cost of providing free coffee. We also determined to sell a cup of espresso for $2, "Pastor Mark's Famous Waffles" for $1 each, and juice for children at $0.50 per box. In time, my plan was to recoup the money I had spent on equipment, eliminate $2,400 from the budget, and find a better use for that money, say, to provide partial scholarships to students wanting to go to summer camp.

Lessons Learned

So in 2017, that's what we did. We opened our café for business while still providing free coffee for those who wanted it on Sunday mornings. Over the next twelve months, we learned and experienced a few things—some expected, some unintended. For example:

- Our café space transitioned from a transactional space (where people simply filled their coffee cups) into a place of warmth and relational connection (where people lingered before and after worship services).

- Our people greatly appreciated the effort. In fact, more than a few commented that they were now arriving to church on time knowing they could get something to eat upon arrival and not otherwise have to stop for food along the way.
- Volunteers were mobilized and energized in a new way, one that enhanced our hospitality and made a wonderful impression on those visiting Mosaic for the first time.

To be clear, not everything went according to plan and we continually made adjustments. As with any business:

- We saw that some products sold better than others, so we adapted our inventory.
- We adjusted prices and ran specials to attract customers.
- We became more intentionally focused, recognizing early on that success could not be presumed; it would have to be earned.

After one year we evaluated our efforts. Had we been successful in recovering the $2,400-a-year we were spending to provide free coffee and/or generating even more income from the café? Unfortunately, we had not achieved our goal. Yes, we were making enough money to cover the additional costs we had incurred to provide food items, but we were still losing approximately $200 a month.

Of all the lessons we learned, perhaps the most pertinent were these:

- Monetizing an existing service requires a shift in mindset among all those involved in the operation. It cannot be run like a "ministry" but must be run like a "business."
- Like many new businesses we opened strong, with lots of interest that soon leveled off and required ongoing stoking, as with a fire, to maintain sales. This required us to identify and empower a general manager, something better done on the front end.
- Similarly, we recognized early on that optimism alone will not sustain earnings. A recent study in the UK found that "overly optimistic business owners earn 30 percent less than those with below-average optimism. . . . The study, published in the *European Economic Review*, found that overly optimistic thinking may be one reason half of businesses in the UK fail in the first five years."[12]

In other words, you can't simply have a good idea and hope for the best. Research, preparation, focused attention, and a sustained effort over time is needed to see positive results. "So, if optimism doesn't help entrepreneurs, what does? Having resilience and knowing how to work well with others are important. Part of succeeding involves believing in your product, but it also means having others believe in it too."[13] Looking back, we had a good idea but not necessarily a plan. Indeed, a strategic business plan should be developed, vetted, and put into place ahead of any attempt to monetize an existing service.

Pursue Economic Competence

As pastors of a multiethnic and economically diverse church, we continually encourage the pursuit of cross-cultural competence. Whether in "becoming proficient in the idiosyncrasies of language or the ins and outs of customs and traditions different from our own, cross-cultural competence, once acquired, allows us to interact in a more informed and effective way with others of varying ethnic or economic backgrounds."[14]

Just as important today is the need for pastors to become economically competent—that is, to understand the way that economics powerfully shapes a local church, for better or worse. For instance, in our own urban setting, many households are sustained by two or more incomes. Therefore, relatively few of our members have discretionary time to serve as volunteers throughout the week, as might be expected in more affluent churches in the suburbs. This affects our philosophy of ministry, how we staff the church, and requires us to prioritize spending in order to compensate.

Similarly, I (Harry) remember when my daughters made significant money in high school by babysitting for parents in an affluent church, so that these parents could attend weekly small group meetings without their children. My daughters made $30 or more an evening from these people. In other words, to participate consistently in a small group, parents were expected to pay somewhere between $120 and $150 per month. At Mosaic, however, this is unrealistic. In addition, since many of our families put their children in day care throughout the week, asking parents to spend even more time away from them for small group interaction does not work in our context—one in which (at the time of this writing) some 26 percent of children are being fostered or have been adopted and 23 percent are living with a single biological parent. Therefore, our small groups are designed with children in mind, whereby they can attend the gathering with their parents who will collectively supervise them or pay (as a group) for someone else to do so.

These types of culturally informed realities have forced us through the years to rethink church economics. They have also required us to understand why, how, and where to engage people in ministry, adjust accordingly to changing circumstances, and be creative in terms of how to fund our work. For instance, virtually every day of the week people from the community are served in one way or another in our building. Many come to us from across the street and the apartment complexes in which they live or from the homeless encampments that surround the church. If we even moved a couple of miles farther west our tithes and offerings would likely increase, but we wouldn't be able to so readily serve marginalized people in our community. Thus, we have learned to generate income beyond tithes and offerings in order to sustain our work in a difficult context.

We should also recognize that those who are affluent have as much to learn from those lacking resources as is true the other way around. Our worldview is shaped by race, class, and culture as well as by experiences, personality, and preferences. Worldviews skewed one way or another—by affluence or need—are equally dangerous to the health and well-being of a church. Indeed, we all need and have much to learn from one another.

People with a worldview skewed by affluence typically:

- Believe they know what others should do to overcome financial challenges, having been successful or privileged in ways that those affected by poverty have not experienced.
- Ignore the financial needs of people in their own communities while being generous with people lacking resources in more appealing, international locations.
- Embrace politicized sound bites such as "People living on the streets or in poverty should take personal responsibility for their financial situation and show initiative."
- Refuse to acknowledge systemic causes and effects that have led to entire groups of people being disadvantaged in terms of gaining economic ground.

People with a worldview skewed by poverty typically:

- Embrace a survivalist mentality, being unable to think or plan for tomorrow. Today, they'll tell you, has enough trouble of its own.
- Experience what others consider to be minor inconveniences as major obstacles, such as when getting the flu triggers a domino effect: work is missed, employment is jeopardized, the paycheck is lighter, and bills pile up.
- Display an entitlement mentality in their personal interactions, having been accustomed to social programs in which it has been easy to receive continuous assistance without having to give back in return.

- Know that it is easier to receive assistance than to try to get beyond it, or are unable to do so for one reason or another.

In either extreme, these worldviews can be destructive to the health and well-being of any church, especially churches pursuing unity and diversity for the sake of the gospel. Therefore, in all contexts and situations, twenty-first-century leaders should pursue both cross-cultural and socioeconomic competence in order to understand the strengths and weaknesses of each worldview as well as to discover best practices for helping people hear, understand, and validate the perspectives of diverse others.

One Way or Another

Had we continued selling biscuits, waffles, and more out of our café on Sunday mornings, I'm certain we could have further developed our plan and in time realized our goal of generating $200 a month in net profit. In fact, "Innovative managers . . . constantly look at opportunities that many times are disguised as problems and put together a task force backed by the appropriate resources to explore improvements."[15] However, just as we began our second year of selling products from the café, owners of a local start-up approached us about renting the space and serving customers throughout the week. Their business, River City Nutrition (RCN), was a smoothie and juice bar offering meal replacement shakes, nutritional counseling, and calorie-burning teas. In support of small business and, again, to leverage assets, we agreed to do so. When RCN moved into the space, we curtailed our own sales on Sunday mornings without losing any of the additional gains we had realized by opening the café for business. The one-year contract with RCN provided us with $750 a month for the first six months and $1,000 a month for the second six months. In addition, we negotiated to receive a percentage of net profits on a sliding scale, although due to sluggish sales we never realized any such income beyond the rent.

In other words, through partnership with RCN our plan to monetize an existing service was met—just not in the way we had envisioned. In fact, net profit from the RCN contract not only covered the $200 a month we had hoped to generate but also an additional $400 a month or so, nearly $5,000 over the course of the next year.

Considering Possibilities

Since virtually every church in the US provides coffee to people on Sunday mornings, we have chosen so far to help illustrate our thinking with this in mind. Of course, there are any number of existing services that churches might be able to monetize, depending on context, once the concept is understood and embraced. Here are an additional five possibilities to further inspire your thinking.

Janitorial Services

At Mosaic, we budget and spend $22,800 annually ($1,900 a month) to ensure that our building is clean from week to week. In 2016 and 2017, the first two years we occupied our new facility, we simply contracted the work. Later, we began to imagine a scenario in which we would use that money to help someone from the church start their own janitorial service company. In this way, it might employ people from the church and/or from the community who need a job. By securing additional contracts outside of Mosaic, we reasoned that if $22,800 could be generated in net profit over the course of a year our facility could be cleaned (essentially) for free by the company who, by prior agreement, would pay workers to do so from profits earned outside the church. Thus, Mosaic would save a significant amount of money (tithes and offerings) annually and repurpose it for more direct ministry impact.

Monetizing such an existing service takes research, preparation, focused attention, and a sustained effort over time to see results. Perhaps your church might explore this possibility by enlisting an entrepreneurial leader to determine feasibility.

Parking Lots

Even before Mosaic purchased the former Kmart, we arranged with its owner to open Little Rock's first food truck court, the University Market at 4Corners. The concept was simple. Various food trucks parked in a semicircle on a designated area of the parking lot, weekdays between 11:00 a.m. and 2:00 p.m., and each truck paid the church $50 a week to do so. In addition, some trucks were allowed to operate in the evening and on weekends.

As with our café, we launched with a bang. Twelve trucks and over four hundred people were involved in our grand opening. The market was big news in Little Rock at the time and the

CHANGING THE CONVERSATION

By MARTIN ROBINSON, Principal and CEO, ForMission, Birmingham, United Kingdom

Rowheath Pavilion Church is located in the heart of a community better known for its connection to Cadbury's chocolate. Bournville Village was created by the Cadbury family as a holistic community expression of what it means for an employer to provide housing, recreational facilities, parks, schools, and shops in addition to employment for their workers. It was an inspiring vision of what community could look like, and it was fired by a family of evangelical Quakers who had a passion for the gospel and for social justice. Originally built in the late nineteenth century as a village to the south of the city of Birmingham, it has now become absorbed as a suburb within that same city.

Our church began as a small group meeting in a home. As this emerging community of believers developed, we searched for larger accommodation in a school or community center. Rowheath Pavilion had been part of the Cadbury vision for the community. It opened in 1922 to provide recreational amenities for Cadbury workers, consisting of a park with a lake, playing fields, and a building in which people engaged in sports could change, shower, and then meet each other in a social setting.

In the late 1980s we rented space in Rowheath Pavilion for Sunday morning worship services. By the middle of the 1990s the church had grown, but the community center was in difficulty. Our church was offered a long-term lease for the whole building and most of the land that surrounded it for just £1.00 a year. That sounds like a bargain, but like many apparent bargains, it was not all that it seemed. The building was not in great shape,

[and] neither were the grounds. The previous charity that had assumed responsibility for the building had been in huge deficit. Taking on the lease was a huge risk, one that we prayed about and eventually embraced.

But being in the neighborhood, having responsibility for a community facility, was not the same thing as having a relationship with the community. In reality, we were an invisible presence with no significant connection to the community in which we were located. So the question became "How can we change the conversation with our neighborhood, or indeed, how could we begin a conversation in the first place?"

Our first action was to ask the neighborhood what they thought were the biggest needs in the area. The answers surprised us. The top perceived need was for a children's playground. The second was for a place where people could meet their friends, and the third was a desire for amenities for young people. Our church committed to address all three of these needs.

First, we led a community initiative to raise funds to construct a children's play area next to the Pavilion building. Our second move was to open a community café as close to the play area as we could. Third, we employed a youth worker to connect with youth both in the local high school and at the community center. These actions radically shifted the conversation we were probably not having with the community.

Today, Rowheath Pavilion has a turnover of around £650,000, employs around thirty people both full-time and part-time, uses the services of many volunteers, opens from 9:00 a.m. until 11:00 p.m. seven days a week, and is used by close to three thousand people every week. In the midst of this renewed relationship with the community, the church is growing numerically and is learning how to serve the neighborhood in which it is located.

next day it was featured prominently on the front page of the *Arkansas Democrat-Gazette*. Since we had already put the property under contract for future purchase, we were also able to work with the city at that time to rezone it as a Planned Commercial District (PCD). This means that we have been granted certain allowances or exemptions on the property. For example, in those days food trucks in Little Rock were not allowed to set up tables and chairs. As part of our PCD, however, they were given permission to do so.

Though we had a focused leader and were upholding our end of arrangements, over the next few months we experienced some difficulties in doing steady business with food trucks. First, the entire concept of a food truck was relatively new to our mid-sized city, which presented certain challenges we had not anticipated, such as general acceptance of this business enterprise. We also found that food trucks came and went with regularity, as owners might be operational one week and out of business the next week. In addition, too many trucks in our market at any one time created competition. Even though we tried never to have two trucks with similar products on the lot simultaneously, a large crowd did not always translate into sales for trucks parking on our property. In other words, we couldn't guarantee to the owners an equitable distribution of customers. Thus, if a better opportunity presented itself on any given day, an owner would not bring his or her truck to the market at the time we had otherwise announced through social media that it would be with us. This created confusion in our customer base and also contributed to a slow erosion in numbers. As these declined, the remaining food trucks began to look elsewhere for business, and within six months or so the market was closed.

Nevertheless, we still believed our parking lot had the potential to generate revenue. Thankfully, our next attempt to monetize it was more successful, as it is to this day.

As mentioned in chapter 6, Fun Time Shows (FTS) brings their carnival to town twice a year (spring and fall) and sets it up on our property. It has been a win for the community and a win for the church. In fact, FTS is now baked into our PCD, meaning we do not have to secure an operational permit from the city every time they come to town. In total, the church receives some $16,000 annually for hosting the carnival on our parking lot, the equivalent of one month's mortgage payment. That is significant money, at least for Mosaic. In fact, I know of a church that was once making $1 million a year by renting its strategically located lot for weekday parking downtown in a large city!

Since most churches have a parking lot and must pay to maintain it, a parking lot too should be considered an existing service that might be monetized. Perhaps, like Mosaic and other churches throughout the country, you and your church can find a way to do so.

Graphic Design and Printing Departments

While the first two examples have been taken from our experience at Mosaic, this next one stems from my previous experience working in a megachurch. There, we had a large graphic arts department dedicated to producing materials for the entire church as well as for its individual ministries. It had computers, printers, and dedicated space, all of which was paid for by the church, as were the salaries of designers employed on staff. Accounting for all related costs, including necessary equipment, paper, ink, maintenance, employee salaries and benefit packages, and the fair market value for the space (its retail price to rent per square foot), I imagine the church was spending well over $200,000 a year to support this department. Of course, this money was budgeted and paid for by tithes and offerings.

In such a church, a graphic arts department is an existing service. How might it be monetized?

For example, if one person on the team was designated or another person hired even part-time to pursue business outside the church, income could be generated. How? Since operational overhead is already being paid for by the church, the department could be monetized to generate income against (say) a $200,000 annual budget. Graphic designers already employed by the church might leverage the department to increase their own earnings via the gig economy and by agreement return a percentage of the money they make in so doing back to the church . . . or the church could allow them to do so in lieu of a larger salary. Alternatively, graphic designers from the community could pay the church rent to use the space and equipment outside of office hours in order to incubate a business. One way or another, by monetizing this existing service a church might be able to recoup a portion of what it's already investing annually to house a graphic design department. The money saved or earned by doing so could then be redirected to other areas of its ministry or perhaps to help erase a budget deficit. Even a 10 percent ROI annually would be significant.

At-Risk Youth Programs

In certain cities, local government will provide grant money to nonprofit programs working with young people considered to be at-risk. "[*At-risk youth*] is a general term for a range of circumstances that place young people at greater vulnerability for problem behaviors, such as substance abuse, school failure, and juvenile delinquency, along with mental health disorders, such as depression and anxiety."[16] Therefore, a church with a student ministries program that draws at-risk young people has an existing service that could perhaps be monetized.

Imagine a church whose youth pastor was funded entirely by tithes and offerings yet whose efforts served many at-risk students. A separate nonprofit established by the church, as discussed in detail in chapter 4 of my book *Disruption: Repurposing the Church to Redeem the Community*, could apply for a grant from local government to help fund such work. Assuming the grant was awarded, the nonprofit could then pay a portion of the youth pastor's salary and/or some programming costs, thereby saving the church money. In this way, the church could help itself by creating another stream of income and by monetizing an existing service, in this case, its student ministries.

Senior Adult Day Care Centers

Many churches provide nonprofit or for-profit day care for children. Given an aging membership, relational connections, and existing facilities, something similar might be done for a growing population of senior adults. According to Senior Living, "An adult day care service is a professionally managed environment for older adults. . . . In addition to providing a safe and secure setting . . . during the daytime hours, an adult day care also offers specialized activities. These often include physical therapy, social engagement, and health care services that are tailored toward this age demographic."[17]

To be clear, "Adult day care centers are required to have state licenses or are regulated to a degree to provide certain types of services. This is especially true of those centers that receive public funding, which most do."[18] In fact, "Many states [now] have non-Medicaid assistance programs; sometimes referred to as nursing home diversion programs. These programs are intended to prevent or delay the placement of elderly state residents in nursing homes. To that end, these programs will pay for or help with the cost of daytime supervision in a group environment. This is especially true for individuals with Alzheimer's or dementia."[19] In other words, public funding is available to support such efforts. If obtained, whether through a nonprofit or a for-profit entity, some of it could be paid to the church in the form of rent and thus reduce the amount of its spending to pay a monthly mortgage.

Both to serve others in the name of Christ and to secure some measure of ROI in pursuit of long-term sustainability, churches already working with senior adults could help themselves and the community in this way—if not by creating a quality day care center, perhaps by organizing a revenue-generating enterprise to provide in-home love, service, and support to this growing segment of our population. Since "many seniors are shut-ins, either unwilling or unable to leave their houses because of a physical disability, or mental obstacles such as agoraphobia or social anxiety," opportunities exist for good people under the right circumstances to bless them individually, and in the process, collectively, on behalf of the church, to leverage assets and monetize existing services.[20]

Eyes Wide Open

Everyone understands that it takes money to establish and sustain a local church. Lots of it! This is why denominations and networks tend to steer church planters toward more affluent or upwardly mobile communities in which live people who can more readily afford to underwrite such efforts. The result? The fastest-growing and largest churches in the United States tend to be in the most affluent suburbs of our cities. I (Harry) am convinced this breaks the heart of God. Sure, God is at work in those churches and places, and pastoring a large church is no easy task. But we should ask why so many pastors believe they are called to plant, grow, or develop churches in these areas— ironically, churches that can afford to pay them well. Is the Lord really sending so many of them to upwardly mobile communities while people in more difficult contexts, including the urban environment, remain largely neglected? Or, perhaps, are some "callings" more informed by pragmatic finances? In the future, church planters should be encouraged by denominations and networks to go where the needs of people in a city are greatest, without fear of financial failure, and be equipped to minimize financial risks through the application of smart economics.

Daniel Cook writes, "While there are hundreds of ways that churches can bring in additional income, it is imperative to do this with a solid business plan utilizing business minds . . . and in some cases outside consultants."[21] So where might you begin in terms of considerations? Michael Feuer suggests, "Take your biggest issue, envision solutions, commit to the most logical or innovative fix, and then set a time frame for completion. Put one of your best and brightest in charge. And don't be afraid to fail. Even if you succeed only one out of three or four times, the math likely still works. By accepting 'that's the way it is' as a mindset (docile acquiescence) you'll always follow, but never lead."[22]

As for strategy, Cook suggests the following five steps for pastors and churches looking to move forward:

1. Develop goals and objectives that are both short and long term. Methodologies should be included to get your congregations on board.
2. Identify the income strategy that best fits your ministry, facility, and community.
3. Select revenue source(s) to pursue. Research each of these and gain expertise either internally or through consultants.
4. Write and execute a revenue source (business) plan.
5. Evaluate the results quarterly while continually striving for improvement.[23]

Cook concludes, "The easiest money you will ever raise is the money you will no longer need by operating your existing facilities wisely from a standpoint of reducing costs on energy, maintenance and janitorial expenses,"[24] or as we have been saying, by generating rental income and monetizing existing services.

These principles are important to understand and critical for churches to apply, especially those looking to start new businesses . . . the seventh and final directive to discuss.

IDEAS AT A GLANCE

- It is common for pastors and churches to spend money freely and later justify it as a "ministry expense" or an "investment in people." In business, casual or careless expenditures are not tolerated. Yes, we should be generous in budgeting to freely serve others yet we should also develop plans to recoup expenditures wherever possible, as is done in business.

- In terms of where to begin, "Take your biggest issue, envision solutions, commit to the most logical or innovative fix, and then set a time frame for completion. Put one of your best and brightest in charge. And don't be afraid to fail. Even if you succeed only one out of three or four times, the math likely still works."[25]

QUESTIONS FOR REFLECTION

1. What existing services in your church might be monetized?

2. How might some of your leaders have fallen into "docile acquiescence" where church finances are concerned? How can you inspire a change in their perspective?

3. In hindsight, what might you have done to ensure that Mosaic's café space was profitable from the start? List at least three steps that could have been taken to do so.

8

Start New Businesses

The Seventh Directive

In 2016, the Barna Group, in partnership with Thrivent Financial, produced a report titled *Church Startups and Money*. To no one's surprise, among four general implications highlighted by the research was that "money equals stress. The greater the financial burden on a church planter, the greater his or her emotional burden."[1] From our experience, this is true not just for church planters but for pastors of existing churches as well. No one tells you at the outset, but in time you learn: The need for money is a constant, and the pressure on pastors to raise ever more of it is one that never goes away. The question is, What are we to do about it?

Conventional wisdom suggests there are at least three ways that pastors can generate income for their churches and thus lighten their own emotional burden:

1. Grow numerically.

   So much has been written to date both to encourage and to discourage size as a measure of success in the church that there's not much left to say. For instance, one church-growth consultant wrote, "The power of the church to do good in the world is controlled and measured, in substantial part, by its size. A tiny church might be a pinch of salt in society, but a larger church has a greater chance of salting it thoroughly."[2] In most instances, it is true: the larger a church, the larger its income (that is, its tithes and offerings). Therefore, seeking to increase attendance is often the first line of attack in dealing with budget deficits and the need for more money in a church. Nevertheless, we take exception with the consultant's last point. As I've written before, "I've seen what a determined [multiethnic and economically diverse] church can do regardless of its size and demographics, without a quality facility to call its own, and without what many others consider the income necessary to make a significant impact on its city."[3]

2. Cultivate generosity.

   The Barna report asks, "What would happen if we considered training in generosity an important aspect of discipleship?"[4] According to Rachel Cruze, "Ultimately, you want to win with money so you can become an outrageous giver! . . . Generous people are not only the happiest people on the planet, but they also truly live more fulfilled lives"[5] This is one aim of Dave Ramsey's Financial Peace University, the course that will teach you "how to save an emergency fund, get out of debt, save for retirement and college, pay off the house, build wealth, and give generously."[6] In his biblically based course The Legacy Journey, Ramsey says, "There's really only one way to heal a selfish and greedy spirit: generosity. . . . Giving is the antidote for selfishness. It's the hallmark character quality of those who win with money. . . . If you live like no one else, later you can live—and give—like no one else. That's the goal!"[7] With this in mind, let's be honest. In part by cultivating a spirit of generosity, pastors hope to help people learn to better manage money so they can and will give more of it to the church.

3. Avoid mistakes.

By strengthening systems of financial accountability and oversight, a church may reduce spending and increase the confidence of its members regarding stewardship, leading to increased tithes and offerings. In a post titled "10 Common Church Financial Mistakes," Josh Henry, the founder and CEO of Simplify Church, suggests, "It may be time to strengthen your financial systems especially if you are making one or some of these most common mistakes."[8] Among the mistakes he lists are these:

» not having enough or any checks and balances
» not using or understanding accounting software
» not having a budget forecast or paying ongoing attention to the budget

What strikes me most about his list, however, is that only one of the ten items suggests our topic: "not having an understanding about how to generate income." On this point, he writes, "The difference in budgeting and finance between a church and other types of business is that the church has very little control over its income. Churches for the most part are solely dependent on the faithful gifts of their donors."[9] He then offers pastors two ideas to generate income:

- Remain attentive to giving by preaching on it to remind donors "of the ways their money is being used to benefit the church and its ministry."[10]
- Make it easy for people to give by "providing options for people to give to the church electronically outside of the service."[11]

In other words, when Henry speaks of generating income he's thinking about traditional giving to a church and not necessarily about ways a church itself can do so beyond tithes and offerings. The fact is that relatively few churches are thinking like this today, that is, entrepreneurially. We predict, however, that an increasing number of churches (if not a majority of churches) will not only be thinking entrepreneurially but acting entrepreneurially in the years ahead in order to generate necessary and sustainable income.

So, sure: grow numerically, cultivate generosity, and avoid mistakes. But is that it? Is that all we can do to reduce financial pressures, strengthen the budget, and increase the effectiveness of a local church? Among the many questions posed in the Barna report, only once is the question asked, "What would happen if we went back to the drawing board on funding . . . churches? What would it look like to make a whole new model?"[12]

By now you realize that is exactly what we are saying. A whole new model and approach to church development and funding is needed, with impact and sustainability in mind. In addition to growing numerically, cultivating generosity, and avoiding mistakes, churches must learn to create multiple streams of income by leveraging assets, monetizing existing services, and/ or by starting new businesses. In fact, "Non-profits are increasingly entering into traditionally commercial spaces in an effort to increase revenues as they face simultaneous challenges of uncertain philanthropic funding, diminishing governmental funding, unstable fundraising revenues, and increased competition for limited resources."[13] This too is fueling the coming revolution in church economics.

It Can Be Done

Contrary to common yet erroneous beliefs, a nonprofit (for our purposes, we'll speak of a church) may start, own, and/or control a for-profit business. "This kind of corporation is referred to as a 'for-profit subsidiary corporation.'"[14] In fact, "Prior to the Revenue Act of 1950, which introduced the unrelated business income tax (UBIT), charitable non-profits were exempt from federal income taxes on all income, including earned income, regardless of whether such earned income derived from related or unrelated businesses."[15]

From a legal perspective, the primary concern is fair competition—that is, making sure the playing field is level when nonprofits compete with for-profit businesses in the marketplace. "When the UBIT rules were enacted, the rationale was to eliminate unfair competition by imposing a tax on a nonprofit's net income generated from unrelated business activities."[16] Thus, the UBIT "allows for the taxation of businesses operated by nonprofits, the profits from which [are] unrelated to the nonprofits' official tax-exempt mission."[17] Put simply, if a church owns a business it must pay related taxes like anyone else owning such a business. Therefore, taxes should be considered part of the cost of doing business (making money). Thus, paying UBIT should not be feared or avoided by churches starting, owning, or otherwise benefitting from a for-profit business.

That said, there are certain businesses that churches can start that would not be taxed by the government. "If a nonprofit can characterize its income-generating businesses as substantially related to its exempt purpose, the net income from such businesses will not be taxed."[18] "To determine if a business activity is *substantially related* requires examining the relationship between the activities that generate income and the accomplishment of the organization's exempt purpose. The activities that generate the income must contribute importantly to accomplishing the organization's exempt purposes to be substantially related."[19]

What are some substantially related income-generating businesses that might not be taxed? Among others, Daniel Cook suggests that churches consider the following:

- a children's day care center
- a senior day care center
- an after-school center for youth

NET INCOME SUBJECT TO UBIT

"501 (c) (3) Tax Guide for Churches and Religious Organizations"

Churches and religious organizations, like other tax-exempt organizations, may engage in income-producing activities unrelated to their tax-exempt purposes, as long as the unrelated activities aren't a substantial part of the organization's activities. However, the net income from these activities will be subject to the UBIT if the following three conditions are met:

- the activity constitutes a trade or business,

- the trade or business is regularly carried on, and

- the trade or business is not substantially related to the organization's exempt purpose. (The fact that the organization uses the income to further its charitable or religious purposes does not make the activity substantially related to its exempt purposes.)

Even if an activity meets the above criteria, the income may not be subject to tax if it meets one of the following exceptions: (a) substantially, all the work in operating the trade or business is performed by volunteers, (b) the activity is conducted by the organization primarily for the convenience of its members, or (c) the trade or business involves the selling of merchandise substantially all of which was donated.

In general, rents from real property, royalties, capital gains, and interest and dividends aren't subject to the unrelated business income tax unless financed with borrowed money.[20]

- a business focused on young moms or providing a mom's day out
- spring or summer camping or a campground
- an indoor children's play zone
- meals prepared and served, such as on Sunday mornings or Wednesday nights[21]

Why and How

In starting new businesses, churches must keep the "why" in mind. We should start new businesses not simply to generate income, as if mercenary or self-serving, but to bless the community by encouraging small business and entrepreneurship, creating jobs, revitalizing abandoned or underutilized space, helping to reduce crime, generating tax revenue for local communities and more, as ambassadors of Christ (2 Cor. 5:20), like good and faithful servants (Matt. 25:23). Ultimately, all such effort should result in others seeing our good works, glorifying our Father in heaven, and (we pray) coming to know him as we do (Matt. 5:16).

We should also avoid becoming overly optimistic as previously discussed. Starting a new business is not the same as starting a new ministry within the church. For instance, if we try and fail to start a vibrant Wednesday night Bible study only time, effort, and perhaps a bit of credibility are lost. If we try and fail to start a business due to poor planning, unrealistic optimism or any number of other factors, these things will be lost too, but so will money—potentially lots of it—that cannot otherwise be recovered. Thus, it's important to understand that businesses do not turn a profit simply because entrepreneurial types devote some time and attention to operations but because they are fully invested, daily engaged, and persistently devoted to ensuring profitability. In other words, successful businesses require focused and effective leadership.

According to the Small Business Administration (SBA), "Starting a business involves planning, making key financial decisions, and completing a series of legal activities."[22] Toward that end, the SBA provides a number of free services and support. Pastors and churches might look to the SBA for practical guidance and assistance in starting for-profit new businesses.

In addition, churches should consider hiring or otherwise designating someone specifically to develop for-profit business enterprise. Traditionally, churches employ executive pastors or business administrators to provide for financial oversight and accounting, and it's important to do so. Your church might go beyond this, however, and hire someone with proven success in business and the acumen not only to provide for oversight and accounting but also to help the church generate sustainable income by:

- attracting small business, pursuing outside investment, and public or private grants
- assisting with contract negotiations, tenant management, and the development of community partnerships
- accounting for UBIT, billing, and legal services[23]

Furthermore, such a person could encourage, equip, and accelerate entrepreneurship within the body. They might also take the lead in donor development, allowing senior leadership to focus more of their efforts on direct pastoral work and to avoid any potentially conflicting motivations where a member's checkbook is concerned. In fact, with someone in this position the work of encouraging generosity and donor development could be more evenly distributed over the course of twelve months instead of crammed into a hurried and worried appeal for funds only at the end of the year.

At Mosaic, Harry and I essentially share this role. He focuses on cultivating generosity in the body while I focus on securing revenue beyond tithes and offerings as well as new business development. Recently, our elder board discussed taking such a next step and hiring someone to serve the church as we are suggesting, whether in a part-time or full-time capacity. Typically,

TEN STEPS TO START A NEW BUSINESS[24]

1. Conduct market research.
2. Write a business plan.
3. [Determine how to] fund the business.
4. Pick a location for the business.
5. Choose a [legal] structure.
6. Choose a business name.
7. Register the business.
8. Get federal and state tax IDs.
9. Apply for licenses and permits.
10. Open a business bank account.

when churches hire someone they have funds in hand. With this role in mind, however, a church might be willing to take an initial risk by guaranteeing a person's monthly income until such a time as it's recouped and ultimately generated through his or her work of donor development. In this creative way, a church might fast-track someone into the position and begin benefitting sooner rather than later from entrepreneurial investment.

An Alternative Approach

To start new businesses, pastors might also leverage relationships and connections with potential investors to form for-profit corporations from which their churches would also benefit. At the time of this writing, this is an approach we're exploring at Mosaic. While it's too early to tell if we'll be successful, we do believe this is a viable option, as do a couple of early adopters who have bought into the concept. Let me explain.

The building we own is approximately 100,000 square feet. As discussed previously, 10 Fitness currently rents about half of the facility from us and we have built out an additional 32,000 square feet for our own use as a church. This leaves 18,000 square feet or so that is still undeveloped. It's our intention to do three things with this remaining space: develop 15,000 square feet to enhance our event center, install and monetize a commercial kitchen, and in the remaining 3,000 square feet open a commercial coffee shop for the community. The entire cost is approximately $1.8 million.

Typically, a church will seek to solicit and raise such money by way of a capital campaign, encouraging tax-deductible gifts, and we will do so as well. To expedite plans, however, we are also considering the possibility of leveraging relationships and bringing investors together to form a holding company. "A holding company is a parent corporation, limited liability company, or limited partnership that owns enough voting stock in another company to control its policies and management. The company does not have any operations or active business itself; instead, it owns assets in one or more companies."[25] In other words, investors would not put money directly into the businesses—the event center, the kitchen (again, which we intend to monetize), or the coffee shop. Rather, they would put money into the holding company and each own a percentage of it. The holding company, led by a CEO, would then use the money to develop the three businesses (or investment types) on behalf of the investors. Each of the businesses would be run by a general manager.

Since all of these businesses would be on church property, grouping them under one holding company makes sense. In addition, it's a good strategy for investors to pursue because spreading out investments reduces risk. "All investments carry a certain degree of risk. There are different investment types and the returns can be unpredictable from one year to the next. By spreading [their] money across different investment types, [investors] reduce [their] overall risk—this is what is known as diversification."[26] Stated another way, by investing in the holding company investors mitigate risk. They would not be putting "all their eggs in one basket." For instance, if the coffee shop is underperforming and the event center is doing well overall profit and loss will be better balanced.

As mentioned, we are still only discussing the possibility. We'll soon see if it becomes a reality. If so, the easiest and least complicated way for Mosaic to generate income would be to collect rent from each of the three businesses rather than become a percentage partner in the holding company, though we've been told this is an option. All such discussion has been included here to stretch your imagination and to help you further consider what might be possible in your own context.

That said, there is one more reason why proceeding in this way could potentially help a church. Say a member of your congregation has $30,000 available and is wanting to do something with it. First, she determines to be generous and gives $5,000 to the church. Next, she considers how to invest the additional $25,000, looking for ROI. As things stand, a local church cannot attract, leverage, or compete for the $25,000 since, as a nonprofit, it cannot provide ROI to the investor. However, assuming there was a holding company such as we've described, a for-profit business owned by the church, or perhaps by introducing her to another investor seeking to bless the church through profitable business enterprise, both goals could be accomplished. The church could compete for her investment dollars, and investors wanting to help the church could get ROI. It's a win-win scenario.

We have no illusions. You may be able to develop or apply such thinking better than we can, given our unique ministry context . . . and we encourage you to do so, if and wherever possible. Indeed, so much is coming—things foreseen and yet unknown—that will require continued adaptation, peer-to-peer learning, and exchange of information as pastors lean into the coming revolution in church economics as, in fact, some are already doing.

Examples

Here's one final adage: "Don't just take my word for it, listen to what others are saying." With this in mind, consider the following four examples, shared by pastors of like mind who today are leveraging church assets, blessing communities, and helping to fund sustainable mission by starting new businesses.

Land Development

Greg Wigfield • Destiny Church, Leesburg, Virginia

It was not my plan to become a pastor. I first became a businessman who later embraced a call to plant a church in the area where I live and work, in the wealthiest county in the nation, just outside Washington, DC. As of this writing, I am in my twentieth year of pastoral ministry at Destiny Church while still owning and leading multiple businesses, including the National Church Group—an insurance agency serving more than three thousand churches throughout the United States as well as other nonprofits, camps, and schools. Given the success of my companies, I have never had to receive a paycheck from Destiny, and the church benefits from this arrangement. After many years of experiencing God's blessing in business and ministry, however, I shoulder a deep concern for the future of the local church in America.

Evangelical ineffectiveness coupled with severe financial decline is becoming a real problem, though many pastors fail to recognize the coming crisis. Alan Hirsch states that we have a strategic problem in that the majority of churches in the US are using a model of church designed to reach 40 percent (and declining) of the American population. This leaves around 60 percent (and climbing) of the American population outside the reach of the local church.[27] John Maxwell says that in today's US culture, 53 percent of people will *never* attend a church![28]

John Dickerson too, writing in *The Great Evangelical Recession*, suggests that over the next fifteen years, the American church will lose approximately 70 percent of its income from people sixty-five and older who are dying at a rate of one thousand people per day.[29]

Since 70 to 90 percent of church income is allocated for staff and building expenses, Millennials are choosing to direct their financial giving and volunteer support to causes with better ROI.

Over the past couple years, the business side of my brain has collided with my heart for reaching people with the gospel. I now realize that present approaches to church funding are ineffective. For instance, our church is housed in temporary facilities. Nevertheless, our rent and related expenses exceed $25,000 per month. That is not sustainable and why (through one of my companies) I decided to purchase land, develop a piece of historic property, and move our church out of rented facilities.

Once completed, the $20,000,000 campus will include the following:

- Destiny School of the Arts, which is already operating on the property; 175 students (preschool through elementary) are currently enrolled, giving us access to hundreds of parents and grandparents as well
- an event venue to house concerts, weddings, a community theater, as well as family-friendly musicals and plays

- for-profit businesses, including a café, a restaurant, a landscaping company, and a cleaning service
- a corporate retreat center as well as Destiny Church

Ultimately, our goal is to net $25,000 a month from the development in order to negate what the church is currently paying in rent and related expenses. In time, we hope to generate even more income from the development to cover the cost of church staff.

Imagine a church with such ROI! Imagine your church pursuing such a course for similar reasons, no matter the scale. I believe these things can and should be accomplished. More than that, I believe such an approach will better position us to engage our communities to change the number of people willing, at least, to give Christ and his church a try.

The apostle Paul gave church leaders a strong admonition: "Keep watch over yourselves and all the flock of which the Holy Spirit has made you overseers" (Acts 20:28 NIV). Pastoral leaders must now pay attention to these things and fundamentally shift in their approach to church economics before things get worse. Failure to make such a shift will have devastating consequences for local churches in the future: spiritually, practically, and financially.

Kingdom Risks

Myron Pierce • Mission Church, Omaha, Nebraska

If there is anyone who understands the dynamics of ROI, it's God, our Father. First, he identified a profoundly deep problem, one that was seemingly impossible to solve. Then in his wisdom, he devised a perfect plan, sending his Son, Jesus Christ, to die for our sins. Heaven continues to reap eternal returns from what our Savior did two thousand years ago.

Such understanding informs my approach to ministry in the twenty-first century. It is my contention that pastors and church planters today must be entrepreneurial and think creatively about how to generate income beyond tithes and offerings.

In 2016, I planted Mission Church in north Omaha, Nebraska, in a difficult urban setting where the average income is about $10,000–$15,000 a year. From the outset, I realized an innovative approach would be required to get the church off the ground. However, since I have helped start businesses and plant churches in the past, it was not a big leap for me, because the two go hand in hand. I determined therefore that this church, Mission Church, would need to integrate both business and mission in order to impact the community and the kingdom.

One day, within the first nine months of development and prior to our church's official launch, it began to hail. I looked up and thought, *Money is falling from the sky!* Searching the internet, I learned that the devastating hailstorm in Omaha had caused approximately $80,000,000 in damage. Roofs would need to be repaired. So one week later, a friend and I began a roofing company, hiring three full-time laborers who had spent time in prison and struggled with addiction. For us, business isn't just about money but about kingdom impact. In addition to the impact we had on their lives, our hiring of those men saved our city thousands of dollars—$145,000 to be exact—by keeping them out of jail.

Building on this model, we later employed forty convicted felons for snow removal in the winter of 2017, again impacting our community socially and economically. In time, we launched four other businesses and started three transitional houses for men getting out of prison and an after-school mentoring program for at-risk kids in our neighborhood. Every month we have community events that bring diverse people together.

All such enterprise is helping us generate sustainable income for the church, engage our community, and advance the common good. We even launched a second church recently and seeded it with $30,000 from our profitable businesses.

After taking kingdom risks, our inner-city church is today having kingdom impact in Omaha by leaning into what I call "kingdom economics." The moral of our story is this: dream big, think differently, and take risks. If you are willing to do so, the kingdoms of this world will give way to the kingdom of our God.

Adding Value

Rowland Smith • Third Space Coffee, Colorado Springs, Colorado

A few years ago, my wife and I decided to live a more missional life outside church walls. Choosing to leave the staff of a local congregation where I was paid from tithes and offerings, we wanted to couple our desires to be kingdom agents with a sustainable way to do ministry. Desiring to create a space where the diverse people of our community could gather, we decided to open a coffee shop. For us, it seemed a perfect contextual way to intersect our love for God and our love for our neighbors.

Through the years, I had become aware of several nonprofit, faith-based coffee shop initiatives. As part of church culture, however, dependent on the generous giving of others, I had been conditioned to think this was the way to do things. Initially, we considered establishing the coffee shop as a nonprofit but ultimately decided to open Third Space Coffee as a for-profit business enterprise for the purposes of God's kingdom.

Third Space is known in our community as a specialty coffee shop that offers great product and a great experience. We don't present ourselves as a Christian coffee shop, per se, but operate as a socially motivated enterprise that happens to be owned by missionally minded Christ followers. We strive to provide a glimpse of the kingdom and to be a fragrant aroma by which others experience Christ through the way we live in community with others, bring value to the neighborhood, and provide employment opportunities in our city.

In conjunction with opening Third Space Coffee, we also determined to plant a missional church community that would gather in the shop on Saturday nights. So we launched Ecclesia COS with the help of others who shared our values.

We didn't want to saddle it with a large salary or other expenses usually associated with church plants. Therefore, since Ecclesia meets in Third Space, and I am compensated through the coffee shop as its owner, I can serve as the church's visionary leader without an expectation of salary. This approach—one that allows the church to meet in a space without paying rent and to have a lead pastor without paying a salary—allowed us to launch Ecclesia without the pressures of raising a large amount of money on the front end typically required to fund a church-planting budget.

It was freeing for me, though a paradigm shift, to begin pursuing "business as mission" instead of relying on donations. Combining a well-run for-profit, such as our coffee shop, where people gladly pay for good product and service, with a local church is a model more pastors should consider in the future. Many churches have a deep talent pool and a rich base of experiential knowledge in members who live in the business world. It is time to start combining these talents with church methodology to create a more sustainable model of church economics in the future. In doing so, we can close the gap between secular and sacred space and not only teach about the kingdom but also bring it to bear in tangible ways.

What If . . .

Matt Roberts • The Genesis Project, Ogden, Utah

In 2005, my family and I moved to Ogden, Utah, with a dream to plant "a church for people who don't do church." During the first two years of our journey, we met with people at local coffee shops, in restaurants, and in our home. Soon, a core group developed, and we were worshiping with approximately sixty people regularly. In time, we outgrew every space made available to us, so we signed a lease to rent a small metal building in the middle of Ogden. It was not ideal, but it was ours.

On our first Sunday in the new space, we were overwhelmed: two hundred people showed up! Yet God was moving in the hearts of addicts, strippers, felons, and the irreligious around us. It was beautiful. Thus, the Genesis Project, the name we gave to the church, was born as a place of refuge and hope in our community.

As we would soon learn and experience, creating a church whose heart is set to serve the broken and marginalized is a noble, spiritual work, but one that is challenging to sustain financially. The bubble of that first Sunday soon burst when it was reported to me that the offering that day was just $242.

My heart sank. The lease we had just signed required us to pay $4,500 a month for the space. Unless something radically changed, I knew we were in trouble. The church was not sustainable.

Over the next seven years, our church continued to grow. Yet the growth was always outpaced by our financial need. My faith was stretched and challenged as I tried to reconcile a ministry that was advancing the gospel yet still financially struggling to stay afloat. We began to wonder if there might be a better way to generate income beyond tithes and offerings. We asked ourselves, *What if our church, collectively, took up the call to tentmaking as a body?* It was this thought that led us into a new frontier and era of the work.

In 2015, we dreamed and purposed to build a for-profit event center that would house the ministries of our church. With generous partners behind us, we realized that everything we wanted to be as a church also made for a great hospitality business. So instead of our building sitting empty over 60 percent of the week, we chose to leverage it as an economic engine to help fuel our mission and vision.

Once we advertised that the building could house events, our calendar soon filled with bookings. We found ourselves surrounded by the community inside our own building. Weddings, business meetings, government events, high school dances, class reunions, and family reunions all came to us and happily paid for use of the space. For the first time, we had a facility that was working for us, not against us, and we loved serving brides, families, and professionals with hospitality, all the while creating a sustainable income stream for our church. It was awesome but not always easy.

Undoubtedly, there will be challenges that arise from a non-profit church and a for-profit business sharing space. Sacrifice and accommodation will be required from both entities; there will be new problems and ongoing needs to address. It is not a perfect model, but it is one that has allowed the Genesis Project to move a little closer to being a sustainable church in the future.

Join the Revolution

In response to research concerning "the myths and realities of church planters and finances," Barna suggests that leaders do the following:

- Embrace complexity and the challenge of change.
    It is a remarkable opportunity to lead during this era, but only those who get comfortable with complexity will navigate these times.
- Realize this era of change requires entrepreneurship and new partnerships.
    This is a time to take risks together, to dare greatly.
- Establish practices to see and shape reality.
    In order to lead effectively toward a vibrant future of church funding, we need . . . to examine what is true about our context. . . . If we can't see what is, we won't see what might be.[30]

In addition, churches could establish a volunteer team to serve as economic advisers or as a council to help curate and vet ideas, cultivate the potential for business incubation, and/or establish microfinancing to accelerate new start-ups. The possibilities are endless once you free your mind, dare to dream, and are willing to take risks. Such advice from Barna is not only good for church planters but also for pastors and others in existing churches, that is, for all those charged with navigating the complexities of church economics in the twenty-first century.

Finally, be reminded as with anything good you will need to commit yourself and your church to long-term pursuit of any such goals in order to see them realized. As someone has said, "It takes ten years to become an overnight success." At Mosaic, we have never experienced overnight success in anything we've set out to do or accomplish. For instance, our food-distribution program called The Orchard began more than ten years ago, with us distributing small boxes of a few dry goods when people came to Mosaic for help. Today, the program requires one full-time person, several committed lay leaders, and an additional thirty volunteers to distribute approximately eighteen thousand pounds of quality food every Tuesday afternoon. The Orchard provides three to four days of groceries, including fresh fruits, vegetables, quality meats, and more, once a month to approximately 55 percent of people living in the 72204 ZIP Code.

Developing the church's budget has not been any different. I (Harry) remember the first Sunday we received a $5,000 offering. What a big deal that was for us! Why? Because I also remember a time when Mosaic had less than $10 in its checking account. I also remember the first time our budget exceeded $250,000, then $500,000, and so on. Today, over seventeen years since we first began, our budget is $1.2 million. Thankfully, 30 percent of it is funded outside of regular tithes and offerings.

If it wasn't for our approach, our budget would include another $400,000 or so in annual expenses.

Reflecting on the illusion of overnight success, John Kim, part of Emory University's Goizueta Business School and full-time faculty in the practice of organization and management, provides a few takeaways gleaned from Jeff Bezos, founder of Amazon:

- Amazon was not a "sure thing." It took grit, experimentation, and salesmanship
- Customers are more dissatisfied than you think; you need to constantly invent for them
- Think three years out; the current financial quarter was determined 3 years ago
- Experimentation inevitably results in a lot of "good failure"
- "Bad failure" does exist—screwing up what you should be getting right
- All "overnight" successes take ten years
- Baseball is not a good business analogy. With technology, the returns can be asymmetrical. What if you swung the bat and hit the ball so hard you got 1,000 runs?
- Focus on 2–3 big, 10-year ideas and pour your energy into them[31]

Given our own experience and the experience of others, we've come to believe that in the context of planting a healthy multi-ethnic, socially just, and economically sustainable church, it takes seven to ten years to get from survival to stability and another seven to ten years to get from stability to sustainability. For existing churches, the forecast may be adjusted to five to seven years for each era. In short, as with so many other things in life and ministry, prayer, patience, and persistence, coupled with a passion that arises from calling, are foundational to any success that leaders experience in leveraging church assets, blessing the community, and generating sustainable income. Indeed, such leaders already exist in entrepreneurial churches that are today shaping the future. It is these leaders and churches that will surely bring about the coming revolution in church economics.

We Know . . . We're Pastors Too!

After considering our disruptive approach, some pastors express concern about not being instinctively entrepreneurial. They are unsure if they can or even should move forward in terms of implementation. The good news? You don't have to be the one to do so. As in our church, God has likely already brought gifted and passionate others your way. They have sharp minds and are smart with money. They are focused and credible people with proven expertise and specific experience. These men and women have a heart for your church and truly desire to make a difference beyond the level of mere employees or managers, as we discussed in chapter 5. Trust us. Such people will welcome the opportunity to run point for you in one or more of the ways we have discussed. They're just waiting for you to ask.

Therefore, whether you are a creative and intuitive entrepreneur or not at all business-minded, we encourage you to do the following:

- Embrace the concepts.
- Establish strategic plans.

- Enlist others to the cause.
- Empower them to lead.
- Enjoy seeing God produce spiritual, social, and financial ROI through you and your church in the years ahead.

Toward that end, we remain one in Christ with you.

IDEAS AT A GLANCE

- New models and approaches are needed for churches and nonprofits regarding funding. Increasingly, they are "entering into traditionally commercial spaces in an effort to increase revenues [due to] uncertain philanthropic funding . . . unstable fundraising revenues, and increased competition for limited resources."[32]

- Given conventional understanding of its economic system, a local church cannot attract, leverage, or compete for investment dollars. By establishing for-profit business enterprise, it can attract investment beyond tithes and offerings, especially from investors wanting to make ROI *and* help the church. It is a potential win-win.

QUESTIONS FOR REFLECTION

1. What is the default understanding in your church for generating more income?

2. What fears exist among pastors and churches, in general, that might inhibit any discussion of starting for-profit businesses to help fund their ministries? What fears exist in your own church?

3. Who might you ask for help in exploring the possibility of leveraging church assets to start a new business, bless the community, and generate sustainable income?

AFTERWORD

In *The Coming Revolution in Church Economics*, Mark and Harry have shared promising practices and given you every good reason to create a sustainable future for your church. To conclude, they have invited me to address one more important issue for you to consider. It is a technological development that has the greatest potential to disrupt church economics in the future—a technology that results from the exponentially increasing developments in artificial intelligence (AI). The technology is automation.

Imagine you are standing in the pulpit on a Sunday morning in the near future, looking into the faces of your congregants, knowing that nearly half of them have recently lost their jobs. Automation is threatening to turn this imagined scenario into reality. In fact, a 2013 study at Oxford initially indicated that up to one-half of all jobs in the United States are at risk of being fully automated in the next twenty years.[1] More recently, in 2017, the McKinsey Global Institute posited a lesser number: up to one-fifth of the global workforce will be affected by 2030.

That study suggested that up to one-third of the workforce in richer nations such as the United States and Germany will be replaced.[2] Keep in mind that these most recent estimates predict that we are just eleven years away from such a scenario. So in a very real way, the clock is ticking.

Even if the percentage of jobs lost is only 12 percent to 23 percent, as the McKinsey report suggested, these unemployment rates would represent the greatest percentage of job loss in the United States since the Great Depression. In those years, a quarter of all Americans were out of work. Of course, our social and cultural context is quite a bit different than it was in the 1930s. Still, it is worth noting that mass unemployment in Germany during the Great Depression undergirded events that led to World War II. Those who remember history know that once the Great Depression hit, the effects lasted for a decade. The unemployment rate in the United States didn't drop below 10 percent until after the country entered World War II and began mass production of military machines, weaponry, and more that reinvigorated the economy.

A loss of jobs anywhere near the magnitude of what either of these reports suggest would shake the social and economic foundations of our society, create new levels of unimagined poverty, increase income inequality, and heighten racial and ethnic divisions. What makes things worse is that governments and churches around the world are wildly unprepared for such disruption.

Automation Armageddon

As a global expert in emerging technology and the foremost clergy expert on AI in the world, I often get invited to advise scientists, business leaders, politicians, and educators on what is next in regard to emerging technology. In addition, I have had the opportunity to advise policy makers in numerous countries. When the issue of automation arises behind closed doors, the one ethical question everyone asks is "How much should we emphasize to people that this is likely to happen?"

It is a fair question.

People in the know talk about a coming automation crisis as if it is a meteor headed toward earth destined to wipe out much of humanity. The debate among them is whether informing the general population will help or hurt preparation efforts. The danger is that telling people everything about what is projected to happen might cause panic, and that panic might snowball to create a very real and negative impact on financial markets and government preparedness.

Of course, it is possible that yet unknown or still emerging technology may be created to offset or even counter massive job loss and subsequent percentages of unemployment. After all, the problem is not that there won't be any jobs left for us to do. In fact, there will be plenty of new opportunities for work, particularly in fields related to technology. The problem is that workers won't be able to be retrained fast enough to avoid severe suffering and the accompanying economic downturn. However, betting on some unknown technology to fix a coming "Automation Armageddon" doesn't sit well with most insiders. There are so many other issues surrounding AI, ranging from questions of human identity to machine learning to weaponized escalation among governments. These too represent a clear and present danger to the future of church and global economics.

Given emerging technology and what most certainly lies ahead, I can assure you that there are many good reasons to heed the counsel that Mark and Harry have provided. Moreover, I suggest they have only scratched the surface in consideration of what is yet to come. In the next twenty-five years, we will witness and be impacted by the greatest proliferation of technological development ever before seen in human history. The speed at which it is coming will only increase with time.

A Technological Reckoning

I believe that not unlike the church prior to the Great Depression of the 1920s, the modern church has largely forsaken the gospel to perpetuate the institution. One hundred years later, we have constructed edifices far too large and expensive to maintain. In so doing, we have forsaken the weightier teachings of Jesus focused on discipleship and helping the poor. For example, we have turned many churches into nothing more than glorified entertainment clubs that numb the self-inflicted pain of pride and greed. Instead of helping people develop a new hermeneutic of life and ultimate meaning in this technological age, we perpetuate insufficient theology that leaves people feeling helpless and without hope. We then wonder why so many churches are stagnant if not declining today in the United States.

If predictions about the future of automation hold true, it will devastate the current economic system of the church. Today, "More than half of US churches are under 100 members,"[3] and many of these already struggle to pay their bills. If those churches lose even more income due to automation in the coming years, it could force a massive number of local churches to close their doors, and the repercussions will be marked by unprecedented social impact. During the Great Depression, local churches served as the glue that held society together and helped those most in need make it through the darkest of times. If even 10 to 15 percent of existing churches in the United States do not survive what lies ahead, who will come to the aid of growing numbers of people negatively affected in ways too numerous to count?

With such things in mind, I believe a technological reckoning is coming—one that will significantly affect local churches. In fact, automation may be just what is needed to bring about a change in our perspective and actions going forward. But we shouldn't wait for things to happen. We must today engage what is happening to shape the future of things to come.

The Opportunity

Advancing automation may be the local church's greatest chance to once again become spiritually vibrant and socially relevant. People in mass numbers are going to need personal, emotional, and practical help. They will need to see and benefit from sacrificial behavior in order for humanity to flourish. Christianity offers a rich history of how to provide such things effectively.

In spite of all the aforementioned challenges, I still remain a techno-optimist. I think that humanity is called to be stewards of all matter. The development of automation provides us with an opportunity to reconsider our own humanity, how we should then live, and what we must do. Personally, I don't think our ultimate vocation is to be laborers who unnecessarily toil. I believe that we have all been called, individually and collectively, to redeem the world.

As a result, the present questions we face are simple and direct. Do we want to participate in Christ's redemptive purposes in the world? Do we want to minimize suffering in the world? If so, as a Christian community, we need now to begin a dialogue in consideration of automation, the coming impact of AI on the church, and the related implications for the world. In fact, nearly five years ago, in an article titled "Churches Need to Prepare for When Robots Take People's Jobs," I proposed four ways that a local church could begin its preparation.[4] In spite of this, a very small fraction of pastors and churches are having such discussions—potentially less than 2 percent.

So what is stopping us? If local churches devote time to preparing for the impact of automation on society in the future, what is the worst thing that could happen? If we rethink conventional wisdom and retool approaches for future ministry, and yet scholarly predictions of massive unemployment don't materialize, have we really lost anything? At the very worst, we will have engaged the topic of emerging technology and considered AI, automation, and ultimately what it means to be human. More practically, we will be better prepared to advance the gospel and the common good in a rapidly changing economy, community, nation, and world.

May I suggest that this is reason enough to lean into the concepts and hope presented in *The Coming Revolution in Church Economics.*

Rev. Dr. Christopher Benek
pastor and CEO, CoCreators, Miami, Florida

NOTES

Foreword

1. *Disruptive innovation* is a term first coined by Harvard's Clayton M. Christensen.

Introduction

1. Mark DeYmaz, *Building a Healthy Multi-ethnic Church* (San Francisco: Jossey-Bass, 2007), xxvi.

2. Barna Group, *Church Startups and Money: The Myths and Realities of Church Planters and Finances* (Ventura, CA: Barna Group, 2016), 10.

3. According to the CDC, "an estimated 240,200 children were born out of wedlock in 1961." The 6 percent (more specifically, 5.6 percent) figure is derived from dividing this number by the number of reported live births in the US in 1961; namely, 4,268,326. Learn more at https://www.cdc.gov/nchs /data/vsus/vsus_1961_1.pdf bottom of page 1–14 or https://www.infoplease .com/us/births/live-births-and-birth-rates-year.

4. While the term was in use prior to 1942, it became commonplace in the 1980s as a descriptor for Generation X, "who according to a 2004 marketing study, 'went through its all-important, formative years as one of the least parented, least nurtured generations in US history.'" Learn more at https:// en.wikipedia.org/wiki/Latchkey_kid.

5. Rube Goldberg was an inventor, sculptor, author, engineer, and cartoonist who poked fun at the way human beings tend to utilize modern machines and technology to complicate rather than simplify life. Learn more at http:// www.rube-goldberg.com/wiki/ and http://www.rube-goldberg.com/.

Chapter 1 Understanding the Need

1. Learn more about Brotherhood Mutual at https://www.brotherhood mutual.com/.

2. Mark DeYmaz, *Disruption: Repurposing the Church to Redeem the Community* (Nashville: Thomas Nelson, 2017). Learn more at https://www.mosaix .info/disruption.

3. Learn more about Ron Blue at http://masteryourmoney.com/rb-bio graphy-1.html.

4. Tami Luhby and Tiffany Baker, "What Is Middle Class, Anyway?" CNN Money, December 1, 2016, http://money.cnn.com/infographic/economy/what -is-middle-class-anyway/index.html.

5. Luhby and Baker, "What Is Middle Class, Anyway?"

6. Kerri Anne Renzulli, "This Is How Much Debt the Average American Has Now—at Every Age," *Money*, April 13, 2018, http://time.com/money/523 3033/average-debt-every-age/.

7. Renzulli, "This Is How Much Debt."

8. Mark Gongloff, "The Sad State of America's Middle Class, in 6 Charts," *Huffpost*, December 6, 2017, https://www.huffingtonpost.com/2015/01/20 /middle-class-charts_n_6507506.html.

9. Tami Luhby, "What Being Middle Class Means to Me," *CNN Money*, June 1, 2015, https://money.cnn.com/gallery/news/economy/2015/05/28/middle -class-americans/index.html.

10. Holly Hall, Eden Stiffman, Ron Coddington, and Meredith Myers, "Philanthropy Surges 5.4% to Record $358 Billion, Says 'Giving USA,'" *Chronicle of Philanthropy*, June 19, 2015, https://www.philanthropy.com/interact ives/giving-usa-2015.

11. "Giving USA 2018: Americans Gave $410.02 Billion to Charity in 2017, Crossing the $400 Billion Mark for the First Time," Giving USA, June 13, 2018, https://givingusa.org/giving-usa-2018-americans-gave-410-02-billion-to-char ity-in-2017-crossing-the-400-billion-mark-for-the-first-time/.

12. "Giving USA 2018."

13. Bob Smietana, "Churches Continue to Face Budget Shortfalls," LifeWay Research, March 10, 2016, https://lifewayresearch.com/2016/03/10/churches -continue-to-face-budget-shortfalls/.

14. Karl Vaters, "4 Assumptions Pastors Can No Longer Make about Church Giving Patterns," *Christianity Today*, January 5, 2018, https://www .christianitytoday.com/karl-vaters/2018/january/4-assumptions-pastors -church-giving-patterns.html?paging=off.

15. Vaters, "4 Assumptions Pastors Can No Longer Make."

16. In 1995, *Harvard Business Review* published an article titled "Disruptive Technologies: Catching the Wave," by Joseph L. Bower and Clayton M. Christensen. "Disruptive innovation," wrote Christensen, "describes a process by which a product or service takes root initially in simple applications at the

bottom of the market and then relentlessly moves up market, eventually displacing established competitors." Learn more at http://www.claytonchris tensen.com/key-concepts. In simpler terms, when research and development in successful companies are more focused on the present than the future, innovative adapters go to work off the grid, discovering what is next, determining how to get there, and, more significantly, defining for others why it matters. For more on the concept of disruptive innovation as it applies to the local church, see DeYmaz, *Disruption*.

17. Thom S. Rainer, "6 Reasons Why Your Church Offerings May Be Declining," *Ministry Today*, March 30, 2016, https://ministrytodaymag.com/life/money /22721-6-reasons-why-your-church-offerings-may-be-declining.

18. Rainer, "6 Reasons Why Your Church Offerings May Be Declining."

19. Rainer, "6 Reasons Why Your Church Offerings May Be Declining."

20. Bridget Hartnett, CPA, PSA, and Ron Matan, CPA, CGMA, PSA, "Generational Differences in Philanthropic Giving," Sobel and Co., LLC, fall 2014, https://sobelcollc.com/sites/default/files/NFP%20Fall%202014%20White paper.pdf.

21. Hartnett and Matan, "Generational Differences in Philanthropic Giving."

22. Abigail L. Coleman, "Gen X Giving—Effective Fundraising Strategies for Gen X Donors," Giving Institute, July 27, 2016, https://www.givinginstitute .org/news/297722/Gen-X-Giving—-Effective-Fundraising-Strategies-for-Gen -X-Donors.htm.

23. Coleman, "Gen X Giving."

24. Hartnett and Matan, "Generational Differences in Philanthropic Giving."

25. Hartnett and Matan, "Generational Differences in Philanthropic Giving."

26. Hartnett and Matan, "Generational Differences in Philanthropic Giving," figure, *The Blackbaud Index*, August 2013, https://www.blackbaud.com /files/5-13-BBIndex.pdf.

27. Jonathan Vespa, David M. Armstrong, and Lauren Medina, "Demographic Turning Points for the United States: Population Projections for 2020 to 2060," United States Census Bureau, March 2018, https://www.census.gov /content/dam/Census/library/publications/2018/demo/P25_1144.pdf.

28. Vespa, Armstrong, and Medina, "Demographic Turning Points."

29. Vespa, Armstrong, and Medina, "Demographic Turning Points."

30. "Older People Projected to Outnumber Children for First Time in U.S. History," United States Census Bureau, March 13, 2018, https://www.census .gov/newsroom/press-releases/2018/cb18-41-population-projections.html.

Chapter 2 Free Your Mind

1. "History of the Nine Dot Problem," Art of Play, August 2, 2016, https:// www.artofplay.com/blogs/articles/history-of-the-nine-dot-problem.

2. Mark McGuinness, "Why Thinking 'Outside the Box' Doesn't Work," September 22, 2008, https://lateralaction.com/articles/thinking-outside-the-box/.

3. Quoted in McGuinness, "Why Thinking 'Outside the Box' Doesn't Work."

4. Chris Brown, "3 Most Misunderstood Bible Verses on Money," Stewardship, June 20, 2016, https://www.stewardship.com/articles/3-most-misunderstood-bible-verses-on-money.

5. Some suggest that Jesus cleansed the temple on two occasions. Learn more at Tim Chaffey, "Why Did Jesus Cleanse the Temple?" Answers in Genesis, August 16, 2011, https://answersingenesis.org/jesus-christ/when-did-jesus-cleanse-the-temple.

6. Randall Niles, "Jesus Cleanses the Temple of the Moneychangers," Drive Thru History, September 12, 2016, https://www.drivethruhistory.com/jesus-cleanses-the-temple-of-the-moneychangers/.

7. Niles, "Jesus Cleanses the Temple of the Moneychangers."

8. Joe Plemon, "Why Did Jesus Drive the Money Changers from the Temple?" Seedtime, September 18, 2013, https://christianpf.com/why-did-jesus-drive-the-money-changers-from-the-temple/.

9. Adam Nicholas Phillips, "Turning the Tables: A Lenten Sermon on Jesus, the Money Changers, and #Selma50," *Sojourners*, March 9, 2015, https://sojo.net/articles/lenten-reflections/turning-tables-lenten-sermon-jesus-money-changers-and-selma50.

10. Mark DeYmaz, *Disruption: Repurposing the Church to Redeem the Community* (Nashville: Thomas Nelson, 2017), 135.

11. Mark DeYmaz and Harry Li, *Leading a Healthy Multi-Ethnic Church* (Grand Rapids: Zondervan, 2010, 2013), 235, iBook.

12. Jeannine K. Brown, "Commentary on Matthew 22:15–22," Preach This Week, October 19, 2008, https://www.workingpreacher.org/profile/default.aspx?uid=2-brown_jeannine.

13. Brown, "Commentary on Matthew 22:15–22."

14. "History of Churches and Taxes," ProCon.org, last updated August 5, 2016, https://churchesandtaxes.procon.org/view.resource.php?resourceID=006554.

15. "History of Churches and Taxes."

16. "How Does Social Security Work?" Peter G. Peterson Foundation, July 10, 2018, https://www.pgpf.org/budget-basics/how-does-social-security-work.

17. *Church Law & Tax Report: Integrating Tax Reform for 2018 and Beyond*, May/June 2018, 1.

18. *Church Law & Tax Report*, 1, 3.

19. Grayson Bell, "It's All in Your Head: Why a Scarcity Mindset Is Keeping You Poor," Debt Roundup, July 16, 2018, https://www.debtroundup.com/scarcity-mindset-keeping-you-poor/.

20. Bell, "It's All in Your Head."

21. J. Jimenez, "John C. Maxwell: 6 Tips to Develop and Model an Abundance Mindset," *Success*, March 4, 2015, https://www.success.com/john-c-maxwell-6-tips-to-develop-and-model-an-abundance-mindset/.

Chapter 3 Stop Begging for Money

1. At the invitation of Dale and Judi Dawson, founders of Bridge2Rwanda, I attended the first P.E.A.C.E. conference at Saddleback Church in 2008, where we heard Rick make this statement.

2. Erin Griffith, "Why Startups Fail, according to Their Founders," *Fortune*, September 25, 2014, http://fortune.com/2014/09/25/why-startups-fail-according -to-their-founders/.

3. According to Wikipedia, "AOL was one of the early pioneers of the Internet in the mid-1990s, and the most recognized brand on the web in the United States. It originally provided a dial-up service to millions of Americans, as well as providing a web portal, e-mail, instant messaging and later a web browser following its purchase of Netscape." "AOL," Wikipedia, last updated April 3, 2019, https://en.wikipedia.org/wiki/AOL.

4. Fillmer Hevener, "A Brief History of Tithing after Christ's Crucifixion," Guthrie Memorial Chapel, 2005, http://www.guthriememorial.org/articles /brief_history_of_tithing.htm.

5. Bob Smietana, "Churchgoers Say They Tithe, but Not Always to the Church," LifeWay Research, May 10, 2018, https://lifewayresearch.com/2018 /05/10/churchgoers-say-they-tithe-but-not-always-to-the-church/.

6. "Tithing Questions," Generous Church, accessed November 22, 2019, https://www.generouschurch.com/tithing-questions.

7. "What Has the Church Said about Tithing?" Catholic Diocese of Wichita, accessed May 28, 2019, https://wichitadio.solutiosoftware.com/stewardship -articles/23851-the-church-s-tradition-of-tithing.

8. Hevener, "A Brief History of Tithing after Christ's Crucifixion."

9. Hevener, "A Brief History of Tithing after Christ's Crucifixion."

10. William Barclay, "The Bible Commands Christians to Tithe," Gospel Coalition, March 28, 2017, https://www.thegospelcoalition.org/article/bible -commands-christians-to-tithe/.

11. As Donald Sanchez writes, "The fallacy *ad ignorantiam*, or an appeal to ignorance, occurs when someone argues that something must be either true or false because it hasn't been proven to be one way or the other. In other words, a particular belief is said to be true because you do not know that it is not true. The issue typically has to do with something that is either incapable of being proven true or false, or has not yet been proven true or false." Donald Sanchez, "Logical Fallacies 101: *Ad Ignorantiam*," Southern Evangelical Seminary and Bible College, May 19, 2018, https://ses.edu/logical -fallacies-101-ad-ignorantiam.

12. Sarah Eekhoff Zylstra, "Your Tithe Doesn't Have to Go to Your Church, Most Leaders Say," *Christianity Today*, September 15, 2017, https://www.christ ianitytoday.com/news/2017/september/your-split-tithe-doesnt-have-to-go-to -church-ministries.html.

13. Brian Kluth, "20 Truths about Tithers," State of the Plate Research, 2013, http://www.ecfa.org/Documents/News/20-Truths-About-Tithers-Exec utive-Summary-eBook-Report.pdf.

14. Kluth, "20 Truths about Tithers."

15. Jayson D. Bradley, *Teaching Your Church to Give: The Guide for Cultivating a Generous Culture* (Redmond, WA: Pushpay, 2018), 9, eBook.

16. See John W. Kennedy, "Big Dream in Little Rock," *Christianity Today*, April 1, 2005, https://www.christianitytoday.com/ct/2005/april/24.42.html.

17. Read more about the financial launch of Mosaic in Mark DeYmaz, *Building a Healthy Multi-ethnic Church* (San Francisco: Jossey-Bass, 2007), 136–38.

Chapter 4 Create Multiple Streams of Income

1. "The Rise in Dual Income Households," Pew Research Center, June 18, 2015, http://www.pewresearch.org/ft_dual-income-households-1960-2012-2/.

2. "The Rise in Dual Income Households."

3. Tami Luhby, "What Being Middle Class Means to Me," *CNN Money*, June 1, 2015, https://money.cnn.com/gallery/news/economy/2015/05/28/middle -class-americans/index.html.

4. Moolanomy, "10 Reasons to Have Multiple Income Streams," *U.S. News & World Report*, November 3, 2010, https://money.usnews.com/money/blogs /my-money/2010/11/03/10-reasons-to-have-multiple-income-streams.

5. Moolanomy, "10 Reasons to Have Multiple Income Streams."

6. Thomas C. Corley, "The Majority of Self-Made Millionaires Have a Similar Approach to Earning Money," *Business Insider*, May 27, 2015, http:// www.businessinsider.com/self-made-millionaires-have-a-similar-approach -to-earning-money-2015-5.

7. "Should Churches (Defined as Churches, Temples, Mosques, Syna-gogues, etc.) Remain Tax-Exempt?" ProCon.org, last updated July 8, 2018, https://churchesandtaxes.procon.org.

8. "Should Churches (Defined as Churches, Temples, Mosques, Syna-gogues, etc.) Remain Tax-Exempt?"

9. Billy Hallowell, "Will Churches Lose Their Tax-Exempt Status in the Wake of the Supreme Court's Gay Marriage Ruling?" *Blaze*, July 3, 2015, https:// www.theblaze.com/news/2015/07/03/will-churches-lose-tax-exempt-status -in-the-wake-of-the-supreme-courts-gay-marriage-ruling. See also "Oberge-fell v. Hodges," Wikipedia, last updated April 12, 2019, https://en.wikipedia .org/wiki/Obergefell_v._Hodges.

10. Erwin Chemerinsky, "The Parsonage Exemption Violates the Estab-lishment Clause and Should Be Declared Unconstitutional," *Whittier Law Review* 24 (2003): 707, 738, http://works.bepress.com/erwin_chemerinsky/255/.

11. Mark DeYmaz, *Disruption: Repurposing the Church to Redeem the Com-munity* (Nashville: Thomas Nelson, 2017), 125.

12. Jimmy Scroggins and Steve Wright, "The Math Doesn't Work: Why the Future of Church Planting Is Bi-Vocational," 9Marks, January 3, 2013, https://www.9marks.org/article/journalmath-doesnt-work-why-future-church-planting-bi-vocational/.

13. J. D. Greear and Mike McDaniel, "Is the Future of Church Planting Bi-Vocational?" 9Marks, August 22, 2014, https://www.9marks.org/article/is-the-future-of-church-planting-bi-vocational/.

14. Scroggins and Wright, "The Math Doesn't Work."

15. Scroggins and Wright, "The Math Doesn't Work."

16. Ed Stetzer, Micah Fries, and Daniel Im, "The State of Church Planting in the US," 2015, https://www.wesleyan.org/wp-content/uploads/Lifeway-2015-ChurchPlanting-Survey.pdf.

17. Stetzer, Fries, and Im, "The State of Church Planting in the US."

18. Stetzer, Fries, and Im, "The State of Church Planting in the US."

19. Carey Nieuwhof, "The Top 8 Reasons Most Churches Never Break the 200 Attendance Mark," Carey Nieuwhof, accessed December 4, 2018, https://careynieuwhof.com/the-8-most-common-reasons-most-churches-never-break-the-200-attendance-mark/.

20. Excerpt from an email exchange with Matt McGue, December 8, 2018.

21. Scroggins and Wright, "The Math Doesn't Work." Not surprisingly, Evangelical Outreach Ministries (EOM) faces a similar problem in attempting to onboard nonwhite staff members through traditional fund-raising strategies. In an article titled "How Support Raising Keeps Parachurch Ministries White," missioeric notes, "The Personal Support Raising model is predicated on your social network connecting you with people who have disposable income they can give on a regular basis. By assuming all staff start from the same place in their support raising, the model fails to take into account the disparity of wealth in this country. [For example,] the Pew Research Center published a study in 2011 entitled, Wealth Gaps Rise to Record Highs Between White, Blacks, Hispanics (see http://www.pewsocialtrends.org/2011/07/26/wealth-gaps-rise-to-record-highs-between-whites-blacks-hispanics). The report analyzed data from the US government and found that 'the median wealth of white households is 20 times that of black households and 18 times that of Hispanic households . . . the typical black household had just $5,677 in wealth (assets minus debts) in 2009; the typical Hispanic household had $6,325 in wealth; and the typical white household had $113,149.' While there are obviously many individual exceptions to the report, this means that when an African-American or Latino minister enters the average home of someone in their community to raise support, the person they meet with will likely have 20 times less wealth than the average person a White minister meets with from their community. Statistically, Asian Americans are just as likely to raise full support as their White counterparts. [However] Asian Americans are 2.5 times more likely to report that their families are embarrassed

of them because of support raising. [It] can be perceived as valuing money over relationship and highlights the white cultural context the model was created and honed in." Missioeric, "How Support Raising Keeps Parachurch Ministries White," Minister Different, accessed November 23, 2019, http://ministerdifferent.com/support-raising-white. See also Samuel Perry, "Diversity, Donations, and Disadvantage: The Implications of Personal Fundraising for Racial Diversity in Evangelical Outreach Ministries," accessed November 23, 2018, https://link.springer.com/article/10.1007/s13644-011-0020-7.

22. Greear and McDaniel, "Is the Future of Church Planting Bi-vocational?"

23. Greear and McDaniel, "Is the Future of Church Planting Bi-vocational?"

24. Learn more about the gig economy at Margaret Rouse, "Gig Economy," WhatIs.com, last updated March 2019, https://whatis.techtarget.com/definition/gig-economy. Learn more about the System D economy at Robert Neuwirth, "The Shadow Superpower," FP, October 28, 2011, https://foreignpolicy.com/2011/10/28/the-shadow-superpower.

25. This approach is fully explained in DeYmaz, *Disruption*.

26. "President Obama Honors the Super Bowl XLVI Champion New York Giants," YouTube, June 8, 2012, http://www.youtube.com/watch?v=JORmlItScHU.

27. The model is fully explained in DeYmaz, *Disruption*.

28. "President Obama Honors the Super Bowl XLVI Champion New York Giants."

29. DeYmaz, *Disruption*, chapter 4 ("Disrupting Witness").

30. DeYmaz, *Disruption*, 21.

31. Learn more at https://www.daveramsey.com/.

32. The term *gig economy* loosely defines "alternative work arrangements." According to *Harvard Business Review*, "There's no denying the growth of the gig economy. Economists estimate that the portion of US workers earning a living as independent contractors, freelancers, temps, and on-call employees jumped from 10% in 2005 to nearly 16% in 2015, and the trend shows little sign of slowing. Advocates of these 'alternative work arrangements'—many of which are enabled by sharing or on-demand apps such as Uber and TaskRabbit—bill them as a way to trade unemployment, burnout, or hating one's job for freedom, flexibility, and financial gains." Nicole Torres, "Are There Good Jobs in the Gig Economy?" *Harvard Business Review*, July–August 2018, https://hbr.org/2018/07/are-there-good-jobs-in-the-gig-economy.

33. Learn more at https://www.nationalservice.gov/programs/americorps/americorps-programs/americorps-vista.

34. Learn more at http://www.theamericanchurch.org.

35. DeYmaz, *Disruption*, 43–44.

36. Spencer Perkins and Chris Rice, *More Than Equals* (Downers Grove, IL: InterVarsity, 2000), 261.

37. Mel Robbins, "Why Trump Is Beating Fox News—and GOP Rivals," CNN, January 28, 2016, https://www.cnn.com/2016/01/27/opinions/donald-trump-republican-robbins.

38. DeYmaz, *Disruption*, xxviii.

Chapter 5 Leverage Church Assets

1. Derrick M. Hall, "Letter from the Team President," *Phoenix*, accessed November 27, 2018, https://www.phoenixmag.com/2018/07/01/arizona-diamondbacks-20th-anniversary-issue/.

2. Hall, "Letter from the Team President."

3. Andy Newman, "Their Mission: Spreading the Word through Business," *New York Times*, November 14, 2005, https://www.nytimes.com/2005/11/14/giving/their-mission-spreading-the-word-through-business.html.

4. Newman, "Their Mission."

5. Newman, "Their Mission."

6. Jo Plummer, "What Is Business as Mission?" Business as Mission, January 14, 2015, http://businessasmission.com/what-is-bam/.

7. Plummer, "What Is Business as Mission?"

8. Plummer, "What Is Business as Mission?"

9. One organization devoted to this very idea is Made to Flourish. Its stated mission is to "empower a growing network of pastors and their churches to integrate faith, work and economic wisdom for the flourishing of their communities." Learn more at https://www.madetoflourish.org, accessed May 11, 2019.

10. See https://bobbbiehl.com/.

11. See https://bobbbiehl.com/product/team-profile-2/.

12. See https://bobbbiehl.com/product/team-profile-2/.

13. See https://bobbbiehl.com/product/team-profile-2/.

14. See http://foclonline.org/sites/foclonline.org/files/biehl_bobb_team_profile_handout_0.pdf.

15. Thom S. Rainer, "Hope for Dying Churches," *Facts and Trends*, January 16, 2018, https://factsandtrends.net/2018/01/16/hope-for-dying-churches/.

16. Robert Schmidt writes, "Ground leases, often called land leases, are simply a lease of the land only. Usually land is leased for a relatively long period of time (50–99 years) to a tenant that constructs a building on the property. A ground lease separates ownership of the land from ownership of the building and improvements constructed on the land. . . . For the land owner, a ground lease provides a stable income stream typically from a creditworthy tenant, while still allowing the landlord to retain ownership of land. Usually ground leases have built-in escalation clauses and eviction rights, which give the land owner adequate rent increases over the term of the lease as well as further downside protection in the event of a default. Another benefit to land owners is that ground leases normally have a reversionary clause, which

transfers ownership of the improvements to the landlord at the end of the lease." Robert Schmidt, "Ground Lease Fundamentals in Commercial Real Estate," Property Metrics, August 4, 2014, https://www.propertymetrics.com /blog/2014/08/04/ground-lease/. For pros and cons of a ground lease, see Michael Dean, "Pros and Cons of Commercial Ground Leases," Wendel Rosen, March 1, 2007, https://www.wendel.com/knowledge-center/publications/2007 /pros-and-cons-of-commercial-ground-leases.

17. Generally, the term *dark property* refers to a building previously leased by a tenant who moves or goes out of business prior to (and often years ahead of) the end of the lease contract. In these cases, buildings can be rented (often long term) on a sublease from these companies at a fraction of their value, saving the church money and providing it with an opportunity to engage and bless the community.

18. Mark DeYmaz, *Disruption: Repurposing the Church to Redeem the Community* (Nashville: Thomas Nelson, 2017), 126–27.

19. Carter and Courtney Reum, "It Takes Money to Make Money—Or Is There Something Else You Should Be Looking For?" *Inc.*, December 13, 2017, https://www.inc.com/carter-and-courtney-reum/it-takes-money-to-make -money-or-is-there-something-else-you-should-be-looking-for.html.

20. DeYmaz, *Disruption*, 110.

21. DeYmaz, *Disruption*, 130–33.

22. Learn more at "Al Worthington," Wikipedia, last updated February 13, 2019, https://en.wikipedia.org/wiki/Al_Worthington. See also Bill Nowlin, "Al Worthington," Society for American Baseball Research, last revised October 1, 2015, https://sabr.org/bioproj/person/db42b586.

23. Learn more about how Mosaix Global Network can help you plant, grow, or develop a healthy multiethnic, economically diverse, and sustainable church at https://www.mosaix.info.

Chapter 6 Become a Benevolent Owner

1. The term "benevolent ownership" was first coined and defined by Mark DeYmaz in *Disruption: Repurposing the Church to Redeem the Community* (Nashville: Thomas Nelson, 2017), 122ff.

2. DeYmaz, *Disruption*, 123–24.

3. DeYmaz, *Building a Healthy Multi-ethnic Church*, 64–65.

4. DeYmaz, *Building a Healthy Multi-ethnic Church*, 64–67.

5. DeYmaz, *Disruption*, 127.

6. DeYmaz, *Disruption*, 117. Part I incidents of crime include homicide, rape, robbery, aggravated assault, burglary, larceny-theft, auto theft, and arson.

7. "The University District Redevelopment Potential Analysis Report," developed by AECOM, a Fortune 500 company and global provider of professional technical and management support services, states, "In the southwest quadrant, new mixed-use structures share parking with a former Kmart store,

the new home of Mosaic Church; this structural re-use keeps an important community anchor within the neighborhood, and promotes a sustainable approach to neighborhood redevelopment. The arrangement of uses is especially conducive to shared parking, with heavy weekday office demand dovetailing with heavy weekend church demand," as cited in DeYmaz, *Disruption*, 130.

8. "May Church Rent to For-Profit?" Don Kramer's Nonprofit Issues, April 1, 2014, https://www.nonprofitissues.com/tothepoint/may-church-rent-profit.

9. Michael E. Batts, *Unrelated Business Income and the Church* (Orlando: Accountability Press, 2011), 34.

10. "Exclusion of Rent from Real Property from Unrelated Business Taxable Income," IRS, last updated October 24, 2018, https://www.irs.gov/charities-non-profits/exclusion-of-rent-from-real-property-from-unrelated-business-taxable-income.

11. See Batts, *Unrelated Business Income and the Church*. This booklet provides an informed, simple, and helpful guide to understanding such legal complexities.

12. Robert W. Wood, "Why Churches Are the Gold Standard of Tax-Exempt Organizations," *Forbes*, September 22, 2015, https://www.forbes.com/sites/robertwood/2015/09/22/lets-tax-churches/#197d3474322b.

13. Kramer, "May Church Rent to For-Profit?"

14. "Should Churches (Defined as Churches, Temples, Mosques, Synagogues, etc.) Remain Tax-Exempt?" ProCon.org, last updated July 8, 2018, https://churchesandtaxes.procon.org/.

15. "Should Churches (Defined as Churches, Temples, Mosques, Synagogues, etc.) Remain Tax-Exempt?"

16. Billy Hallowell, "Will Churches Lose Their Tax-Exempt Status in the Wake of the Supreme Court's Gay Marriage Ruling?" *Blaze*, July 3, 2015, https://www.theblaze.com/news/2015/07/03/will-churches-lose-tax-exempt-status-in-the-wake-of-the-supreme-courts-gay-marriage-ruling. See also "Obergefell v. Hodges," Wikipedia, last updated April 12, 2019, https://en.wikipedia.org/wiki/Obergefell_v._Hodges.

17. Erwin Chemerinsky, "The Parsonage Exemption Violates the Establishment Clause and Should Be Declared Unconstitutional," *Whittier Law Review* 24 (2003): 707, 738, http://works.bepress.com/erwin_chemerinsky/255/.

18. Ann Omley, "The Wise Man and the Foolish Man," Timeless Truths, accessed November 20, 2018, https://library.timelesstruths.org/music/The_Wise_Man_and_the_Foolish_Man/.

19. DeYmaz, *Disruption*, 131.

20. DeYmaz, *Disruption*, 131.

21. Daniel Cook, *10 Tsunamis Impacting Ministries: How Do We Survive What's Coming?* (Ogden, UT: BGW Press, 2015), 272.

22. DeYmaz, *Disruption*, 136–37.

23. In 1987, Skatechurch was the first skateboard ministry to be established by a local church in the US. Learn more at https://www.skatechurch.net/about/history/.

24. "What Is Gentrification?" pov.org, June 17, 2003, http://archive.pov.org/flagwars/what-is-gentrification/.

25. Wood, "Why Churches Are the Gold Standard of Tax-Exempt Organizations."

26. This content was aggregated from a social media account; name withheld.

27. This content was aggregated from a social media account; name withheld.

28. This content was aggregated from a social media account; name withheld.

29. DeYmaz, *Building a Healthy Multi-ethnic Church*, 120.

Chapter 7 Monetize Existing Services

1. This is a play on the more common phrase "the cost of doing business." By definition, "The cost of doing business refers to all the expenses incurred by a firm or a sole proprietor in producing and selling goods or services. The cost of doing business depends on a wide variety of factors, including the cost inputs of goods and services, the costs of compliance with any regulations, interest rates on borrowed funds and taxes. The lower the costs of doing business, the easier it is for the business to operate, hire labor and pay taxes." Eliah Sekirin, "How to Calculate the Cost of Doing Business," bizfluent, September 26, 2017, https://bizfluent.com/how-6763183-calculate-cost-doing-business.html.

2. "Customer Acquisition Cost," techopedia, accessed December 16, 2018, https://www.techopedia.com/definition/16537/customer-acquisition-cost.

3. Molly Galetto, "What Is Customer Acquisition?" NG Data, July 2, 2015, https://www.ngdata.com/what-is-customer-acquisition/.

4. Galetto, "What Is Customer Acquisition?"

5. Michael Feuer, "It's Just the Cost of Doing Business," *Smart Business*, March 23, 2018, http://www.sbnonline.com/article/just-cost-business/.

6. Mark DeYmaz, *Disruption: Repurposing the Church to Redeem the Community* (Nashville: Thomas Nelson, 2017), xxviii.

7. Mel Robbins, "Why Trump Is Beating Fox News—and GOP Rivals," CNN, January 28, 2016, https://www.cnn.com/2016/01/27/opinions/donald-trump-republican-robbins.

8. Robbins, "Why Trump Is Beating Fox News—and GOP Rivals."

9. DeYmaz, *Disruption*, xxviii.

10. DeYmaz, *Disruption*, 6. In addition, Susan Scott says, "Companies derail because people don't say what they are really thinking. . . . The role of the (disruptive) leader is to make the implicit explicit." This was shared

with students in a three-year Doctor of Ministry (DMin) cohort, for those pursuing a degree in Redemptive Leadership and Organizational Development in the Multi-Ethnic Context, under professors Dr. Rod Cooper and Dr. Mark DeYmaz at Gordon-Conwell Theological Seminary, Charlotte, North Carolina, May 16, 2016. Learn more at https://www.gordonconwell.edu/doctor-ministry/Redemptive-Leadership.cfm.

11. "The State of the Church 2016," Barna, September 15, 2016, https://www.barna.com/research/state-church-2016/.

12. Alessandra Malito, "Starting a New Business? Too Much Optimism Can Cost You Your Shirt," MarketWatch, October 5, 2018, https://www.marketwatch.com/story/too-much-optimism-can-cost-you-your-shirt-2018-10-04.

13. Malito, "Starting a New Business?"

14. Mark DeYmaz, *Building a Healthy Multi-ethnic Church* (San Francisco: Jossey-Bass, 2007), 96.

15. Feuer, "It's Just the Cost of Doing Business."

16. "When studying youth at risk, researchers typically focus on the risk factors that contribute to—and the protective factors that serve to buffer against—problematic outcomes. Protective factors can be thought of as either personal factors, such as problem-solving abilities or competence, and perceived efficacy or environmental resources, such as social support in the community or family income." Craig Winston LeCroy and Elizabeth K. Anthony, "Youth at Risk," Oxford Bibliographies, last modified February 22, 2018, http://www.oxfordbibliographies.com/view/document/obo-9780195389678/obo-9780195389678-0112.xml.

17. "Adult Day Care for Seniors," Senior Living, accessed December 16, 2018, https://www.seniorliving.org/lifestyles/adult-day-care/.

18. "Paying for Adult Day Care and Adult Day Health Care," Paying for Senior Care, last updated April 2018, https://www.payingforseniorcare.com/longtermcare/find_adult_day_care.html.

19. Derek Jones, "How to Help a Senior Shut-in," Griswold Home Care, January 21, 2013, https://www.griswoldhomecare.com/blog/how-to-help-a-senior-shut-in/.

20. Jones, "How to Help a Senior Shut-in."

21. From an email exchange with Daniel Cook, January 4, 2019.

22. Feuer, "It's Just the Cost of Doing Business."

23. From an email exchange with Daniel Cook, January 4, 2019.

24. From an email exchange with Daniel Cook, January 4, 2019.

25. Feuer, "It's Just the Cost of Doing Business."

Chapter 8 Start New Businesses

1. Barna, *Church Startups and Money: The Myths and Realities of Church Planters and Finances* (Ventura, CA: Barna Group, 2016), 12.

2. Bob Jackson, *Hope for the Church: Contemporary Strategies for Growth* (Norwich, UK: Hymns Ancient & Modern Ltd., 2002), 23.

3. Mark DeYmaz, *Disruption: Repurposing the Church to Redeem the Community* (Nashville: Thomas Nelson, 2017), xxxiii.

4. Barna, *Church Startups and Money*, 51.

5. "Be Generous for a Lifetime, Not Just a Season," Dave Ramsey, accessed January 10, 2019, https://www.daveramsey.com/blog/a-lifestyle-of-giving.

6. See https://www.daveramsey.com/blog/take-control-of-your-money-financial-peace-university.

7. "The Payoffs of Generous Living," Dave Ramsey, December 8, 2018, https://www.daveramsey.com/blog/the-payoffs-of-generous.

8. Josh Henry, "10 Common Church Financial Mistakes," North American Mission Board, May 4, 2017, https://www.namb.net/replant-blog/10-common-church-financial-mistakes/.

9. Henry, "10 Common Church Financial Mistakes."

10. Henry, "10 Common Church Financial Mistakes."

11. Henry, "10 Common Church Financial Mistakes."

12. Barna, *Church Startups and Money*, 51.

13. Gene Takagi, "Fair or Foul? A Review of Federal Tax Laws Governing Unfair Competition," *Nonprofit Quarterly*, October 28, 2014, https://nonprofitquarterly.org/2014/10/28/fair-or-foul-a-review-of-federal-tax-laws-governing-unfair-competition/.

14. "When Should Churches Form Subsidiary Corporations?" Law Office of Dan Beirute, July 26, 2014, https://uschurchlaw.com/churches-form-subsidiary-corporations/.

15. Takagi, "Fair or Foul?"

16. Takagi, "Fair or Foul?"

17. "When Nonprofits Compete with Businesses: Leveling an Unfair Playing Field," Capital Research Center, November 1, 2013, https://capitalresearch.org/article/when-nonprofits-compete-with-businesses-leveling-an-unfair-playing-field/.

18. "When Nonprofits Compete with Businesses."

19. "Substantially Related," IRS, last updated March 26, 2019, https://www.irs.gov/charities-non-profits/substantially-related.

20. "501(c)(3) Tax Guide for Churches and Religious Organizations," IRS, https://www.irs.gov/pub/irs-pdf/p1828.pdf.

21. From an email exchange with Daniel Cook, January 4, 2019.

22. "10 Steps to Start Your Business," U.S. Small Business Administration, accessed December 2, 2018, https://www.sba.gov/business-guide/10-steps-start-your-business/.

23. DeYmaz, *Disruption*, 133.

24. "10 Steps to Start Your Business."

25. Andrew Bloomenthal, "Holding Company," Investopedia, last updated April 11, 2019, https://www.investopedia.com/terms/h/holdingcompany.asp.

26. " Spread Your Money, Lower Your Risks," Schroders, accessed May 28, 2019, https://www.schroders.com/en/id/mutual-fund-investment/education /investing-a-quick-start/spread-your-money-lower-your-risks/.

27. Alan Hirsch and Dave Ferguson, *On the Verge* (Grand Rapids: Zondervan, 2011), 250.

28. John Maxwell, "Paul's Blueprint for Evangelism," Church of the Highlands, July 26, 2017, https://www.churchofthehighlands.com/media/message /pauls-blueprint-for-evangelism.

29. John Dickerson, *The Great Evangelical Recession* (Grand Rapids: Baker Books, 2013), 81–85.

30. Barna, *Church Startups and Money*, 52.

31. John Kim, "Jeff Bezos: 'All Overnight Success Takes about 10 Years,'" Consultant's Mind, accessed January 6, 2019, https://www.consultantsmind .com/2018/05/19/jeff-bezos-all-overnight-success-takes-about-10-years/.

32. Takagi, "Fair or Foul?"

Afterword

1. Carl Benedikt Frey and Michael A. Osborne, "The Future of Employment: How Susceptible Are Jobs to Computerisation?" September 17, 2013, https://www.oxfordmartin.ox.ac.uk/downloads/academic/The_Future_of_Em ployment.pdf.

2. James Manyika et al., "Jobs Lost, Jobs Gained: What the Future of Work Will Mean for Jobs, Skills, and Wages," McKinsey & Company, November 2017, https://www.mckinsey.com/featured-insights/future-of-work/jobs-lost -jobs-gained-what-the-future-of-work-will-mean-for-jobs-skills-and-wages.

3. Aaron Earls, "Majority of American Churches Fall Below 100 in Worship Attendance," LifeWay Research, February 24, 2016, https://factsandtrends .net/2016/02/24/majority-of-american-churches-fall-below-100-in-worship -attendance/.

4. Christopher Benek, "Churches Need to Prepare for When Robots Take People's Jobs," *Christopher Benek*, accessed April 30, 2019, http://www.chris topherbenek.com/2015/10/churches-need-to-prepare-for-when-robots-take -peoples-jobs/.